DRAGON LADY

DRAGON LADY

The History of the U-2 Spyplane

Chris Pocock

PRINTED IN ENGLAND

This edition first published in 1989 by Motorbooks International
Publishers & Wholesalers Inc,
PO Box 2, 729 Prospect Avenue, Osceola, WI 54020, USA.

Published by Airlife Publishing Ltd.,
Shrewsbury, England, 1989.

Printed and bound in the United Kingdom.

Library of Congress Cataloging-in-Publication Data
ISBN 0-87938-393-3

Motorbooks International books are also available at discounts in bulk quantity for industrial or sales-promotional use. For details write to Special Sales Manager at the Publisher's address.

Contents

Preface

It is now more than a quarter of a century since the U-2 suddenly became, for a while, the world's most famous aircraft. Time moves on and memories fade, but if you ask most people over thirty if they remember the U-2, it is still likely that they will recall the time in 1960 when US pilot Francis Gary Powers was shot down over the Soviet Union and wrecked a summit meeting of the superpowers. Sometimes they will ask: 'They don't still fly those planes, do they?'

Indeed they do. In fact, Lockheed is still building them. Not to the original 1954 specification, it is true, but the U-2 that is flown today differs little in basic design principles from the machine that soared across the Soviet Union from 1956 until 1960 with impunity, in the most successful aerial espionage venture of all time. Moreover, there seems every chance that this unique aircraft will still be flying vital intelligence-gathering missions in 2005, fifty years after the type's first flight.

Since May 1960, the U-2 has attracted more than its fair share of attention. Acres of print were written about the summit crisis that year, and more followed in 1962 when pilot Powers was returned to the US. Later that same year, the aircraft was back in the headlines when it played a crucial role in the Cuban missile crisis. Yet much of what was written was both speculative and misinformed. Even when Powers published his autobiography in 1970, the air of mystery and notoriety which surrounded the aircraft was not entirely dispelled. The US Air Force eventually grew weary of this image, and in an attempt to break with the past, redesignated the U-2 as the TR-1 in 1978.

In this decade, official archives from the period when the U-2 became famous have been made public, thereby prompting fresh articles and books on the 1960 and 1962 world crises in particular, and US intelligence resources of the period in general. There has also been one attempt to chart the aircraft's technical history in detail. Unfortunately, very little official material about the aircraft itself has been released, nor about U-2 operations. Many of the myths and the mysteries have been allowed to continue. This is partly due to the involvement of the Central Intelligence Agency (CIA) in over twenty years of U-2 history. For this was no ordinary military aircraft project run by a branch of the armed services. The U-2 has led a double life, during which many of its most amazing exploits have been conducted under cover by the CIA, in an operation which was not directly accountable to the Pentagon. When Powers was shot down in 1960, for instance, he was officially listed as a 'civilian' working for Uncle Sam, rather than as a regular US military officer.

Many U-2 people are proud of their association with the programme, and anxious that the true story should not be suppressed forever. Unfortunately, the authorities appear to wish otherwise. Former U-2 pilots have been refused permission to talk publicly about their exploits and details of the U-2 flights which actually penetrated the Iron Curtain are still secret, although they took place thirty years ago. The CIA pretends that their official narrative history of the U-2 programme does not exist. Researchers have been told that histories of the individual aircraft will not be declassified 'until 2011'. By that time, no doubt, much of the material will have been lost or thrown away. As recently as 1987, Lockheed was obliged to shred thousands of U-2 documents in its archives, after a government audit revealed that the company's security procedures did not conform to official requirements.

Meanwhile, memories are fading and some U-2 veterans from the early days are beginning to pass away. Without access to the official records, this history of the U-2 programme cannot pretend to be definitive. However, the author believes that these pages contain the most comprehensive account of this extraordinary enterprise that has yet been attempted. Hopefully, the reader will not find any of the more serious misconceptions about the U-2 repeated here. These include the frequently-aired contention that the plane was intentionally flown as a glider with the engine shut down during portions of its flight. Also, that it was capable of flying as high as 100,000 feet, or that it had photographed every blade of grass behind the Iron Curtain during the four years prior to May 1960. In this book, the reader will search in vain for some of the more fantastic stories concerning the aircraft that have surfaced in the past. This author does not believe, for instance, that the Cubans froze in a block of ice the body of a U-2 pilot they shot down, although some officials in Havana apparently told a visiting American group this story in 1976. He doesn't think much of the claim by a Norwegian, who once spied for the Soviet Union, that Powers' aircraft was downed in 1960 by a bomb which had been hidden in the tail before it took off. He also finds it difficult to credit the idea, advanced by Powers and some others, that Lee Harvey Oswald passed crucial information about the U-2 to the Soviets during his three-year sojourn in Moscow from 1959 to 1962.

Nevertheless, there are many incredible tales to be told about this mysterious aircraft and the men who flew it. In the history of the U-2, truth has sometimes indeed been stranger than fiction.

Acknowledgements

I first began to take an interest in this strange aircraft almost fifteen years ago, during a visit to Davis-Monthan AFB outside Tucson, Arizona, which was then the home base of the only US Air Force U-2 squadron. Work on this volume began in 1984, shortly after the publication of *Aerograph 3: Lockheed U-2* by Jay Miller, proprietor of Aerofax Incorporated, Arlington, Texas.

A good friend and fellow author, Jay Miller had invited me to contribute to his book, which was the first serious attempt to produce a technical description and history of the U-2 for public consumption. Jay's volume remains an outstanding work of reference, which all those interested in the nuts and bolts of this unique aircraft would do well to purchase. Its publication created much interest in the U-2 community, as initiates began to realise that an honest attempt to trace the aircraft's detailed history was long overdue. Jay generously made available to me the considerable feedback which was prompted by the publication of his volume, and has continued to provide all possible help and encouragement. To him and Susan, my grateful thanks are due.

I have drawn upon a variety of sources. A mass of written material was consulted, and that which has been previously published is acknowledged after each chapter. However, the most significant contributions to this book have been made by those who designed, managed, flew or maintained the Dragon Lady herself. For reasons which I have already made clear, many of these people had not previously felt able to discuss the U-2 programme with anyone outside of official circles. Some of them were only willing to contribute to my research on the understanding that they remained anonymous. These people are therefore not listed in the credits below, but to them and all the other U-2 alumni who helped, may I say that I consider it a great honour that they should have placed their trust in me.

My thanks go to Fred Carmody, Tony LeVier, Ben Rich, Eric Schulzinger and Richard Stadler of Lockheed-California Company; Bill Bender and Lee Levitt of Hughes Radar Systems Group; Jim Barrilleaux, Dick Davies, Doyle Krumrey, Joddi Leipner, George Plambeck, Walt Prouty and Gary Shelton at NASA Ames Research Center; and to the following serving or retired members of the US Air Force: Tony Bevecqua, Roger Cooper, Larry Driskill, Thom Evans, Deke Hall, Steve Heyser, Don James, Chuck Kern, Tony Martinez, Rich Smith and Don Webster.

Other valuable help and material was provided by Richard Bissell in Farmington; Cheryl Hortel at Edwards AFB; Jim Long in Del Rio; Dave Ostrowski in Washington; Mick Roth in Camarillo; Hank Tester in Las Vegas, and Nolan Tucker in Lancaster; by fellow aviation authors Robert Archer, Paul Crickmore, Bob Dorr, Peter Foster, William Green, Stephane Nicolaou, Lindsay Peacock and Dave Wilton; and by the staff at the British Library, British Library of Economic and Political Science, British Newspaper Library, Science Museum Library and Science Reference Library in London, as well as the Library of Congress and the Defence Audio-Visual Agency in Washington.

The following were gracious hosts or made travel arrangements: Timothy Deason in Del Rio; Peter Lowes and Tony Sackett in London; Roger Giles and the Kok family in Los Angeles; Pete and Ann Cundall and Doug Coker in San Francisco.

Thanks are also due to Sophie Jebb, my editor at Airlife Publishing, and to Alastair Simpson, Airlife's ever-patient joint chairman and managing director.

Finally, I must acknowledge the debt I owe to my wife Meng, who endured my obsessions and idiosyncracies with fortitude during this book's very long period of gestation.

Chris Pocock
Uxbridge, Middlesex, England
April 1989

Author's Note: The Deuce and the Dragon Lady

The Lockheed U-2 aircraft was never officially named, but it has acquired more sobriquets than almost any other flying machine. In the earliest days, it was simply referred to as 'The Article' within Lockheed. Then they took to calling it 'The Angel', since it flew so high. After the aircraft received an official military designation, some of the crews began referring to it as 'The Deuce', and this became a popular epithet within the CIA's detachments. Once the U-2 became public property in 1960, the press took to calling it 'The Shady Lady', or 'The Black Lady of Espionage'.

But of all the nicknames which have been applied to the aircraft, the one which has gained the greatest currency is 'The Dragon Lady'. Very early in the U-2 programme, someone in the US Air Force had the idea of giving the aircraft a codename based on a memorable character from a long-running comic strip. The Dragon Lady was a mysterious oriental who appeared in 'Terry and the Pirates', a newspaper series drawn originally by Milton Caniff, who has been called the Rembrandt of the comic strip artists. When it began in 1934, the series was set along the coast of mainland China, and the Dragon Lady was a pirate warlord who frequently frustrated the law-abiding endeavours of the youthful Terry Lee and his guardian Pat Ryan.

By the time that the US entered the war, Terry had grown up and joined the Army Air Corps in the Pacific theatre of operations. He and the Dragon Lady were now fighting on the same side, but the two still had their differences. The series continued after the war, now drawn by George Wunder, with Terry as a fighter pilot and the Dragon Lady leading an underground group fighting the communist regime in mainland China. She remained distant and mysterious, sometimes defeated but never conquered, and it ill-behoved anyone to take her for granted. She knew how to bite back!

Terry's relationship with the Dragon Lady typified that between the U-2 pilot and his tricky mount. As a codename for the new super-secret spyplane, it was an inspired choice, and proved to be even more suitable a few years later when Chinese pilots began flying the aircraft. The nickname remains popular with U-2 crews to this day.

Dedication

The crews who flew and maintained the U-2 took an unproved airplane, flew it into an unknown world, and gutted it out, day after day, week after week, and month after month, for years. They asked no questions, and explored the high altitude world to provide the answers for those who came later. It was trial and error, one step forward and two steps backward at times. Their combat was with the unknown, where one mistake could result in death. Those of us who lived through those early days, when many of our friends perished, never felt that they died without cause. Their heritage was a fierce determination to make it work, safely . . . and those who followed maintained that fierce determination. They really had 'the right stuff,' many years before the term became fashionable.

Colonel Donald James
4080th Strategic Reconnaissance Wing,
US Air Force, 1957-1965

Chapter One: Project Aquatone

Well, boys, I believe the country needs this information, and I'm going to approve it. But I tell you one thing. Some day one of these machines is going to be caught, and we're going to have a storm.
President Dwight D. Eisenhower, 1955.

They called him an aeronautical genius, an aerospace legend, a great team leader, and no doubt Clarence 'Kelly' Johnson was all these things. He and his tightly-knit team at the Lockheed 'Skunk Works' turned out the P-80, America's first operational jet fighter, followed by the F-104 — the missile with a man in it — when missiles were in their infancy. The U-2 came next, followed by the Mach 3 'Blackbird', which was a decade ahead of anything else aeronautical. Some hold that excellence brings its own rewards, but with Uncle Sam footing the bill for all of these great projects, and plenty of other aerospace companies eager for business, the Lockheed people also had to fight their corner in Washington. If Kelly Johnson had been the 'shy genius' that the press made him out to be, he would have left all the lobbying for contracts to Lockheed's top executives. But Kelly knew a few important people, and knew how to impress others; if not, the Dragon Lady would never have left the drawing boards. And a small group of equally dedicated engineers at the Bell Airplane Co., who thought *they* had won the contest to produce the secret reconnaissance plane, would not have been so sadly disillusioned.

For Kelly Johnson's original design had actually been rejected by the US Air Force, and Bell were proceeding to cut metal on the X-16 prototype. Unknown to them, however, another US government agency was entering the spyplane business at the President's behest. Now the Air Force were being pushed aside, and Lockheed had the inside track with this new customer . . . .

So Johnson was a useful lobbyist as well as an excellent designer of aircraft, but he alone did not conceive the idea of a high-altitude spyplane capable of overflying the Soviet Union with impunity. Neither, for that matter, did Richard Bissell, the Yale and Harvard whizz-kid in the CIA who was brought in to manage what became the U-2 programme. Like most ambitious technical projects of this kind, the seed was sown

The prototype 'Article' in August 1955 at the secret test site, Groom Lake, Nevada; the official designation U-2 was not allocated until some months after the first flight. The aircraft was also known as 'The Angel'. Groom Lake was also called Watertown Strip, and unofficially, 'The Ranch'.

many years earlier, in research laboratories, think-tanks and development offices, to await the right combination of conditions for successful germination.

No-one was more influential in these earlier years than Brigadier-General George Godard, the father of US aerial reconnaissance, who established the Photographic Laboratory at Wright Field before the Second World War. Then there were the people gathered around him, such as Colonel Richard Philbrick and the civilian Amrom Katz, who continued to work at Wright Field after the war in the renamed Aerial Reconnaissance Laboratory. There was also a group of optical research specialists that Godard brought together in 1946 at the Boston University Optical Research Center, including its director Dr. Duncan Macdonald. Finally, the list of credits for the U-2 should extend to a group of notable industrialists and academics who served the President and the Air Force chiefs on various high-level advisory panels in the early 1950s. Men like Dr. James Baker, a Harvard astronomer and wizard with lenses; Edwin Land, inventor of the Polaroid camera and founder of the company of the same name; Allen Donovan from the Cornell Aeronautical Laboratories; and Colonel Richard Leghorn, an airborne reconnaissance expert from Eastman Kodak who flew photo-fighters in the Second World War and was recalled to active duty during the Korean War.

Way back in 1946, Dick Leghorn effectively set the scene for the U-2 programme, when he told a symposium organised by Godard's team at Boston that 'it is unfortunate that whereas peace-time spying is considered a normal function between nation states, military aerial reconnaissance — which is simply another method of spying — is given more weight as an act of military aggression. Unless thinking on this subject is changed, reconnaissance flights will not be able to be performed in peace without permission of the nation state over which the flight is to be made. For these reasons, it is extraordinarily important that a means of long-range aerial reconnaissance be devised which cannot be detected. Until this is done, aerial reconnaissance will not take its rightful place among the agents of military information protecting our national security prior to the launching of an atomic attack against us.'

Leghorn called for higher and faster aircraft which would be invulnerable to interception, but within two years the Air Force cancelled its first-ever attempt to develop a custom-built reconnaissance aircraft. Competitive contracts had been let to Republic and Hughes Aircraft during the war, and although both companies built their designs round the big Pratt & Whitney R-4360-31 radial piston engine, the XF-11 and the XF-12 were very different machines. Republic's XF-12 was a large four-engine design carrying seven crew, multiple cameras and a built-in darkroom in a fully-pressurized fuselage. Howard Hughes, on the other hand, modified his earlier design for a long-range escort fighter into the single-seat, twin-boom XF-11. The movie mogul promptly crashed this sleek-looking machine into a Los Angeles suburb on its first flight in July 1946. Hughes luckily survived, and flew a second prototype in flight trials against the two XF-12 prototypes, one of which was also written off in a crash. But by 1948 both these designs were looking decidedly old-fashioned now that the age of the jet fighter had dawned. Besides, a dedicated reconnaissance aircraft seemed unaffordable in the post-war budget crunch, despite the manufacturers' claims of top speeds in excess of 400 m.p.h. and maximum altitudes of some 48,000 feet.

When, therefore, Strategic Air Command (SAC) began probing Soviet defences with photo, radar and electronic reconnaissance sorties in 1949-50, it had to use stripped down versions of the lumbering B-29, B-36 and B-50 heavyweight bombers, in an attempt to gain altitude and safety. The RB-45C version of America's first jet bomber, the Tornado, wasn't much of an improvement since its higher top speed was offset by a lower ceiling — only 40,000 feet. This was 10,000 feet lower than could be reached by the new MiG-15 jet fighters, which were first revealed in the Korean War. As Soviet air defences expanded, the reconnaissance flights were detected by radar and intercepted by the MiGs. Increasingly, they were confined to brief incursions over border areas. Even so, by 1956 a dozen USAF and US Navy reconnaissance aircraft would be shot at or shot down by MiGs, with the loss of eighty airmen. This didn't please the politicians of course, since they had to handle the diplomatic protests, and therefore insisted on close control of the flights, even closing them down from time to time.

Converting the big SAC bombers for reconnaissance had two advantages, however. One was long range, and the other was space. Their large bomb bays afforded plenty of room for the huge cameras which had now been developed. Jim Baker had produced the first 100-inch focal length precision lens for an aerial camera at Harvard by the end of the war; Dunc Macdonald and others from his research team continued this work in the early post-war years at Boston. If the aircraft were to be denied the opportunity of flying over the target, these big lenses could at least look sideways into enemy territory while the aircraft remained out of danger at high altitude a hundred or more miles away. The trouble was, these camera installations weighed two tons or more, and were very complex pieces of engineering, extremely sensitive to variations in temperature, pressure and aircraft movement. They also required huge windows of very expensive optical glass to be cut in the carrying aircraft, with all the attendant problems of airframe strength and pressurization. The ultimate in this line of development was reached when Macdonald and his team came up with a 240-inch focal length lens which would only fit into a B-36 — even though folding optics were employed, it was fourteen feet tall. Ultimately, however, the success of the U-2 would depend on the return to a somewhat smaller camera which could nevertheless match the larger lenses in resolution, despite being operated from even greater altitude.

By the time that Dick Leghorn returned to the Air Force from Kodak in 1951, to a planning job at USAF headquarters, he wasn't the only one worrying about a surprise Soviet atomic attack. Most of the rest of the intelligence community was sounding alarm bells about the dearth of information on Soviet military progress. America was waging a hard-fought war with the Communist forces in Korea. The Iron Curtain was proving virtually impenetrable by traditional espionage means, and the Soviets gave virtually nothing away, but they had exploded an atomic bomb in 1949 and were well on the way to thermonuclear devices. Stalin had ordered the development of long-range nuclear bombers capable of striking the US. Within a year, a group of German scientists who had been co-opted by the USSR for further rocket research after the war had finally been allowed to leave. They reached the West with disturbing tales of Soviet progress with surface-to-surface missile technology.

Accordingly, in May 1951 the Air Force set up a special study group at Boston on the aerial reconnaissance problem. Codenamed 'Beacon Hill', the study team was chaired by Carl Overhage (who went on to direct the Lincoln Laboratory) and brought Baker, Land and Donovan together for the first time. Most of the same team continued to tackle the problem over the next few years as a special panel of the Air Force's Science Advisory Board. Their first chairman was George Kistiakowsky, who later became President Eisenhower's personal advisor on science. Ultimately, these people became key figures in the decision to build the U-2 through their membership of the so-called Killian Panel, which was set up by the President himself in 1954.

Some of this high-powered debate began to filter down the command structure, and back at Wright Field in late 1952 the USAF began only its second attempt to create a dedicated reconnaissance aircraft. Bill Lamar ran the New Developments Office of the Bombardment Aircraft Branch, assisted by Major John Seaberg. Like Dick Leghorn, Seaberg was a veteran of the Korean War who had returned to military service, having been working as an aeronautical engineer with Chance-Vought. From their office, Lamar and Seaberg had witnessed the introduction of the jet-powered B-45 and B-47 bombers, powered by 6,000 lb. s.t. General Electric (GE) turbojets. A new generation of turbojets promising greater thrust was now being developed, and Lamar and Seaberg began to realize that they might be matched to the right kind of airfoil with very low wing loading to produce an aircraft capable of sustained cruise at over 65,000 feet.

This was very interesting. Such a plane might not even be detected on radar, let alone intercepted, either by fighters or the forthcoming guided missiles. It might therefore make an ideal light bomber. But with SAC now investing in massive production of medium and heavyweight nuclear bombers, it seemed more appropriate to think about reconnaissance. After all, the team in SAC intelligence was tearing its hair out just trying to assemble a target list long enough to justify the 1,000 B-47s that were on order! Apart from Soviet cities, and the perimeter radar and anti-aircraft defences that had been identified by

Clarence 'Kelly' Johnson, designer of the U-2. This brilliant aeronautical engineer contributed to the design of more than forty aircraft during his long career. He was responsible for the wartime P-38 interceptor, and the early postwar F-104 Starfighter. This track record, together with Johnson's unconventional method of working, earned him the contract for Project Aquatone.

the probing flights, they really didn't know where to send the bombers. Where were the rumoured rocket test sites, the atomic bomb production facilities, the new bomber bases? Not surprisingly, therefore, Lamar soon gained approval to contract for design studies, and by March 1953 Seaberg had worked out a set of specifications.

These called for an altitude of 70,000 feet and radius of 1,500 nautical miles carrying a camera payload weighing up to 700 lb. Gross weight would be kept very low by eliminating as much ancillary equipment as possible, such as defensive armament and an ejection seat for the pilot, and by providing only rudimentary navigation and communications aids. If the design could minimize detectability by radar, so much the better; this was probably the first serious request to the aircraft industry for a 'stealthy' aircraft. It was required in service by 1956.

Lamar and Seaberg now took a somewhat unusual decision: they wouldn't circulate their specification to the big airframe manufacturers. A large production run wasn't envisaged, so they reckoned on getting closer attention from smaller companies. They settled upon Bell Aircraft, builder of the first two famous X-plane jets, and Fairchild, best known for the C-119 and C-123 military transports. In addition, they asked the Martin Company to see if they could improve upon their existing B-57, which was actually a re-engined version of the British Canberra bomber, built under licence. All three manufacturers were interested, and six-month study contracts under the codename 'Bald Eagle' were issued in July 1953. Although Seaberg did not originally specify a particular powerplant (and the contractors were even asked to consider turboprops), opinion at Wright Field hardened in favour of Pratt & Whitney during 1953. Pratt & Whitney's J57 turbojet had just gone into production for the B-52 and F-100 as the first 10,000 lb. s.t. jet engine, and seemed to have the edge on the rival offerings from Allison and General Electric. So Pratt & Whitney were therefore asked to take a preliminary look at the modifications needed to keep the J57 running in very thin upper air, and their positive response led to Bell, Fairchild and Martin all nominating it as the powerplant of their choice.

In other respects, the three designs which were submitted to Wright Field in early January 1954 were very different. Martin could do little more than take a standard B-57 fuselage and marry it to a much larger wing and two J57s. Although this aircraft couldn't make 70,000 feet, it would nevertheless be a quick and relatively cheap interim solution. Bell also built their Model 67 proposal round two wing-mounted J57s; theirs was a breathtakingly delicate and spindly machine with a super-slim fuselage for maximum drag reduction and wings with an incredibly high aspect ratio of twelve for maximum lift. They were nearly 115 feet long, nine feet longer than the more conventional airfoil that Martin was proposing. Having decided that a single J57 could do the job, the Fairchild team opted for a much smaller wing, but their M-195 design featured a novel engine installation above and behind the cockpit. Weighing in at a little under 11,000 lb. empty, it was less than half the weight of the Bell, but both design teams had produced uncompromisingly lean machines.

Kelly Johnson knew a few things about airframe weight reduction too, because the revolutionary XF-104 — soon to be named Starfighter — was only two months away from its first flight. Having heard from US pilots in Korea that speed and altitude were of paramount importance if you wanted to hassle with MiGs, Johnson had sat down to design a Mach 2 plus hotrod fighter capable of reaching over 60,000 feet. To achieve this with the jet engines then available had meant aiming for a gross weight as low as 15,000 lb. Through ruthless pruning of systems and the design of the famous very short, thin wing, this had been achieved.

The first of Johnson's friends in high places now entered the picture. Philip Strong was a retired general now advising the CIA on intelligence matters. Strong and his boss, Richard Amory, had spent a frustrating 1953 trying to advance US knowledge of the Soviet missile testing site at Kapustin Yar in the Ukraine. SAC had virtually nothing on it, and didn't seem inclined to risk a reconnaissance sortie so far into the interior. Instead, the British had mounted a flight, sending one of their new Canberra photo planes over the vital target. The RAF plane had been shot at, but fortunately made it to a landing in Iran with some pictures. Surely Uncle Sam could do better than this! Amory was all for the CIA developing its own analysis of technical intelligence, instead of relying on the military. The Agency was already in the process of expanding its Office of Scientific Intelligence (OSI), and it now took steps to expand the CIA's rather inadequate photo-interpretation department by recruiting the excellent Arthur Lundahl, who was teaching the subject at the University of Chicago. Soon, the CIA began scheming for its own intelligence *collection* systems, free from control by the military.

Strong knew Johnson, and travelled out to California to seek his advice on reconnaissance

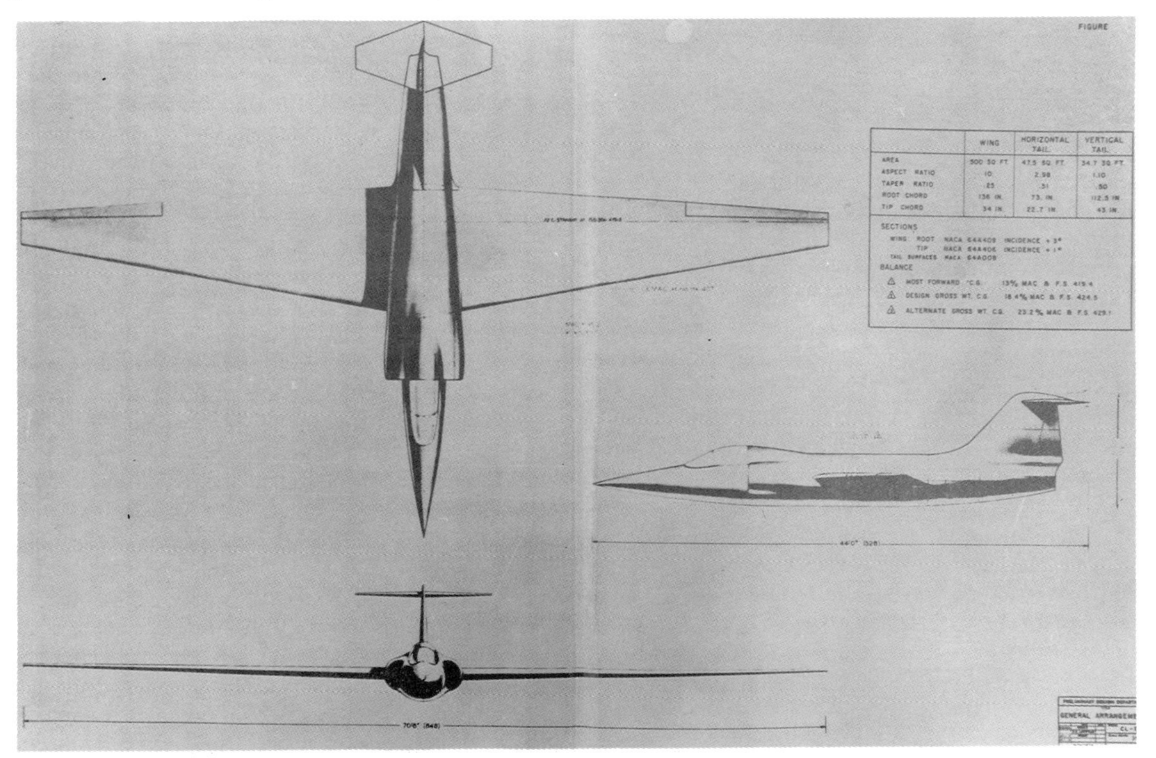

Johnson's CL-282 proposal shown in a preliminary general arrangement drawing from the Lockheed Skunk Works. It married an F-104 fuselage to a new, 500 sq ft wing. The aircraft would have been launched from a wheeled dolley and landed on a single skid which retracted into the lower fuselage. The Air Force rejected the design.

Sceptics doubted whether turbojets could be made to work reliably at the cruising altitudes proposed for the new spy plane. Engineers at Pratt & Whitney promised to adapt the successful J57 fighter engine for the project. This is the definitive -31A version fo the J57 which was fitted to production U-2 models from 1956.

aircraft. Assisted by fellow Skunk Works designers Phil Coleman and Gene Frost, Johnson soon came up with a new design which he designated CL-282. It was essentially the F-104 fuselage and tail married to a new, seventy-foot wing. To save even more weight, they eliminated the F-104's sturdy landing gear entirely: the vehicle was designed to take off from a wheeled dolly and land on a skid built into the belly. Colonel Bernard Schreiver, who was Dick Leghorn's boss in the Pentagon, heard about the proposal and wrote to Johnson requesting details.

In early April 1954, while Lamar and Seaberg at Wright Field were still mulling over the three Bald Eagle submissions, Johnson presented his CL-282 proposal to the Air Force hierarchy in Washington. He briefed Trevor Gardner, the newly-appointed assistant secretary for R&D, a bright and forceful engineer who had worked on the Manhattan Project and later founded the Hycon Company. Gardner was a great believer in the 'red menace', already knew Johnson, and believed that here was the very kind of military-industrial project that the country needed. The enhancement of American security through better technology was Gardner's main theme; he was critical of the scientific advice that the Pentagon and the White House were receiving, and had just succeeded in persuading President Eisenhower to set up a high-powered panel to study the problem of surprise attack — a fear that had haunted many Americans since Pearl Harbor. Gardner told Johnson to submit formally the CL-282 to the Air Force.

While Johnson returned to California to refine his proposal, Gardner helped organize the top-secret panel, which would study the state-of-the-art in strategic weapons, air defence and intelligence gathering. This became known as the Killian Panel after its chairman James Killian, the president of MIT (Massachusetts Institute of Technology), who by August had gathered a group of forty-six prominent scientists together in the Executive Office for a six-month investigation. Specialist teams were set up to study the three categories, and that which concentrated on the technology for intelligence once again included Jim Baker and Edwin Land, who was its chairman.

Meanwhile, at Wright Field, Seaberg had eliminated the Fairchild design and recommended contracts for Martin and Bell. His plan was well received by the Generals in briefings conducted at Air Research & Development Command (ARDC) and SAC headquarters in late April and early May. While the Bald Eagle development contracts were ascending the Air Force chain of command for approval, Lockheed's proposal was travelling in the opposite direction, and it reached Seaberg in mid-May. He was not particularly impressed. Johnson's design offered less range than the Bell proposal, was single-engined, and since pressurization had been eliminated to save weight, the pilot would have to fly the whole mission with an inflated pressure suit. Moreover, Johnson wanted to use the GE J73 turbojet.

The XF-104 was now flying with a temporary Wright J65 engine, but the development and production programme was being built around the J79 engine now being developed by GE for the supersonic B-58 bomber. The J73 was an interim engine derived from the first generation J47 as used on the B-47 and F-86, but it still fitted the XF-104 fuselage contour. Offering just over 9,000 lb. s.t., it was smaller and lighter than the Pratt & Whitney J57, but the people at Wright Field preferred the greater thrust and considerably better fuel economy offered by the latter, with its unprecedentedly high pressure ratio. The J57 was going to be a great engine, and more than 1,000 were being produced this year alone for the B-52, F-100, F-101, F-102, and Navy Skyray and Skywarrior jets.

In 1954, there was still a significant body of aerospace opinion which doubted whether any turbojet could be made to work reliably in the thin air above 60,000 feet. Up there, you were dealing with air that was only a fraction of its density at sea level. Either the engine would overspeed or overtemp, or the compressor blades would surely stall, and then there would be a flame-out. Undeterred, Pratt & Whitney had promised to adapt the J57, even though it would only be able to produce around seven per cent of its sea-level thrust at the highest altitudes. In the specialised J57-P-19 high-altitude version now under development (soon redesignated J57-P-37), Pratt & Whitney was offering a whole range of suitable modifications to the hydraulic pump, accessory box, oil cooler, alternator and constant speed drive. They would virtually hand-build much of the two-spool compressor and three-stage turbine to much closer tolerances, to eliminate weight and minimize pressure losses and other inefficiencies. Great — but it wouldn't fit the CL-282.

So Seaberg rejected Johnson's design in early June, and hearing no more from his superiors, thought that was the end of the matter. Martin received a go-ahead for the RB-57D on 21 June, with an initial order for six, and Bell were contracted for what was now known as the X-16 in September. Bell were asked to produce a prototype within eighteen months, but the interim solution from Martin was expected to be flying within a year. In Burbank, Johnson took a fresh look at the CL-282. In the end, he and his preliminary design team virtually threw out the F-104 fuselage, excepting the cockpit, and Ed Baldwin designed a new one from scratch. The undercarriage was redesigned, cockpit and payload bay pressurization introduced, and a conventional tail adopted. The wing had to be lengthened slightly, but was otherwise unaffected.

Trevor Gardner continued to push the Lockheed proposal, by submitting it to the Killian panel. Allen Donovan visited the Skunk Works and reported back favourably to Land and Baker, who were now assessing the latest camera technology and signals intelligence receivers, which might allow a lighter payload requirement than Seaberg had envisaged.

Through the autumn of 1954, the Killian panel worked up its recommendations, which were to be presented to the President by the following February. (Their chief recommendations would be to start immediate work on intermediate-range ballistic missiles [IRBMs] to be based in Europe as a short-term counter to similar Soviet weapons; to assign the highest national priority to development of intercontinental ballistic missiles [ICBMs]; and to initiate work on an earth-orbiting satellite that could provide overhead reconnaissance.) But Land and his team on the intelligence sub-panel thought the need for a reconnaissance vehicle which could fly deep inside the USSR was now so urgent that they began to circulate their idea more widely within the intelligence community in early November. Capitalizing on their sense of urgency, Johnson had pledged to have his aircraft flying within eight months of go-ahead. He travelled to Cambridge, Massachusetts, where he met Killian at Land's home, and the three of them discussed his proposal in detail. Moreover, Johnson kept in touch with his CIA contacts who had first tipped him off about the spyplane requirement. They still seemed keen on sponsoring such a project. The Lockheed bandwagon was now rolling.

After gaining approval from the US Intelligence Board, Land and Killian took their spyplane proposal to a personal meeting with the President in mid-November. He was very receptive to their basic concept, but not at all keen on the proposed vehicle being flown by Air Force pilots. He reasoned that, if the worst happened, the loss of a 'civilian' reconnaissance pilot over Soviet territory would be less politically provocative. Moreover, it seems that he shared the CIA's suspicions of military intelligence, especially that being provided by the Air Force, which was thought to be exaggerating the Soviet bomber threat in order to justify its own continued expansion. Within a few months another report on the US intelligence community would land on the President's desk from the Hoover Commission, and it too would be critical

Dwight D Eisenhower, the soldier-turned-politician who became President in 1953, and approved the ambitious scheme to overfly the Soviet Union nearly two years later. He took the key decision to hand control of the project to the CIA, and later approved every single overflight.

of the Air Force intelligence set-up. Eisenhower's decision to demilitarize the U-2, taken before the programme was even launched, would be the cause of continuing tension between the Air Force and the Agency for the next decade or more. For the President now decided that he would assign management of the new spyplane project to the CIA. The Air Force would play a supporting role, but the spooks would be running the show.

Land's team now called Seaberg and Johnson to Washington on successive days for intensive briefings. The meetings were held in the Pentagon and conducted by Gardner and General Donald Putt, who had earlier given approval for the Bald Eagle projects while still Commander of ARDC, and had since been transferred to the Pentagon as Deputy Chief of Staff for development. On 18 November, Seaberg was ushered in to the top secret gathering, and outlined the Air Force programme, with the reasons for preferring the RB-57D and the X-16 to the Lockheed and Fairchild proposals. Seaberg was now aware of Johnson's end run on the CIA and Killian flank, so he came prepared with a graph which showed that the three all-new designs were aerodynamically close, and that if the CL-282 were re-engined with the J57 it would be competitive.

Gardner and Land were pushing strongly for Lockheed, and the CL-282 proposal could be altered to accommodate the favoured powerplant. That more or less settled it. Next day, Johnson arrived for what he later described as the most thorough grilling he had experienced since his college exams. At the end of it, the Killian team were satisfied with every aspect of the vehicle's design and performance, and Johnson was lunching with CIA Director Allen Dulles and Air Force Secretary Harold Talbott. Someone said, 'Let's stop talking about it and build the damn plane.' Within two days, Allen Dulles, Talbott and Putt were in the Oval Office seeking the final go-ahead from Eisenhower, together with Secretary of State Foster Dulles (Allen's brother), Defense Secretary Charles Wilson and General Nathan Twining, Air Force Chief of Staff. At this high-level meeting, the CIA's control of the project — now codenamed 'Aquatone' — was confirmed. The civilian agency would foot Lockheed's bill from one of its many secret accounts, but the Air Force was asked to hide the cost of the dedicated high-altitude engines within the much larger military amount allocated for procurement of standard J57 powerplants. Twining was distinctly unhappy about this unprecedented breach of Air Force sovereignty, although for the moment he kept his objections to himself. His SAC commander, General Curtis LeMay, had no such inhibitions when he heard about the arrangement. He hit the roof. How could a highly complex air operation such as the one now being proposed, be put in the hands of a bunch of civilians who knew next to nothing about the whole business? It was obvious that they would have to lean heavily on the Air Force for support, if this project was ever to get off the ground.

Although the team at Bell in Niagara Falls didn't know it, the X-16 was doomed. They were allowed to continue prototype development as an insurance should Lockheed fail to deliver.

The enormous wingspan and incredibly thin fuselage of the Bell X-16 is illustrated by comparison to the F-86 Sabre fighter. Bell's design team elected to use two J57 engines for reliability, and won the formal Air Force design competition for Project Bald Eagle. The X-16 was cancelled a few months before its first flight, after the U-2 had successfully flown.

Martin was also allowed to press on with the RB-57D, which first flew in November 1955, and twenty were eventually delivered to the Air Force in 1956–57 in three versions tailored to specific reconnaissance missions. By then, the Lockheed machine had caught up and overtaken this interim solution. There seemed to be no stopping the Skunk Works when it was in full swing.

The Skunk Works had been created in 1943 when Lockheed got the contract to build the first jet fighter. Johnson managed to cream off the pick of the Burbank factory's engineering talent into an experimental department where design engineers, mechanics and assembly workers would work closely together in a streamlined fashion, free from the constraints imposed by the wider company bureaucracy. The department was completely independent of the rest of the company for purchasing and all the other support functions. Walled off from the rest of the plant, and only accessible to the select few staff, it was also a very secure method of building secret prototypes — it built the XP-80 in just 143 days. What on earth were Kelly and Co. up to in there? A popular wartime comic strip featured a hairy and eccentric Indian who regularly stirred up a big brew, throwing in skunks, old shoes and other unlikely raw material. With Johnson's secret shop situated right next to the plastics area, it smelt like a similar deal, and so the Skunk Works name caught on.

Johnson had promised to build his spyplane in eight months. The clock started ticking on 9 December 1954, when Gardner paid a formal visit to Burbank to confirm a contract worth $22 million for twenty aircraft with Lockheed chairman Robert Gross and engineering Vice-President Hal Hibbard. Johnson was already at work: a matter of hours after Eisenhower had given his approval, Gardner had telephoned Burbank from the Pentagon with an informal go-ahead. Also present when that call was placed was Colonel Ozzie Ritland, who had been assigned to run the Air Force part of the project, and the CIA's nominee Richard Bissell, who had rushed over from the Agency downtown headquarters after being informed of the project by Allen Dulles. Bissell turned out to be an excellent choice; a brilliant economics professor from Yale and MIT who had joined the CIA as special assistant to Dulles the previous February, he had an informal and unorthodox style of management which suited similar spirits like Gardner and Johnson. An advisory group of scientists and engineers, including luminaries from the Killian panel, contributed guidance and appraisals throughout the development programme. Bissell liberated Johnson and his team from the paperwork tyranny of technical specifications so beloved by the mandarins at Wright Field. Instead, the CIA agreed a simple set of *performance* specifications with the Skunk Works. The relationship was so informal that progress payments worth millions of dollars were sent as private cheques to Johnson's home address.

Project Aquatone soon gained another secret codename — 'Idealist' — but neither gained much currency within the Skunk Works, where the emerging aircraft was known as simply 'The Article' or (less prosaically) 'The Angel' until a decision was made a year later to allocate the deliberately misleading military designation in the U-for-Utility series. So what was so special about this 'Angel' that had captured the imagination of so many in Washington?

It was probably the clever blend of innovation and conservatism. In some respects, the U-2 was a conventional design, were it not for the very fine manufacturing tolerances and the extraordinary measures taken to save weight. The first U-2s weighed in empty at just 12,000 lb. Construction was mainly of aluminium, but there was a decided lack of stringers and other forms of structural stiffening, and the skin was wafer-thin — 0.02 inches. So this was a fragile aircraft, especially if mishandled — the structural limits were as low as plus 1.8g and minus 0.8g under some flight conditions. The fuselage was just under fifty feet long, with bifurcated intakes feeding the single J57. Like the rest of the airframe, these intakes had to be designed and constructed with intricate care to coax and shepherd the thin high altitude air into the compressor face or around the airframe. Skin panels were flush riveted and the rudder and elevator were aerodynamically balanced with virtually no gap at the hinge point.

Ahead of the engine was a pressurized area for reconnaissance systems known as the 'Q-bay'. One handy innovation was that the payloads for the Q-bay were mounted on interchangeable hatches. When placed in position, they formed part of the aircraft's lower fuselage skin. The cameras and other delicate reconnaissance equipment could therefore be built up on pallets and mounted on the hatches, before being offered up to the aircraft. Another large detachable hatch fitted the top of the Q-bay — when both were removed the aircraft looked quite peculiar, with this big gap between the bulkhead at the rear of the Q-bay and the cockpit/nose section, with the latter supported only by the fuselage mid-side spar structure. Certain non-photographic sensors could also be carried in the nose. Later, more space for payloads such as electronic countermeasures (ECM) and a datalink was found in small cavities fore and aft of the tailwheel.

The cockpit was extremely cramped, with a very large control yoke which seemed to the casual observer to be more suited to a large transport aircraft. The otherwise conventional instrument panel was dominated by the hooded display of a combined driftsight/sextant. The optics for this were situated in little glass bubbles which protruded from above and below the fuselage, just forward of the cockpit. This notable device (also included on the X-16 proposal) would serve the pilot as a primary navigation aid, and he could also use it as a sort of downward-looking periscope to check if there were enemy fighters below. There was no ejection seat, and pilots in trouble would even have to open the canopy themselves before baling out. A limited amount of space was available for payload and avionics in the nose. The tail was conventional, with large, unboosted control surfaces. It was attached to the fuselage by just three ⅝-inch bolts.

The most noticeable fuselage feature was the landing gear, a bicycle and outrigger arrangement like that of the B-47, with one important difference. The outriggers which kept the aircraft stable for taxi and take-off were designed to drop out as the aircraft left the ground. These tiny compressed rubber wheels on sprung steel legs were known as pogos. For recovery, pilots would have to keep the aircraft steady until the end of the landing run, when one or other wing would drop until down-turned endplates on the wingtips made contact with terra firma. The ground crew would then rush forward and re-insert the pogos for taxying. This was certainly an improvement on the original dolly-and-skid idea for the CL-282 — who knows what problems that would have caused in operational service — but a landing would still call for considerable skill and effort by the pilot, considering that none of the controls were boosted, and he would be fighting that huge wing lift.

That long, narrow wing! This was the key to sustained high altitude flight, with a high aspect ratio (10.67) and very low wing loading which needed to be around or below thirty pounds per square foot. This bestowed good cruise efficiency and low induced drag. But unlike the heavier Bell X-16 and Martin RB-57D, the Lockheed design achieved the necessary low loading at an eighty foot span. Bell had already encountered problems in designing the X-16's 115-foot wing, which had been modified to try and eliminate a bending instability known as aeroelastic divergence. The RB-57D wing was even longer (122 feet), and was later afflicted by serious structural problems. The U-2 wing had a three-spar construction, but the conventional rib stiffeners were replaced by an unusual latticework of aluminium tubing.

One crucial characteristic of the U-2 airfoil was its very high camber, which was meant to give the best possible lift/drag ratio for high-altitude cruising efficiency. But this amplified the airframe's pitching moment, especially at the higher airspeeds, and resulted in heavy balancing loads on the horizontal tail. As very many U-2 pilots were later to find, once this balance was upset, the aircraft could be only seconds from break-up. By means of an innovation known as 'gust control', the large trailing edge flaps and ailerons were enlisted to reduce wing and tail loadings, thereby eliminating the need for structural strengthening and its concomitant weight penalty. When flying at higher speeds, or in turbulent air, the wing control surfaces could be raised to the gust control position (four degrees for the flaps and ten degrees for the ailerons). The effect was to compensate for the high camber by moving the wing centre of pressure forward, thereby reducing wing and tail loads. This gust control device was pioneered on the U-2 (although it was not the exclusive that the Skunk Works claimed — it was also featured on the losing Fairchild design). Later, it would be utilized extensively in transport aircraft.

Without the Type B camera, shown here on its transport stand, the U-2 overflights could never have been successful. It was designed by Edwin Land of Polaroid fame, with thirty-six-inch focal length lenses supplied by Harvard astronomer James Baker. The large twin magazines on top carried enough film for an eight-hour flight.

Virtually all of the fuel was carried in four integral wing tanks, which fed a fuselage sump tank. Standard military JP-4 fuel was not suitable. On the long cruise missions at extreme altitudes, where the outside air temperature might be as low as minus seventy degrees centigrade, JP-4 in the wing tanks would be sure to freeze. Moreover, there was a risk that the fuel would 'boil off' due to the low atmospheric pressure; although the wing tanks were pressurized by bleed air from the engine compressor, the sump tank was not, and wing tank pressure would in any case be lost if there was an engine failure. Lockheed turned to Shell Oil, who came up with a special low vapour pressure kerosene which was designated LF-1A by Lockheed and JP–TS (for Thermally Stable) by the military. The abbreviation LF stood for 'Lighter Fluid' — the fuel wasn't very different from that with which you lit a cigarette! Internal fuel capacity on the U-2 was 1,345 gallons, but a further 200 gallons could be carried in two slipper tanks, one attached to each wing leading edge approximately ten feet outboard. The pilot had no indication of how much fuel had been used from each tank. He had to rely on a simple subtractive-type fuel counter which showed him how much was left (in gallons), backed up by a warning light which came on when only fifty gallons remained.

While the Skunk Works team of only fifty engineers worked hard to produce an airframe prototype at Burbank, important development work on a new camera was going on elsewhere. As we have seen, the trend in strategic reconnaissance cameras had been for ever-larger devices with longer focal lengths. Unfortunately, longer focal lengths meant smaller ground coverage, so either you sent up a number of these devices together in a fan arrangement with the lenses fixed to point in varying degrees from the vertical — a very heavy rig — or you mixed the so-called 'target analysis cameras' with panoramic types to give the wider ground coverage needed for the 'search mission'. When Seaberg wrote his Bald Eagle specification at Wright Field, he envisaged the carriage of two cameras of six-or twelve-inch focal length producing a nine by nine-inch negative for search, and two thirty-six-inch focal length cameras with nine by eighteen-inch format for target analysis. But the U-2 went to work with just one reconnaissance camera performing both roles (augmented by a small tracker camera), and the payload weight thereby saved could be devoted to better aircraft performance or to electronic reconnaissance devices.

When Jim Baker and Din Land sat down on the Killian panel to consider the state-of-the-art in reconnaissance photography, there were some recent developments which looked promising. Eastman-Kodak were working on new emulsions, coatings and a mylar base which would offer thinner but much improved films. Then there were the new lightweight reflective optics, whereby mirrors were used to preserve the long focal lengths and other characteristics desirable of a long-range camera, but enabled them to be packaged into a more manageable size. Another was the panoramic scanning camera pioneered by Colonel Richard Philbrick, who had come up with two solutions to the problem of ground coverage. The first of these was simple enough — you mounted one large camera on a turning crank and thus aimed it to the left and right of the vertical. The other involved sweeping the camera axis by means of mirrors or prisms to build up the picture on a continuously moving strip of film, thus building upon Godard's pioneer wartime work with shutterless strip cameras for low-altitude fighter-reconnaissance aircraft. The Boston team had been working on these developments since the war and, with help from Baker and the Perkin-Elmer company, a forty-eight-inch panoramic camera which promised to do a good solo job in strategic reconnaissance aircraft was now becoming available.

But it still weighed a great deal and there was seemingly no way for Johnson or his rivals to provide such a payload. They were working on a Q-bay capacity of around 750 lb. Realising that high-resolution photography would be the supreme requirement for Project Aquatone, with maximum coverage of secondary importance, the Killian panel proposed a secret new development effort. With CIA funds now flowing, the result was available before the U-2 was long into its flight test phase — and this was the now-famous 'Type B' camera. It weighed less than 400 lb. empty and had only a thirty-six-inch focal length lens, yet it brought back pictures of unprecedented clarity from nearly twice the height of its predecessors. Reconnaissance people measure picture clarity in terms of photograph resolution, the number of lines per millimetre (l/mm) that can be distinguished by a camera system. Twelve l/mm was considered a good performance only a few years earlier, but the Type B now turned in results of around thirty l/mm. From 60,000 feet, it could resolve objects on the ground as small as 75 cm/2½ feet across. This was achieved by synthesizing the best of the previous research, drawing on the emerging computer technology for design, and by insisting on the strictest quality control.

Din Land worked on the camera design and Jim Baker perfected a new lens. Drawing on the Boston developments, they chose a single reflecting mirror and adapted the turning crank and continuous moving strip ideas to a stop-and-shoot principle; the camera would still be able to cover horizon-to-horizon through seven separate small windows in the Q-bay hatch. Baker used a new IBM CPC computer to work out all the lens dimensions, before turning the fabrication work over to Perkin-Elmer. He insisted that the finished product be returned to his Harvard laboratory for final grinding and examination and then personally fitted the lens in each Type B camera, as it became known. Focusing and

Kelly Johnson chose Tony LeVier, the Lockheed test pilot who had flown the prototype F-104, to make the maiden flight of the U-2. LeVier also played a key role in the selection of Groom Lake as the secret test site. This photograph of him in a U-2 cockpit was taken more than a decade later.

exposure were automatically controlled. The new Kodak film would feed simultaneously in two nine-inch rolls to provide the useful stereo effect for interpretation, travelling in opposite directions to keep the camera centre of gravity steady. Every time the shutter clicked, the two films were exposed as a paired eighteen-inch negative, and the film was so thin that the two magazines could each carry more than a mile of it — enough to produce 4,000 pairs of images. A production contract for some twenty-five cameras went to Trevor Gardner's old R&D firm in California, Hycon Corporation. Big as the B-camera was — it weighed about 500 lb. all told, including a full film load for an eight-hour flight — there would still be some room left over in the Q-bay for the thirty-five millimetre tracker camera and electronic equipment. The tracker would continuously scan from horizon-to-horizon so as to provide the photograph interpreters with an accurate track of the entire flight path of the aircraft.[1]

Johnson beavered away in Building 82 at Burbank with his hand-picked team of eighty engineers. During recruitment, he had warned them this project was so secret that they might have an eighteen to twenty-four-month gap in their work records. Dick Boehme was his chief engineer on the project, and Art Viereck was head of manufacturing. By the turn of the year, they had the configuration fixed. Not long after, Tony LeVier was up at Lockheed's Palmdale plant, working as chief test pilot on the XF-104, when Johnson summoned him to the Skunk Works and asked if he would fly a new plane. 'What plane?' 'I can't tell you unless you agree to fly it!' said the designer.

LeVier agreed. His first job was to find a secret site for the flight tests — even with the security constraints operating on all new military aircraft at this time, Johnson had been told by the government that neither Palmdale or Edwards AFB would do. Off went LeVier with Dorsey Kammerer, Johnson's favourite mechanic, in the Lockheed flight test department's V-tail Beech Bonanza for an airborne survey of possible sites in the surrounding desert. The security was ridiculously tight — the pair had to tell everyone at Burbank that they were off on a hunting trip to Mexico, and dress accordingly! Only when they were out of sight of the factory could they turn the plane around and head north to search the most likely areas. After two weeks of photographing and mapping, the pair came up with three possible sites, and Johnson eventually chose the top one on their list. Over thirty years later, it remains America's most secret test site, off limits to all but a select few, who work on the stealth fighter and other 'black' projects. Like Edwards, it is a dry lake bed in rolling desert terrain. Unlike Edwards, it is miles from any town, and could only be sensibly reached from Burbank by air.

However, the handy thing about Groom

Lake, Nevada was that the main Atomic Energy Commission (AEC) nuclear test site was only a few miles away over the mountains. In consequence, a massive area around here had already been cleared and fenced off, with restricted airspace overhead. There were few people about anyway — Las Vegas was the nearest town and it was some hundred miles to the southeast. LeVier flew Johnson and Bissell up to see for themselves; the next time he saw the site a few months later, the tarmac runway was down, two hangars were up and there were accommodation trailers all over the place. Bissell had drafted in a large team of construction crews from the AEC, who had worked round the clock to get the secret base ready. It became known as 'The Ranch'.

In the small hours of a late July morning, the fuselage, wings and tailplane of the secret 'Article' 001 were mounted separately on trailers and covered with tarpaulins, before being wheeled out of the Skunk Works and onto an Air Force C-124 cargo plane. After some delay, the lumbering transport lifted off from Burbank and headed for the secret site. In the next week, the Lockheed team completed final assembly and systems check-out under the eye of a growing government contingent. They planned to fly the aircraft off the dry lake, rather than the runway, so LeVier spent his time clearing debris from the lakebed in a pick-up truck, assisted by fellow test pilot Bob Matye. There was plenty of it — the site had been a gunnery range during the war, and there were spent shells, rocks, sagebrush, even half a steamroller to be cleared away.

The first taxi trials were set for 1 August. The third run turned into a first flight of sorts when, much to LeVier's surprise since he had already chopped the power to idle, the aircraft left the ground at around seventy knots and flew for about a quarter of a mile with the ground observers in hot pursuit. This aircraft's wing had all the lift expected . . . and more! Because the lake bed had no markings to judge distance or height (despite an earlier recommendation from LeVier that this be done), the test pilot eventually made heavy contact with the ground in a ten-degree left bank. The aircraft bounced back up, then settled down again under semi-control and eventually skidded to a halt a mile further on. The brakes were lousy and a tyre had blown. The team were out again the next morning for another taxi test. This time, LeVier pushed the control yoke hard forward as the speed increased and thereby succeeded in keeping the main gear in contact with the ground throughout. The control surfaces seemed to be working as predicted, but the brakes were still causing concern. However, there was plenty of room to stop on the lakebed, so LeVier reported the Article ready for a proper first flight.

Another day passed while final preparations were made to the Article. August 4 dawned

The prototype U-2, heavily shrouded for security reasons, is unloaded from an Air Force C-124 at The Ranch. All the subsequent aircraft were transported here in similar fashion, for final assembly and first flight. Later, those aircraft allocated to the CIA squadrons for overflights of the Soviet Union were dismantled and loaded aboard C-124s again for the journey overseas.

A view of the prototype U-2, serialled 001 and with US Air Force roundels, taken during early flight tests at The Ranch. Behind the aircraft, the dry lakebed stretches into the distance. The unique pogos are in position beneath the long wings, which are also protected during landings by the downturned skid visible at the wingtip.

unsettled, with thunderstorms threatening, but Johnson was determined to keep to his very tight schedule, so in mid-afternoon they wheeled her out again. At 3.55 p.m. the strange-looking aircraft with the shiny metal finish soared into the sky at planned operational gross weight (17,000 lb.) on eighty-five per cent r.p.m. 'It flies like a baby buggy!' exclaimed LeVier to Matye and Johnson, who were following in a T-33.

He soon changed his mind when he brought it in to land after an uneventful twenty minute check-out of the flight regime up to 150 knots and 8,000 feet. Some days previously, LeVier had debated the landing technique with Johnson, who insisted that the Angel be recovered tail up, with the main gear touching first, to avoid stalling the wing. The test pilot reckoned it should be brought in to a two-point landing — 'just like any taildragger', he said. He had heard too many bad stories about B-47 landings, when a dangerous porpoising effect could follow a nose-heavy landing, and lead to structural failure. Johnson prevailed, but on his first landing attempt LeVier ran into exactly the problem he had predicted. Despite partial flaps, fuselage speed brakes extended and almost idle power, the aircraft had the flattest of glide angles and when he attempted to grease the main gear at about ninety knots, the porpoising started.

LeVier applied power and went around, only to encounter the same problem again. And again. And again! By now, the thunderstorms were closing in and Johnson was getting very anxious. He even contemplated a belly-landing, but LeVier saved the day by trying his taildragger theory, holding the nose up into a flare and executing a perfect two-pointer just as the aircraft entered a mild stall. The pogos had been locked in position for this first flight, and even at sixty knots with the gear on the runway they were still airborne. The aircraft started to porpoise again, so LeVier activated the gust control to reduce lift and the Article finally rolled to a halt at 4.35 p.m. Ten minutes later, a downpour flooded the lake with two inches of water! This didn't dampen the spirits of the Skunk Works team, who spent the evening celebrating with beer-drinking and arm-wrestling contests.

The next day, LeVier took it up again to verify his landing technique. On 8 August the Article made its *official* first flight in front of Bissell and other government representatives. It was a spectacular sight, for the power-weight ratio was so great that the take-off roll was just a few hundred feet and the aircraft had to be pulled into a steep climb as soon as it left the ground lest it exceed the structural limits. After one pass over the spectators, LeVier went on climbing to 30,000 feet as planned, and Matye struggled unsuccessfully in the T-33 to catch up. The hour-long flight ended with another low pass and a reasonable landing. The top brass from Washington were very impressed!

Over the next three months, things went very well out at The Ranch. LeVier eventually got his lake bed runway markings. With Ernie Joiner running the flight tests, pilots Bob Matye and Ray Goudey made their first trips in the aircraft, and neither they nor LeVier came up with any significant criticisms. They did recommend a few cockpit changes, such as the provision of a sliding canvas sunshade inside the canopy. To help with landings, small fixed strips were added to the wing leading edge, just outboard of the root. These provided a little bit of stall warning as the aircraft floated over the threshold, and were specially tailored to individual aircraft, each of

A scene from early flight tests at The Ranch. While one Lockheed engineer leans inside to prepare the cockpit, another helps the heavily-clad test pilot ascend the ladder. He is wearing the early partial-pressure suit and helmet, with the attached oxygen bottle providing a temporary supply until he can be hooked up to the aircraft system. Another ground crew member carries two cushions to place in the pilot's seat well.

which exhibited slightly different stall characteristics. The flight envelope was extended to the maximum Mach 0.85 and to 50,000 feet, which meant that the pilots had to don the uncomfortable partial pressure suit to protect themselves against the consequences of a cockpit depressurization. LeVier made a successful first planned deadstick landing, before bowing out of the programme in October upon his promotion to Director of Flying at Burbank. He would never fly the Angel again.

All through the spring and summer of 1955, General Curtis LeMay had fought a rearguard action to have the new spyplane reassigned to the Air Force. The irascible SAC commander, a famous wartime bomber hero who had acquired the nickname 'Iron Pants' for his rigid discipline and control, was finally confronted by Dulles and Bissell at a tense meeting in Colorado Springs. After an unsuccessful appeal by the military to the President, CIA management of the U-2 project was confirmed by a formal agreement between Allen Dulles and General Twining. But SAC was given the responsibility of training all the pilots, and the agreement also defined the considerable support that the Air Force would provide to the project headquarters and proposed detachments. Blue-suiters would staff the operations, navigation, logistics and weather forecasting functions, and provide maintenance for all the support aircraft. Bissell's deputy in Washington would be an Air Force Colonel. As far as the military's Bald Eagle competition was concerned, the Martin RB-57D was allowed to proceed, but the X-16 project was curtailed.

Throughout the year, a dedicated team of Bell engineers led by Richard Smith had made steady progress with the X-16, and were ahead of schedule. First flight was due early in 1956. They knew nothing about the U-2 — they hadn't even been officially told what use the Air Force was going to make of the X-16, which had been numbered in the experimental category for extra security. Unofficially they knew, of course, and were confident of success. After all, they had beaten Fairchild to the contract by offering the extra safety margin of twin engines, together with superior range and altitude — they were expecting 3,300 nautical miles and a maximum 72,000 feet. In late summer 1955, things seemed to be going well for the X-16, and a production contract for twenty-two aircraft was issued by Wright Field. Just hours after the Bell management signed this contract one day in early October, an order to terminate the project reached the plant. It was quite a blow for a relatively small company. Dick Smith, Richard Passman, Mel Zisfein and the others on Bell's design team were pretty sore about it for many years thereafter.

1 Three possible camera systems were studied by Baker's team for carriage by the U-2. The 'Type A' designation was given to various existing cameras in the Air Force inventory, such as the KA-1 and K-38. Like the 'Type B', the 'Type C' was a new development, but it was never used. It had a 180-inch focal length lens and was designed for technical intelligence of the sort which required even greater resolution. Jim Baker's attempt to persuade Kelly Johnson to devote the necessary extra space in the U-2 for the 'Type C' camera was unsuccessful.

Chapter Two
Bissell's Air Force

All of the pilots engaged in this difficult enterprise were most carefully selected. They were highly trained, highly motivated, and as seemed right, well compensated financially. But no-one in his right mind would have accepted these risks for money alone.
Allen Dulles, Director of Central Intelligence, May 1960.

What kind of a deal was this? When the wing operations officer had called them on weekend leave in New York, insisting they present themselves back at Turner AFB, Albany, Georgia in the commander's office on Monday morning at 08.00 sharp, Marty Knutson and Carmen Vito figured they were for some kind of high jump. Neither had exactly been model young officers during their tour with the 31st Strategic Fighter Wing, but they had good flying records. You needed to be a good pilot to survive flying the F-84, a real dog of a fighter-bomber which had killed thirteen pilots at Turner alone that year. Now the pair were sitting in Colonel Horton's office, looking slightly the worse for wear after a long overnight drive preceded by the usual rabble-rousing and bar-cruising, and he was offering them a mysterious special assignment. It was top secret, and if they wanted to take up the offer, there would be a further interview with some unidentified civilians tomorrow.

Both were single, both were set to quit the military in favour of an airline career, but both were tempted by the offer. Knutson thought they were being recruited for some kind of spaceship enterprise. They expressed interest. Another Colonel from Air Force Headquarters was present — they would later know him as Leo Geary, who had replaced Ozzie Ritland as the officer responsible for arranging the Air Force part of the U-2 operation. He and Horton agreed to send them forward for the interview. Knutson and Vito became the first two recruits for Operation Overflight.

Nearly all the early U-2 pilots came from the two SAC bases at Turner and Bergstrom, Texas, where the command would soon be phasing out its Strategic Fighter Wings. There were the 31st and 508th wings at Turner and the 12th and 27th at Bergstrom. They were therefore all facing reassignment, so the sudden disappearance of a number of them would probably go unnoticed amid all the upheaval. Their experience with the F-84, with all its lousy handling and engine failure problems, would stand them in good stead for the Dragon Lady, where 'good stick-and-rudder men' would be needed. Therefore, SAC checked personnel records before turning the likely candidates over to the Agency for exhaustive interviews, medicals and background checks. It was all done in suitably cloak-and-dagger style, and the pilots from SAC were mightily impressed! The mysterious civilians from 'the government' hired downbeat hotel and motel rooms to conduct the initial interviews. The first group of pilots from Turner were interviewed in the Albany Hotel, an insalubrious venue which was virtually a bordello!

Later, they would all be assigned false surnames and ID cards, there would be trips to Washington, more briefings in hotel rooms, lie-detector tests, and so on. Then to Wright-Patterson AFB for the altitude chamber test, a ride on the centrifuge and an uncomfortable exposure to 'The Furnace', a hot room where they cooked you and then put your feet in ice! They called it stress testing. After this, a mysterious journey round middle America would end at Albuquerque and the Lovelace Clinic. Here the recruits would undergo the most rigorous medical ever devised, lasting a whole week and consisting of all sorts of indignities.

White-coated sadists would pump cold water down your ear canal, administer electric shock treatment, and stick pipes up every available orifice. The Project Mercury astronauts would later undergo the same treatment, but the U-2 pilots were the true pioneers at Lovelace. They felt like guinea pigs — which they were.

Never mind, reasoned the erstwhile spy pilots, at least the money was good. Fifteen hundred dollars per month while training, with an extra thousand for overseas duty. This was nearly four times what one of these young lieutenants or captains could earn in the Air Force, and although they had to formally resign from the service, they had a guaranteed return upon completion of the CIA contract. What is more, they could rejoin with no time lost toward military retirement benefits, or toward prospects of promotion. The biggest problem was that they couldn't brag about what they were doing. Wives, parents and girlfriends were all supposed to be told as little as possible.

There were the inevitable few washouts from the pilot screening process, but eventually some twenty-five expectant flyers made it to The Ranch in three groups for checkout in the strange aircraft. By the time the first group arrived, the 'Angel' or 'Article' had become the U-2, the second aircraft to be assigned the Air Force's U-for-Utility designation. This subterfuge condemned the Dragon Lady to permanent cohabitation in the aviation reference books with a motley bunch of Cessnas, Pipers and Beechcraft, but it suited the cover story that was being dreamed up by the wizards in Washington. They would soon bring the National Advisory Committee for Aeronautics (NACA — the predecessor to NASA) into the picture, so that the U-2 could pose as a weather research aircraft. For the moment, the cover story was that the aircraft was investigating certain flight handling data of interest to the F-104 programme.

By the end of 1955, five Lockheed test pilots had taken the aircraft into the upper atmosphere, and confirmed most of the expected performance. LeVier picked Robert Sieker and Robert Schumacher to join Ray Goudey and Bob Matye, and they now soared almost daily past the world record altitude of 65,890 feet established only a couple of months earlier in the UK by Wing Commander Walter Gibb flying a specially modified Canberra. There were no

Lockheed pilots Ray Goudey (left) and Bob Matye (right) soon joined the flight test programme at The Ranch. Together with Robert Sieker and Bob Schumacher, they proved the viability of the U-2 concept over the skies of Arizona and Nevada in the closing months of 1955. Matye was the first of many U-2 pilots to experience a flame-out at high altitude.

triumphant announcements from Groom Lake, though. On his third high-altitude flight, Bob Matye had a flame-out and thereby proved that the pressure suit, regulator and emergency oxygen system all worked. Pressure suit design was still in its infancy, however, and there would be many modifications to this essential gear as the U-2 programme progressed. Without it, a pilot flying above 63,000 feet who experienced complete loss of cabin pressurization (most likely through an engine flame-out) could be dead within a minute, as his body fluids would literally boil away.

Flameouts were a frequent occurrence in these early days, as Lockheed's pilots explored the flight envelope, and Pratt & Whitney engineers strove to perfect the high-altitude J57. The biggest problem was the fuel control. Since the engine could not be restarted in the thin upper air, a flamed-out U-2 would have to descend to 35,000 feet or lower in order to get a relight. The initial -37 version of the J57 also had a propensity for dumping engine oil into the cockpit via the ventilation system. A greasy black film would accumulate on the cockpit transparency, and things got so bad that the pilots were forced to take along a swab mounted on the end of a stick, in order to wipe the windshield clean! Clearly, the engine's reliability had to improve before operations over the Soviet Union began. If an aircraft had to descend to get a relight, it could easily fall prey to the MiG fighters.

Moreover, there was no guarantee that relight would occur, so on their way down to relight level, pilots would also be looking for an airstrip on which they could deadstick the aircraft *in extremis*. According to the book, the Dragon Lady could glide for 240 nautical miles from 70,000 feet, taking seventy-three minutes to do so. The first group of Agency pilots had a number of scares: one flamed-out high above the Grand Canyon and pointed the nose towards Nellis AFB while descending for a relight. He could see the base only eighty miles away . . . but it didn't seem to be getting any nearer. As usual, this pilot had encountered virtually no winds at the cruising altitude, but now he was descending into a strong jetstream, which was almost pushing him backwards. He now realised that those gliding figures from the book were for still air conditions only. Soon he was down to relight level at 35,000 feet. Here he had been advised to hold the airspeed steady at 160-180 knots, get the engine RPMs up to fifteen per cent, throttle to idle, and press ignition. Nothing happened. So he tried again. Still nothing, though he could smell the fuel. Nellis was getting further and further away. He descended through 20,000 feet, still trying to get a relight. Through 15,000 feet . . . at least the jetstream had eased off now, and the aircraft was making forward progress again. The rim of the canyon loomed closer, with the pilot now sizing up a grass airstrip at the bottom. He tried one last time to relight, the engine coughed, the exhaust gas temperature (EGT) started rising, and finally she fired! The pilot made it safely back to The Ranch.

On two other occasions in early 1956, a Dragon Lady flamed out and failed to make it back to base. During one famous incident, a pilot suffered repeated flameouts and relights, eventually gave up, and deadsticked it into Kirtland AFB at Albuquerque, New Mexico. The airborne drama was so protracted that word reached Colonel Geary at his desk in Washington, giving him time to organize a security clamp-down at Kirtland *before* the Dragon Lady glided in. They had the aircraft surrounded by armed guards and wheeled straight into a hangar before most people noticed. A second flameout resulted in a landing at Palm Springs airport, but the few onlookers that day figured it was just another glider. A C-124 and recovery crew was despatched to collect it within the hour. The incident did make the local paper, however.

Pratt & Whitney were working on the definitive, -31 high-altitude version of the J57 at the factory but it hadn't yet reached The Ranch. Besides offering increased reliability, the -31 would also provide a useful extra boost — it was rated at 11,200 lb. take-off thrust, a gain of 700 lb. over the -37 which translated to a little extra in payload or altitude capability. In any case, all the U-2 engines were soothed and coaxed and babied by Pratt & Whitney technical representatives Al Hannon and James Herbein to an extraordinary degree. Like the physiological support people responsible for the pressure suits, their one hundred per cent conscientious effort was vital to the overall success of the operation.

A reliable autopilot was another absolute essential for trouble-free high-altitude operation. The usual flight plan called for a rapid climb to

Another scene from the early flight test programme, as ground crew and flight test engineers crowd around the top-secret aircraft. Engine intake is covered by a wire mesh to prevent the ingestion of foreign objects during ground runs. The Q-bay lower hatch is removed, allowing one of the team to stand beneath the aircraft and reach up into this area, which appears to be empty.

60,000 feet, whereupon the aircraft would be established in a cruise-climb at a standard speed schedule for the rest of the trip. As fuel burned off, the aircraft would slowly climb over the next several hours to as high as 75,000 feet, depending on a number of variables such as aircraft take-off weight and outside air temperature. But as the U-2 passed beyond 60,000 feet, the precious margin between the aircraft's indicated stalling speed and its never-exceeded speed eroded quickly, so that by about 68,000 feet in an aircraft that had taken off fully-fuelled there could be as little as ten knots difference between the two.

In other words, the slowest the aircraft could go was approaching the fastest it could go! This was 'coffin corner', a condition that afflicted all high-altitude flyers, but in the U-2 with the engine going full tilt and that high-camber wing, the condition was particularly acute. Sometimes a pilot would turn and the inside wing would be in stall buffet while the other was in Mach buffet. Recovery from a stall could be difficult (not enough elevator response), and if the pilot exceeded Mach buffet by more than four knots, the aircraft was liable to come apart. Certainly, the aircraft gave the pilot a reasonable enough indication of the onset of buffet, but it wasn't particularly easy to judge which of the two types of buffet it was. The only thing to do was to make sure the trim was constantly adjusted and to keep an eagle eye on the airspeed indicator to make absolutely sure that the Mach and IAS needles didn't cross.[1] As the aircraft burnt more fuel and became lighter, the angle of attack decreased, thereby slightly increasing the slender margin of safety.

However, it was clearly asking too much to expect a pilot to manually ride this roller-coaster for hours on end, while also monitoring the driftsight, taking navigation fixes, operating photo and SIGINT systems,[2] and so on. So the idea was to fly the aircraft on an autopilot with Mach hold above 60,000 feet, keeping to a rigid cruise-climb schedule. Even then, the pilot could not relax. The airspeed schedule was not a constant Mach number, so the autopilot trim had to be frequently reset. Then there was the possibility of autopilot malfunction, or of flying through a sudden temperature change, which could cause the system to over-react and input too much pitch. Lockheed turned to the Lear company for the autopilots, but these didn't reach The Ranch until later in 1955. Numerous test flying hours were then spent trying to perfect the Mach hold, but Lockheed was satisfied with its performance by the time the first group of Agency pilots were ready to deploy. Nevertheless, there were to be many autopilot-related accidents and incidents over the ensuing years. Lower down, the aircraft really had few bugs at all, as Ray Goudey was fond of demonstrating. Returning to The Ranch from some particularly boring systems test in a U-2, he would swoop down and corral the desert cattle! Antics like this were fine as long as the pilot had remembered to balance the fuel in the wings; otherwise, a wing-heavy condition might develop with fuel sloshing down those long main tanks until the wing dropped and the aircraft spun into oblivion.

Early flights were kept within a 200-mile radius of The Ranch. As confidence in the new plane increased, the Lockheed pilots ventured further afield, flying triangular patterns up to 1,000 miles from the base. In December 1955, one of these sorties was staged during a visit to Lockheed by Defense Secretary Charles Wilson. Old 'Engine Charlie' — a former boss of General Motors — was still a bit sceptical about the U-2. So they wheeled him round the Skunk Works, flew him up to The Ranch, and installed him in the control tower with a direct r/t line to the high-flying pilot. He had already been aloft for more than eight hours, but told the Secretary he was prepared to stay up another hour and a half. Wilson was convinced, and shortly thereafter the full production decision was made. Lockheed soon had contracts for fifty U-2s; they were drawn up on the incentive sliding fee basis: the lower the total cost, the higher would be Lockheed's fee. Johnson returned some $2 million to the government on that first contract; he credited the hands-off approach adopted by the CIA for this. As a yardstick of progress, he and Bissell had agreed simple performance figures, rather than the detailed technical specifications which a contract with ARDC at Wright Field would have entailed. Later, there was another bonus for Uncle Sam — a further five U-2s were created out of spare fuselages and other parts largely paid for under the first contract.

Strangely enough for such a top-secret project, most of the airframes were not actually built within the secure confines of the Skunk Works' Building 82, or even elsewhere within the sprawling Burbank complex. Here, they were turning out P2V Neptune patrol planes and T-33 jet trainers as fast as the military could order them, and there simply wasn't the extra production capacity available. However, Lockheed ran a satellite plant ninety miles north of Burbank at Oildale, just outside Bakersfield. This plant had produced C-121 subassemblies, and there was surplus local labour available from the farming communities. So the Dragon Lady was actually put together in a humble, tin-roofed warehouse known simply as 'Unit 80', with a few experienced Lockheed people augmented by a bunch of cotton-pickers! The satellite operation was under way by January 1956, and there were over 400 people on the Oildale payroll by the end of 1956. Twelve months later, the job was done and they were all gone. Lockheed sold the property and it survives virtually unchanged to this day as a warehouse for Firestone tyres.

The whole U-2 came together at Oildale, was checked for fit and systems operation, then disassembled and loaded onto a flat bed truck. Securely covered in tarpaulin, it made a short journey down to the local airport for loading into a C-124 transport for the flight to Groom Lake. Many of the Oildale workers had little idea where this was, or what the plane's true purpose was.

Neither, for that matter, did some of the Air Force transport crews who were called upon to haul these mysterious cargos to and from The Ranch. They were kept as much in the dark as possible. The place was so secret, they would be told to get airborne at night and head for a certain area on the California-Nevada border before contacting a certain UHF frequency — 'Sage Control'. A voice would reply out of the ether advising them not to acknowledge further instructions, then give them descent clearance into an uncharted area with no airfields shown. They obviously had radar down there, but the C-124 crews began getting a little edgy when a final approach controller came on air ('Delta') and told them to lower gear and flaps. As they ploughed on through the blackness getting lower and lower, Delta would eventually flick the runway lights on and clear them to land. Met by a group of seemingly anonymous civilians, they would be loaded or unloaded and on their way within an hour or two, with strict instructions to say nothing about the whole experience.

The Ranch was actually operated by an ad-hoc training group set up by SAC and commanded by Colonel Bill Yancey. His group included four experienced instructor pilots (IPs) to train those selected by the CIA to fly the U-2 operationally. This was the theory; in practice, the IPs had minimal experience of flying the U-2 themselves and, of course, they couldn't climb aboard the single-seat U-2 to accompany a new pilot on his first flight. Moreover, since development flights by the Lockheed test pilots continued in parallel with the training programme, new aspects of the U-2 flight regime became apparent nearly every day. It was trial and error all the way.

The Ranch was no kind of place to spend a vacation. As the training contingent from SAC moved in, the pressure on accommodation grew. There were a couple of barrack blocks, but the majority lived in trailers. According to almost universal verdict, the food was excellent. The mess hall doubled as a pool room and movie theatre. Alcohol was freely available, and with not much else to do and nowhere to go, a great deal of it was consumed. Some unofficial amusements were organized; the first Agency pilot class scrounged gunpowder and woodshavings to make mini-rockets out of cigar tubes. They went off with a tremendous swoosh, but one day they scored an unintentional near-miss on a C-131 in the pattern, which brought the scheme to an abrupt halt.

There was one road off the site, leading towards the AEC area, but nearly everyone commuted to The Ranch by air. A daily shuttle service was started by the Air Force, using a C-54 transport. SAC decided it was better from a security point of view to let everyone disperse for their leisure time in the Los Angeles area, rather than have them descend on the Las Vegas bars and casinos en masse.

At 07.00 on the morning of Wednesday 17 November 1955, the regular C-54 left Burbank for The Ranch with ten people from Lockheed and the Agency on board, plus five crew. In poor weather, it smashed into the top of Mount Charleston near Las Vegas, killing them all. If it had been only thirty feet higher, the lumbering transport would have cleared the 11,000 foot peak. It took three days for rescue parties to reach the wreckage; they were accompanied by an Air Force colonel who removed briefcases and top-secret U-2 documents from the charred bodies of the victims. In an official statement, they were described as civilian technicians and consultants, and the Air Force did not object when outsiders jumped to the conclusion that the deceased were all scientists working for the Atomic Energy Commission, probably engaged on some top-secret nuclear weapons project out on the Nevada test range. A larger number of Skunk Works people would have caught the flight had they not had hangovers from a party organized by the Flight Test Department the previous night. Following this tragic accident, Kelly Johnson insisted on using the company DC-3 to transport his own people to and from The Ranch, but the Air Force shuttle was later resumed.

By early April 1956, just eight months after its first flight, the U-2 was ready for deployment. The need to know what was happening behind the Iron Curtain seemed greater than ever. The Soviets gave tantalizing glimpses of military progress each year at the May Day parade in Moscow, and at an annual air display held above the capital's Tushino airfield in July. Two new and obviously long-range jet bombers had been revealed on May Day 1954 — the previously unheard-of Types 37 and 39. By the time of the 1955 Tushino display, these were known in the West as 'Bison' and 'Badger', but the intelligence people figured they were still a year or so away from series production. Western military attachés invited to the show were therefore astonished to witness a flypast of no fewer than twenty-eight Bisons in three groups, first ten, then nine, and later another nine. There were also a number of medium-range Badgers, and seven examples of the third bomber then under development — the Tupolev Tu-95 'Bear'. Somehow, the Soviets had seemingly accelerated development of a long-range strike force capable of delivering nuclear warheads on American soil. The display also revealed large numbers of the new Soviet twin-engined long-range all-weather fighter and an obviously supersonic single-seat fighter (later confirmed as the Yak-25 'Flashlight' and MiG-19 'Farmer' respectively).

If this was what the Soviets were prepared to reveal, what else remained hidden away? The dreaded missiles, for sure. Air Force Intelligence now had a huge radar installed near Samsun on Turkey's Black Sea coast, and a listening station post at Karamursel near Istanbul in the same country. These were picking up missile tests from Kapustin Yar which were increasingly frequent now. Encouraged by former Air Force Secretary Stuart Symington and fellow cold-war warriors, the legislators on Capitol Hill were becoming alarmed. They called in the top military brass to testify. Twining, LeMay and other generals told them that the Soviet aircraft 'were approaching us in quality and surpassing us in quantity'. Before long, there was talk of a 'Bomber Gap' between the superpowers, to be followed by a 'Missile Gap'. As a result of the ensuing consternation, the Air Force gained an extra $928 million in funds, largely to expand the B-52 force and accelerate missile development.

In an attempt to clear away the mutual suspicion and maybe thereby halt the accelerating arms race, Eisenhower's advisors had come up with an 'Open Skies' proposal the previous summer. As a first step toward potential arms control agreements, why not let both sides overfly each other's territory and take as many pictures as they liked? Eisenhower had proposed the idea at a summit meeting with Soviet leaders Bulganin and Khrushchev in Geneva during July. It was rejected. Then came the balloons.

For five years, the Air Force had been secretly working on a system to carry cameras aloft into the jetstreams which periodically ran right across the Soviet Union from west to east. A balloon fabricated with the new polyethylene film could be capable of flight at constant pressure altitude. A 600 lb. gondola could be suspended beneath it carrying an automatic ballasting system, automatic cameras and a parachute system, which would deploy after a number of days and set off a transponder to aid in recovery of the reconnaissance payload. From the outset, the ARDC team at Wright Field had been sceptical, especially about temperature and pressure control, and the recovery system, but pressure from Air Force Intelligence saved the project and advanced it into flight tests across the US. In March 1955 it was turned over to SAC and renamed Project Genetrix, or Weapon-System 119L.

The balloon effort was later portrayed as a tentative, covert affair which wasn't very well organised. In fact, it was a regular, multi-million dollar military operation which SAC took very seriously indeed — they even created an entire new Air Division to run it. There were five launch sites in Western Europe ranging from Norway to Turkey, ten tracking sites, and six recovery bases in the Far East. They tried five different camera models, each producing nine by nine-inch negatives from paired six-inch lenses, each set at an angle of thirty-four degrees from the vertical. Up to forty balloons per day would be launched. There was a squadron of C-119 transports to haul all the equipment around and to 'snatch' the gondola and parachute as it floated down. A convenient cover story was constructed around the simultaneous launch of meteorological balloons codenamed 'Moby Dick' from Okinawa, Hawaii and Alaska. But the Soviets weren't to be fooled, and less than a month after the first launches on 10 January 1956, a diplomatic protest note reached Washington, and an elaborate display of some fifty of the balloons plus equipment was put on in Moscow. Either the Soviets had shot them down using fighters, (they could be tracked on radar, and some floated along at less than 40,000 feet) or they had simply floated down ahead of schedule. Only some forty gondolas were recovered out of over 500 balloons eventually launched. The CIA had been opposed to the project all along, and Allen Dulles advised Eisenhower to scrap it forthwith. By early March SAC had wound it up, but insisted that worthwhile coverage had been obtained. There were, after all, 13,800 exposures which covered about eight per cent of the Soviet landmass, but the Generals had to admit that resolution was poor and detailed interpretation impossible.

The way was clear for a U-2 deployment, and the British agreed to host the first detachment. The Agency had built up a close working relationship with its British counterpart, MI6. This was the Anglo-US 'special relationship' in action, and it extended way beyond published agreements such as the NATO treaty, into the realm of intelligence-sharing and (from July 1958) nuclear weapons collaboration. There had been a secret 'UK/USA' agreement since 1947 to share signals intelligence facilities and analysis, and British expertise in the ELINT field was particularly admired by US military intelligence. There had been some joint MI6/CIA covert operations in eastern Europe; and the shared operation of US Air Force reconnaissance aircraft operating from British bases. These were sometimes flown by all-British Royal Air Force crews. So the British seemed a natural first partner in the great new project. The existing US Air Force base at Lakenheath in Suffolk was chosen by Bissell, and in mid-April two aircraft were duly delivered by C-124 transport and re-assembled in a remote hangar on the far side of the base. The 'First Weather Reconnaissance Squadron (Provisional)' was about to go into action.

This designation had been devised as part of an elaborate cover story which required the reluctant help of NACA boss Dr. Hugh Dryden. On 7 May a press release was issued in his name, announcing that the first U-2 aircraft were flying from Watertown Strip in Southern Nevada in a new research programme. Capable of reaching 55,000 feet, the aircraft would gather research data about the upper atmosphere to help plan future jet airliners which would routinely be flying at the higher levels. Jetstreams, clear air turbulence, cosmic ray particles, the ozone layer, convective clouds — all would be studied by the new type, a few of which had been 'made available' to NACA by the US Air Force. More detailed releases followed, giving full details of the instrumentation being carried aloft, and announcing an extension of the programme to the UK and 'other parts of the world'. Most of this was true — the U-2s did carry NACA research packages on some 200 training flights over the next several years. A full-time U-2 project officer had been assigned by NACA, and one of their technicians did once visit the deployment bases in Japan and Turkey, but the research agency had no control over where the flights were routed and received all of its data second-hand.

The WRSP-1 designation was strictly for public consumption. To the initiates, the unit was known simply as 'Detachment A'.[3] Like the other two units that would soon be formed, Det A was a strange hybrid composed of CIA employees, serving Air Force officers, and contracted civilians. An Air Force Colonel would serve as commanding officer, with one of the Agency people as his executive officer. There were mission planners (for the most part, flight-rated navigators) and a surgeon from the Air Force, while the communications, intelligence and security types were provided by the

The elaborate cover story which was supposed to disguise the true nature of the U-2 programme relied on the co-operation of NACA, The National Advisory Committee for Aeronautics, which became NASA in 1958. NACA issued an official statement in May 1956 identifying the U-2 as purely a weather research aircraft. But it wasn't until February 1957 that this first official picture of the aircraft, suitably adorned with NACA markings, was released to the public.

Agency. (The security officer on the very first deployment was one Edmund P. Wilson, who would later achieve notoriety as an illicit arms dealer to Libya and end up serving a fifty-year gaol sentence for conspiracy to murder.)

Then there were the pilots, who were 'retired' Air Force officers who had been 'sheep-dipped' (as the phrase went) to disguise their suspended military status. The CIA paid their salaries, but on their ID card they were described as civilian employees working for Lockheed and under contract to NACA. There were also, of course, genuine contractor personnel attached to each unit for maintenance purposes. Some were from the engine, camera and SIGINT system manufacturers, but the majority were from Lockheed, which was responsible for the physiological support division (PSD — which provided the pilots' pressure suits) as well as aircraft maintenance.

The stay at Lakenheath was short-lived. In April, Bulganin and Khrushchev paid a goodwill visit to London, arriving at Portsmouth on a Soviet Navy cruiser. In direct contravention of government orders, British Intelligence despatched a frogman to discover details of the ship's construction which were hidden below the water line. He was never seen alive again; a torso washed up on a nearby shoreline over a year later was tentatively identified as the missing frogman, Lionel 'Buster' Crabbe. The Soviets soon publicized the incident. An embarrassed Prime Minister Anthony Eden fired the head of MI6, Sir John Sinclair, and told Parliament that the secret service had acted 'without the authority or knowledge of ministers'. Fearing a repetition of the affair, the PM withdrew permission for the U-2s to operate from Britain for the time being. The boys of 'Det A' weren't entirely displeased to be moving on. It was a cold English spring and some of them had been billeted off-base in a local pub (the 'Bird In The Hand' at Mildenhall), where they had a hard time adjusting to rigid mealtimes, freezing cold bedrooms, and rationed bathtimes.

They went off to the US air base at Wiesbaden near Frankfurt in West Germany. Another two aircraft arrived from the US and there were more test flights — the engine idle had to be adjusted to suit European conditions. The first overflight was assigned to twenty-six year-old Carl Overstreet, who wasn't even there at the time but on a short leave trip back home. The other five elected Hervey Stockman to tell him when he returned. In the White House, Eisenhower provisionally agreed the first package of overflights, but insisted on being consulted before a final go-ahead. The weather over the western Soviet Union remained cloudy, but it was clearer over the satellite countries. Pilot and aircraft were therefore prepared for a flight over East Germany and Poland, and on 20 June Overstreet soared into the morning sky praying

Nikita Khrushchev, the mercurial son of a coal miner who rose to become First Secretary of the Communist Party and supreme Soviet leader throughout the four-year period of the U-2 overflights. 'As far as we were concerned', he reportedly said, 'this sort of espionage was war — war waged by other means!' Khrushchev's intemperate language and frequent boasts of Soviet military prowess set much of the tone for the cold war years of the late fifties.

that he wouldn't screw up and that the weather would be fine. He flew as far as Warsaw before turning back to overfly Berlin and Potsdam; anti-government riots had broken out in East Germany four days earlier, and Soviet tanks were busily engaged below him in restoring order. It was a successful flight.

Another pause ensued, largely because a US Air Force delegation led by General Twining was in Moscow during the final week of June, having been invited to attend this year's set-piece display at Tushino. It didn't seem prudent to send a U-2 over Soviet airspace while the delegation was there. Twining's team were treated to another massed fighter flypast, comprising over one hundred 'Farmers' and 'Flashlights', but the bomber flyby was muted, with only four 'Bears', three 'Bisons' and nine 'Badgers' in evidence. The best was saved for last — a flypast of new fighter prototypes, including a new version of 'Flashlight', three Sukhoi deltas and two new MiGs — forerunners of the MiG-21.

On Monday 2 July the weather was clearing over European Russia, and Bissell sent a request to begin Soviet overflights to Eisenhower's personal assistant, General Goodpaster. Permission to go was granted, with a ten-day limit set, after which the Agency had to submit a full report. At project headquarters on the fifth floor of the Agency's leased building at 1717 H Street, the mission planners refined the overflight routes. At 18.00, Bissell chaired a major review. Everything looked good. Just before midnight, the go-signal was transmitted over secure communication lines to Wiesbaden, which was six hours ahead of Washington time. It was therefore 06.00 on US Independence Day, the fourth of July, when Hervey Stockman took off for the first penetration of Soviet territory. He flew over East Berlin again, across northern Poland via Poznan and into Belorusskaya as far as Minsk. Then came a left turn to roll out heading north to Leningrad. And now the Soviets began to intervene. Glancing down through the driftsight, Stockman saw MiG fighters rising to try and intercept. Contrary to American hopes and expectations, they had managed to detect the flight! Stockman flew steadily on, putting his faith in Kelly Johnson and all the other experts who had assured the U-2 pilots that they couldn't be reached at altitude by the Soviet fighters. Reaching Leningrad, he turned back down the Baltic coast of Estonia, Latvia and Lithuania and eventually made a safe landing back at base. He had been airborne for eight hours and forty-five minutes. Some time after the debrief, an ELINT technician who had analysed the tapes told Stockman that a radar lock-on had been recorded on his plane.

The next flight was planned to go all the way to Moscow. When asked to justify such a daring venture, the mission planners in project headquarters told Bissell: 'Let's go for the big one straight away. We're safer the first time than we'll ever be again.' Take-off was scheduled for 05.00 the next morning, 5 July, and according to the procedures that had been established, Washington was supposed to send out a single codeword twenty-four, twelve and two hours before take-off to confirm the weather was good and the mission was still on. Otherwise, everyone at the detachment could relax. This they duly proceeded to do when no signal came through at 17.00 on the Wednesday evening, and all the pilots except Carmen Vito went off to join the various Independence Day parties going on around base. Then the go-signal arrived — two hours late — so Vito was chosen for the mission by default and packed off to bed for some rest.

With all the noise of celebrations going on outside, the pilot didn't get much sleep. Soon it was time to get up and take the standard high-protein, low-residue meal. (The lavatory arrangements of the partial pressure suit were rudimentary, consisting of a bottle for liquid wastes which could be connected up in flight after much tugging at zips and hoses — a ten-minute process for most pilots. Many had no need for the bottle even on long flights: the rubberized, airtight suit caused constant skin perspiration, and much body moisture was eliminated this way rather than through the kidneys. There was no provision for solid waste — a messy, uncomfortable and embarrassing business if the pilot couldn't restrain himself.) Then it was time to put on the long underwear, seams to the outside so that the tight, uncomfortable suit wouldn't impress them indelibly onto the occupant. Now two technicians from the Physiological Support Division (PSD) helped him into the suit itself for the pre-breathing session. In the early days, this was a two-hour process to purge the body of nitrogen, so that the pilot wouldn't experience the bends if he had to make an emergency descent. By the early sixties, it was realised that individuals vary in the rate at which they eliminate nitrogen, and that one hour of pre-breathing was usually sufficient for most pilots. This breathing under pressurization was no easy matter, since the normal process of inhaling with effort and exhaling without thought was reversed. The suit was so restrictive of movement that a fellow pilot would be detailed to perform certain preflight checks at the waiting aircraft, and to help the mission pilot strap himself in and connect all the hoses, wires and so on.

After final briefings from the navigator, weather forecaster and detachment commander, Vito and his accompanying entourage emerged into the early morning light and moved towards the waiting silver aircraft — Article 347 again. He climbed aboard and was strapped in. Engine start and systems check was normal. From the end of the runway, the pilot glanced towards the tower, awaiting the final clearance to go, which in the conditions of strict radio silence being observed, would take the form of a light signal. At 5 a.m. precisely, the light shone, and Vito was away.[4]

His route took him further south than Stockman had flown, over Kracow in Poland, then due east to Kiev, before turning north towards Minsk. There was considerable cloud cover, but it cleared away as Vito turned towards Moscow, his preplanned course virtually following the railroad from Minsk to the Soviet capital. Down below, the Soviet air defence system was going berserk. They had him tracked all right, and the MiGs were soon on his trail. All other traffic was grounded as the Soviets vainly tried to intercept the intruder. A couple of the fighters apparently crashed during the process. James Killian happened to be inspecting a US listening post run by the super-secret National Security Agency (NSA) in West Germany at the time. He watched as the baffled operators tuned in on the Soviet panic; the Soviets apparently thought they were being overflown by a plane way above 60,000 feet, but how could this be? Despite the technicians' high-level security clearances, Killian didn't enlighten them on the U-2 project.

Vito flew on, across the rolling farmland

U-2 photograph of a snow-covered Soviet air force base identified by the CIA's photo-interpreters as Engels Airfield. Thirty-two large bombers (apparently Myasishchev M-4 Bisons) line a taxiway, picked out perfectly by sharp early-morning sunlight. Another thirty aircraft are visible in this picture, but the overflying spyplanes found few other bases to match this one. The U-2 therefore helped to prove the notion that the USSR was opening up a 'bomber gap' to be ill-founded.

seemingly divided into a mosaic pattern by the stone walls. Nearing Moscow, a new danger awaited. He could make out the herringbone pattern of an early Soviet surface-to-air missile site, housing SA-1 rockets. He counted three such sites as he turned over the capital, but apparently no missiles were fired that day. Heading back home via the port cities along the Baltic coast, Vito met the cloud cover again, and it was overcast all the way back to Wiesbaden, where the weather was poor and he had to let down with a GCA. Underlining the U-2's obvious detectability, he was picked up while still way over East Germany and at altitude by the radars operated by Canadians as part of their First Air Division contribution to NATO. But the aircraft — Article 347, the same one that Stockman had flown the previous day — had performed flawlessly again, and so had the camera.[5] Another mission was successfully flown across the satellite countries and European Russia on the following Monday.

On 10 July the Soviet Ambassador in Washington delivered a formal, public protest against the flights 'from West Germany by a US Air Force twin-engine plane'. Of course, the US denied it, disingenuously claiming that no *military* plane had violated Soviet airspace. In the White House, however, the President was concerned about the provocation the flights were clearly causing, and he determined to exercise far greater control over U-2 flights in future; there would be no more ten-day, do-as-you-please permissions. From this day on, every single overflight was to be approved by Eisenhower himself.

Even so, the President cannot have failed to be impressed with the initial results from U-2 overflights. After each sortie, film from the 'B' camera was quickly unloaded, developed and duplicated. One set of duplicated film was rushed by special aircraft to Washington, where it was processed and interpreted in a new set-up organised by Art Lundahl of the CIA's Photographic Intelligence Division. (The other set of film was retained at the detachment base, in case the first should be lost or damaged in transit.) Some way across town from CIA headquarters, at Fifth and K Street in one of Washington's sleaziest and most rundown neighbourhoods, Lundahl had leased the upper floors of an auto repair shop and moved in the equipment and people needed to process the huge quantity of material brought back by each flight — enough to cover all four lanes of the entire Baltimore to Washington highway, as Ludahl pointed out to his boss, Richard Amory. This secret location was code-named 'Automat', and would grow to employ over a thousand people.

Lundahl's team prepared prints of the Kremlin and the Winter Palace at Leningrad taken on the first series of overflights, and Allen Dulles brought them to the President. Thanks to the combined efforts of Automat, camera-makers Hycon and film-makers Eastman-Kodak, they were spectacular. The intelligence analysts meanwhile scanned the huge 'take' and began to take serious issue with the Air Force's gloomy predictions about Soviet bomber strengths. The first few flights had overflown most of the known bomber bases, as well as the production factory near Moscow, yet had found no evidence of a massive Soviet build-up. This flatly contradicted the latest National Intelligence Estimate (NIE), just issued, which suggested sixty-five long-range 'Bisons' and 'Bears' in service now, and predicted 470 for mid-1958 and 800 by mid-1960. Taken together with economic analysis of the Soviet industrial base, which concluded that their productive capacity was quite limited, and new doubts about the Bison's powerplant efficiency, the CIA concluded that the 1955 flypast of twenty-eight Bisons had been a ruse, and that the small formation had circled round out of sight for a second fly-by to fool the observers. The NIE was revised, although the Air Force continued to take the gloomy view.

The most disturbing aspect of the first overflights was the fact that they had been tracked by Communist Bloc early warning and height-finding radars. Moreover, the Soviets were now fielding some large and powerful acquisition radars associated with the surface-to-air missile sites which had been detected around Moscow. Soviet target tracking and missile guidance techniques were apparently still in their infancy, however, so this fixed-site SA-1 system was not yet considered a serious threat, any more than the current Soviet fighters were. Nevertheless, the flights had been found out, and protests issued. As Nikita Khruschev subsequently recalled in his memoirs, mysteriously smuggled to the West and published in 1970, 'From such (American) behaviour, we drew one conclusion — *Improve rockets! Improve fighter planes!*'

Back at the Skunk Works, Kelly Johnson

When Kelly Johnson heard that the early U-2 overflights had been tracked on radar by the USSR, he tried to make the aircraft invisible to radar by adding some very early 'stealth' technology. In this photograph, radar-absorbent material known as 'Echosorb' is being added to the entire lower half of the aircraft. Just visible beneath the coating is the grid pattern of a Salisbury Screen, another radar-foiling device.

went to work on the detection problem, by trying to reduce the radar cross-section of the U-2. Lockheed soon came up with two different schemes. One involved stringing wires of various dipole lengths from the tail to the wing, from the wing to the nose, and elsewhere. This was an attempt to scatter radar energy striking the aircraft in as many directions as possible, in the hope of reducing the amount returned to the transmitter/receiver location. The pilots who test-flew this modification hated it; the U-2 flew badly, and the wires whistled in the wind, sometimes breaking loose and flapping against the cockpit or fuselage.

The other modification comprised the addition to the aircraft's lower surfaces of a metallic grid known as a Salisbury Screen, which was then covered in 'Echosorb', a microwave absorbent coating based on black foam rubber. The grid was designed to deflect the incoming radar energy into the absorber. Lockheed test pilots spent hour upon hour flying the modified aircraft up and down the range as radars were trained upon them to measure the effectiveness of the devices. The aircraft were nicknamed 'The Dirty Birds' by the Skunk Works.

The Dirty Bird project was not a success. There was the added weight and drag, of course, when all of Lockheed's previous efforts had been directed towards their elimination. Furthermore, in the case of Echosorb, the slightest leak or spillage of fuel or hydraulic fluid would corrupt the surface. Worse still, the coating prevented the dissipation of heat from the engine through the airframe skin. Under these conditions, the hydraulic pump was prone to fail. In any case, it turned out that the radars weren't to be fooled. At some frequencies, the Dirty Bird modifications appeared to work, but at others, the aircraft's radar signature was actually enhanced rather than suppressed. This was no good — Soviet air defence radars operated on a range of different frequencies.

Before long, Kelly Johnson concluded that the only way to make an aircraft 'stealthy' was to design the required characteristics in on the drawing board. This he proceeded to do with the A-12 and SR-71 series, the supersonic successors

Although Soviet interceptors couldn't reach the U-2 at cruising altitude, a flame-out could force the aircraft down to around 40,000 feet for a relight. As an added precaution to avoid visual detection, the CIA's aircraft were painted a matt midnight blue to match the colour of the surrounding sky. This U-2A is returning to one of the operational detachments from a training flight some time in the late fifties. In black and white photographs such as this, the aircraft appears to be painted black.

to the Dragon Lady. As far as the U-2 was concerned, the best that could be done was to camouflage it with an all-over coat of midnight blue paint (which appeared more like matt black under certain light conditions, or in photographs). At the U-2's cruising altitudes, the sky all around was this same midnight blue shade; if a fighter pilot did manage to get up there, he might spot an unpainted U-2 against this background, especially if a glint of sunlight were to be reflected from the shiny metal finish.

During a Dirty Bird test flight with Article 341 — the Dragon Lady prototype — on 4 April 1957, airframe heat built up and Bob Sieker suffered a flame-out at about 72,000 feet. As his pressure suit inflated, the clasp which secured the bottom of his faceplate failed to hold against the pressure and became unlocked. Lockheed had their first fatality in nearly two years of U-2 flight testing. Without oxygen, Sieker would only have had about ten seconds to react to the failure before losing consciousness, and there was virtually no chance of being able to resecure the faceplate against the seventy pounds pressure coming out of the suit. The aircraft entered a flat spin and descended, but Sieker apparently revived and attempted a bale-out as the aircraft neared the ground.

The wreckage of Sieker's plane was eventually located about ninety miles from The Ranch in a long valley near Pioche, Nevada. A maximum security cordon was thrown around the modified aircraft. After three days of intensive air search by the military, the crash site was finally discovered when one of Sieker's flight test colleagues, Herman 'Fish' Salmon, (who wasn't actually part of the U-2 programme) hired a twin-engine Cessna from Las Vegas and scoured a new area. As in many subsequent Dragon Lady accidents, most of the aircraft landed in one piece, rather than breaking up at altitude. Sieker's faceplate was found loose in the cockpit. The pilot's body was about fifty feet away, which suggested that he had departed the aircraft just before impact. His parachute, however, had not had time to deploy. Since Sieker was known to take food up on long U-2 test flights, word got around the programme that he had caused his own death by loosening the faceplate to eat (there was a faceport installed in the helmet through which one could poke a straw and thus take liquids). His colleagues recall, however, that his candy bar was found unopened in the leg pocket of his flight suit, and that the faceplate clasp was made to fail again at pressure during subsequent ground tests. Johnson designed a lock to keep the faceplate clasp from slipping, and re-evaluated the decision not to provide U-2 pilots with an ejection seat. In this case, it might have saved his test pilot's life.

1 The ASI was set up so that the maximum allowable airspeed pointer would move to indicate the danger speed — 240 knots or Mach 0.8, whichever was less.

2 Following the US convention, SIGINT is used in these pages as a general term covering all kinds of electronic reconnaissance. SIGINT equipment is colloquially known as ferret gear. Where more specific references are made, these employ the following nomenclature: TELINT for telemetry intelligence; COMINT for communications intelligence; and ELINT for electronic intelligence, such as that concerning radar.

3 Similarly, it was to be followed by WRSP-2 (actually 'Detachment B') in Turkey and WRSP-3 (actually 'Detachment C') in Japan. The flying base back home (initially at Groom Lake, later at Edwards AFB) was officially designated WRSP-4. In this account, the author has preferred to use 'Det A' etc, since this was the style adopted by those within the programme. It should additionally be noted that the detachment at Incirlik became known locally as 'Detachment 10-10, TUSLOG (Turkey-US Logistics Group)'.

4 Due to the time difference, it was still the dying hours of 4 July in Washington, which later led some participants mistakenly to recall that this first deep penetration of the Soviet Union had taken place on the US Independence Day.

5 U-2 aircraft were identified by a dual system, depending on whether they were being operated by the CIA or the Air Force. The CIA used the Lockheed construction number, actually referred to as the 'Article Number', and this series began with the prototype numbered 341. When Air Force serials were issued in late 1955 (Fiscal Year 1956 allocation: 56-6675 through 56-6722), these ran parallel to the Article numbers but starting at number 342. A handy method of deducing the last three digits of the serial number from the article number was to add 333. Article 347, the first aircraft to overfly the Soviet Union, was therefore also Air Force serial number 56-6680. This aircraft is now displayed with the latter identity in the Smithsonian Museum, Washington DC.

Chapter Three
Mayday

'I'm very proud of my participation in Operation Overflight. While I might wish that many of the things that followed had never happened, I have the satisfaction of knowing that I served my country — and, I believe, well.'
Francis Gary Powers, 1970.

As he flew steadily westward over the Mediterranean, Frank Powers[1] periodically leant forward to peer through the long rubber sighting cone which protruded over a foot from the top of the instrument panel. This hood helped to shade the six-inch viewing scope through which a U-2 pilot was provided with a pretty good view of the terrain below. Using a hand control to his right on the console sill, he could rotate the scanning head of the driftsight through a full 360 degrees in azimuth, and elevate it to an almost horizontal position, thereby gaining complete visual coverage beneath the aircraft. A switch next to the hand control provided a four times magnification, if required. This driftsight was a primary navigation tool, whereby the unfolding terrain could be checked against the flight plan maps, to make sure the plane was on course and that cameras and other reconnaissance systems were turned on and off at the required times. It also provided the pilot with a valuable lookout for the opposition, in the form of contrailing fighters or the tell-tale smoke puffs of a missile launch.

Today, though, Powers was more concerned with ships. He had been instructed to look for concentrations of two or more vessels, whereupon he was to operate the 'B' camera (via a master switch right next to the driftsight control). The Agency weren't searching out the Soviet navy today, however. The target was the British, French and Israeli navies, for this was late September 1956 and Uncle Sam wanted to know more about the military preparations its two NATO allies were making in response to Egypt's annexation of the Suez Canal, and whether they were colluding in a possible Israeli attack across the Sinai. Having reached the island of Malta, Powers reversed course and returned east to the airbase at Incirlik, near the town of Adana in southern Turkey.

The second group of former Air Force pilots to check out on the U-2 had gone through The Ranch between late May and early August. By mid-September, seven pilots plus planes had

Francis Gary Powers, the pilot who earned lasting fame for himself and the U-2 when he was shot down over the USSR on 1 May 1960. Although known to all the sundry in the U-2 programme as Frank Powers, his first wife Barbara insisted on the use of his formal names, by which he therefore became known to the world after the shootdown. He was one of the most experienced flyers in Operation Overflight, having served at the important Det B in Turkey throughout.

been installed at Incirlik as Detachment B. The US Air Force already had reconnaissance aircraft based there, since it was relatively close to Soviet military test facilities on the Black Sea coast. It was also the nearest they could get to Kapustin Yar, although this was still some 950 miles to the northeast. Bissell sent his colleague Jim Cunningham to negotiate Turkish permission to base the U-2s at Incirlik.

Rather than heading north across the Soviet border, as they had expected, Powers and the other U-2 pilots found themselves pitched into the Suez Crisis. Following that first flight across the Mediterranean, 'Det B' was in action nearly every day for the next two months, soaring high over Cyprus, Egypt, Syria and Israel. Before the fighting started, U-2 photo intelligence had determined that France was secretly arming Israel with many more fighter aircraft and other equipment than it cared to admit, and had revealed the exact composition of the British and French invasion fleets being assembled in Toulon, Marseilles, Malta and Cyprus. The all-seeing cameras were overhead the Sinai as Israel pressed home its attack on the Mitla Pass during 30 October. The next day, they were overhead Cairo West airbase to capture on film the RAF attack which left much of the place a smoking ruin. As soon as they were processed, the CIA is said to have transmitted the key pictures to their friends in MI6 and RAF intelligence in London, prompting a return cable which read 'Warm thanks for pix: quickest bomb damage assessment we've ever had'. Over this period, the U-2 pilots in Turkey reckoned they provided US intelligence with a better overview of the war than was available to many of the battlefront commanders. They were probably right.

While the big 'B' camera performed over Suez, other U-2s of 'Det B' were outfitted for an even more esoteric black art, that of electronic reconnaissance along the Soviet border. Despite the relatively small payload, a surprising amount of electronic equipment found its way onto the U-2 over the years. Although much of the U-2's avionics was supplied off-the-shelf from Air Force stocks, the Agency made its own contracts with the suppliers of the SIGINT equipment. Much of it was supplied by Ramo-Woolridge and was designed by a brilliant scientist, Dr Albert 'Bud' Wheelon. Ramo-Woolridge later became TRW; Wheelon later became the CIA's Deputy Director for Science and Technology. The SIGINT equipment for the U-2 was designated in a simple numerical sequence, unlike the complex alphanumeric designations favoured by the military. Thus, the very first assembly of recorders and antennae was designated System 1. In the mid-eighties, the sequence was still in use on the modern TR-1, and had reached System 29. In the intervening years, a huge variety of electronic warfare equipment was added, removed or substituted on the U-2, including state-of-the-art SIGINT gear, passive radar warners and active jammers. Some of this equipment found its first airborne application on the U-2, and was subsequently adopted by the military for use on Air Force and Navy aircraft.

On most missions, including all the overflights, a fair amount of ferret gear could be taken along without removing the big camera. Receivers, amplifiers and recorders were squeezed into various nooks and crannies: in the nose, on the Q-bay hatch; beneath the cockpit side consoles; in small compartments aft of the engine intakes and below the fuselage. Unlike the converted bombers or patrol aircraft flown by large military crews on ferret flights, the U-2 had no extra crew members to monitor the signals, tune receivers and operate antenna drives. The interface between the Dragon Lady pilot and the SIGINT gear was usually a simple on-off/antenna switching control, situated on the right console. Receiver scanning and tuning therefore had to be automatic, or alternatively preset on the ground if the particular target frequency was already known. Every signal picked up by the receivers had to be preserved for later analysis on the ground. Usually, this was accomplished by three tape recorders of three channels each, with enough tape for eight hours continuous operation, feeding across the recording head at two and a quarter inches per second.

Finding space for antennae was quite a problem, considering the nature of the beast. For some frequency bands, flat helical or small parabolic types would suffice, and these would be fitted behind flush-mounted fibreglass panels. The antennae for System 6, a wide frequency range ELINT receiver, were accommodated in this fashion: a pair to cover C-band in the nose, plus two other pairs covering L-band and X-band mounted on the lower Q-bay hatch. Others had necessarily to protrude from the airframe in order to do the job. This was true for System 3, the COMINT receiver designed to cover

VHF frequencies, which used a seven-foot long scimitar antenna extending over a foot below the rear fuselage. This was encased by a moulding of fibreglass, which became known as 'the canoe' (not to be confused with a fairing designed some years later for the top of the fuselage, which received the same nickname from a later generation of U-2 crews).

What a harvest of SIGINT some of the overflights must have brought back, especially when the Soviet air defence system was frantically trying to bring the aircraft down. Typically, the U-2 carried Systems 3 and 6, and could therefore monitor a wide selection of frequencies used by Soviet ground-to-air communications, early warning radars, medium-range search and height-finding radars, surface-to-air missile acquisition, target-tracking and guidance radars, and airborne intercept radars. Not only this, but the various listening posts run by the National Security Agency (NSA) around the Soviet borders would also electronically capture much of the commotion caused when a U-2 was penetrating 'denied territory', as Communist Bloc airspace was coyly termed by the intelligence community.

For the unit at Incirlik, however, Soviet missile tests were as much a focus of interest as the air defence system. The pace of testing from Kapustin Yar increased considerably in 1956, and the first 700-mile range rockets were deployed along the Baltic Coast facing Western Europe. Designated SS-3 and codenamed 'Shyster' by NATO, the deployment sites were reconnaissance targets for the first U-2 overflights out of Wiesbaden. These early Soviet missiles retained the liquid oxygen/alcohol propulsion system from the old German V-2 rockets, but now they were stretching the range and/or payload through new storable liquid propellants. Western intelligence allocated the designation SS-4 'Sandal' to the development, which looked as if its range might reach a thousand miles.

The listening posts in Turkey could detect the preparations for launch as Soviet radio traffic to and from the site intensified, and capture some of the telemetry that the missile returned while inflight. The powerful GE AN/FPS-17 radar near Samsun could track the missile once it rose above the radar horizon and headed east from Kapustin Yar to impact in the remote desert areas beyond the Aral Sea. But the stations were too far away and below the radar horizon to capture some vital signals relating to the early stages of flight — the so-called 'first burn'. Since the Soviets were using radio command guidance on these early missiles, and had established a series of relay stations below the projected flight path, there was also the need to pick up these ground-level trans issions. It now became known that another launch site had been established further east beyond the Aral Sea for the test of long-range missiles, and although the big Turkish radar was upgraded to detect objects at 3,000-mile range (provided they were flying high enough), US intelligence was faced with an increasing number of blind spots.

The Air Force sent modified versions of the B-47 bomber codenamed 'Tell Two' and packed with TELINT equipment to Incirlik, but even when flying at its maximum height around 43,000 feet, these also could not monitor the vital first stage. Flying at 70,000 feet and above, though, the U-2 could bring home the goods — without crossing the border into Soviet territory. So the Agency detachment would keep an aircraft equipped with System 4 (a special TELINT package) on full alert, ready to go, waiting for the word from a listening post that a Soviet test was imminent. On these flights, as well as others which gathered ELINT, the pilots of 'Det B' would fly north to cruise around the Black Sea, remaining over international waters. Or they would head east, crossing Lake Van before setting course parallel to the Soviet border and following it all the way across Iran and into Afghanistan. Frank Powers described missions flown at night, on cue from the ground stations, during which the pilots would see the missile launched and the sky lit up for hundreds of miles in spectacular fashion. The border surveillance missions became something of a milk run, but the pilots had strict instructions not to violate Soviet territory during them; accurate navigation was therefore all-important throughout these long excursions.

Powers flew the first actual overflight of Soviet territory from Turkey in November 1956, one of two relatively shallow incursions that month in which the main interest was Soviet air defence bases. Overflights were planned in great detail over a number of days, first at project HQ in Washington prior to Presidential approval being sought, and then at the detachment. There were always specific targets — there were no 'let's go see what we can find' missions. There

Famous U-2 photograph of the main launchpad for missiles at the test site subsequently revealed as Baikonur Cosmodrome by the USSR. Lacking such illumination from Moscow, which kept the name and location secret until 1961, the CIA named it Tyuratam after the rail junction nearest the site. This vital intelligence target was revisited a number of times on U-2 overflights, including one occasion when the 'B' camera caught an SS-6 missile erected and ready for launch.

were five main categories of Soviet targets for U-2 flights. These were the bomber force, the missile sites, the air defence system, including fighters, radars and SAMs, the submarine yards, and the atomic energy programme. In the latter case, border flights would normally suffice to monitor the nuclear tests, but overflights were staged in order to confirm and photograph uranium ore mines and processing sites, weapons production facilities, and stockpile sites.

A vital role was played by the programme's flight-qualified navigators, as they worked out the various computations to achieve a successful flight covering the maximum number of targets, and drew up the maps and instructions which the pilot would take along. The flight line was colour-coded: red for portions of the flight where the course had to be followed exactly so that a particular objective was covered, blue for less important sections, and brown for emergency abort routes back to base. The course consisted of straight-line sections of variable length followed by so-called 'ten-cent turns': since it was risky to bank the U-2 steeply at altitude in the coffin corner regime, gradual turns had to be planned. The navigators discovered that the optimum bank-angle could be laid out on the map by tracing round a dime coin! Notations on the map instructed the pilot when to turn the 'B' camera or SIGINT systems on and off. The much smaller tracker camera mounted on the rear of the Q-bay hatch functioned continuously.

However, there were certain sections of a photo overflight where the pilot might be given discretion to alter course if the primary objective ahead seemed to be covered in cloud, or if a slight diversion seemed promising. On just such an occasion in early 1957, the pilot of a U-2 flight that was following the railway line southeast from Aralsk in an attempt to locate the new missile testing site noticed construction off to one side in the desert.[2] He altered course, and there it was, at the end of a fifteen-mile spur leading from the main line! In an apparent attempt to mislead Western intelligence, the Russians would eventually name this site as the Baikonur Cosmodrome after a small town 200 miles away to the northeast. But as he pored over the U-2 photographs in Washington, Automat's chief information officer Dino Brugioni decided to name it after the spot where the spur line left the main railroad, marked as Tyuratam station on the Second World War-vintage German map of the area which was still the best available.

The name stuck, and Tyuratam was captured on film only a few months before the Soviets announced in late August the successful test of their first intercontinental range missile. It had flown downrange all the way across Soviet Asia and landed in the sea beyond Sakhalin Island. Less than six weeks after that, on 4 October 1957, this huge rocket was used to launch the Sputnik satellite, the first-ever object to go into space orbit. In the US, this event was received with mounting alarm, almost hysteria, since the missile gap theorists were quick to point out that a nuclear warhead trained on the American heartland could be easily substituted for the harmless satellite. Despite the highest national priority being assigned to the US ICBM programme following the Killian Panel recommendations, Convair's Atlas ICBM had made its first test flight only three months earlier, and was still two years away from deployment. Shorter-range Jupiter and Thor IRBMs had flown earlier, but had since encountered a number of embarrassing failures.

Not only that, but this Soviet A-booster (designated SS-6 'Sapwood' by the West) turned out to be three times more powerful than the Atlas. With the benefit of hindsight, the A-booster would be characterized as a missile of 'brute force and ignorance', wherein the Soviets had made up for their relative lack of progress in rocket propellants by simply strapping on four additional boosters, and they needed the huge power to compensate for their slowness in developing smaller nuclear warheads. For the moment, though, the launch of Sputnik was perceived as a national disaster akin to Pearl Harbor in the US, and led to much criticism of the Eisenhower administration. The official intelligence estimate that the Soviet Union would not have ICBMs ready for operational use until 1960-61 was revised. It was now concluded that they might have a hundred by then!

Taking their cue from the cold war warriors like Senator Stuart Symington, the American media suggested that the Sputnik launch had caught the US on the hop. This view seemed to be echoed by the Gaither Committee, another high-level think-tank set up to advise the President on national security and civil defence, which reported in late 1957. This was far from the case, however, since US intelligence had been predicting a Soviet satellite launch in 1957 for over a year. Moreover, a U-2 actually photographed the mighty SS-6 sitting on the pad at Tyuratam before its first test flight on 3 August. The question at issue was how quickly the other side could deploy an operational weapon system, and whether a period would ensue during which the US had not yet deployed the ICBM but the Soviets had fielded enough missiles to hold the US strategic deterrent of SAC bombers at risk on its own airfields.

As far as the U-2 overflights and border flights conducted in 1957-58 could detect, there were no immediate preparations being made for deployment of the SS-6. None of the early SS-6 tests demonstrated full range, and the pilots of 'Det B' witnessed a number of aborts and spectacular failures from their lofty observation perch. Moreover, the pace of test firings from Tyuratam ground to a halt in 1958, although it was generally reckoned that between twenty and thirty such firings would be necessary to perfect

The mighty first-generation Soviet ICBM designated SS-6 'Sapwood' by Western intelligence. It had four strap-on boosters which doubled its take-off thrust, making it three times more powerful than the Atlas, America's equivalent. News of its development sparked near-hysteria in the US, and fear of a 'missile gap' grew. As far as the CIA could tell from its U-2 overflights, however, the fear was much exaggerated.

the propulsion and guidance systems. It seemed that the programme had run into difficulties. At least that's what the CIA analysts concluded. But as one of them later remarked, 'To the Air Force, every flyspeck on the film was a missile'. And the cold war warriors also had an answer to the perceived lack of testing: since the SS-6 was really just a scaled-up version of the earlier MRBMs, the Soviets had obviously done with testing already and were preparing to deploy!

An intensive series of U-2 overflights might have been able to settle the matter, but was never mounted. As it eventually turned out, there were less than forty penetrations of denied territory during the entire four-year period of Operation Overflight, and they became progressively less frequent as time went on. After it was all over, a folklore developed that the U-2 had covered 'every blade of grass' behind the Iron Curtain, prompted by a misleading comment by Allen Dulles, but this was very far from reality. At the detachments, crews grew restless as months passed without such a flight; out of the twenty-odd pilots who flew the U-2 for the CIA in these four years, most had two or less overflights to their credit, while some had none at all.

Bissell's team continued to devise overflight plans, after consultation over potential targets with an *ad hoc* committee made up of representatives from the CIA, Army, Navy, Air Force and NSA. Then it would be over to the White House, where the highest counsels in the land would gather to deliberate each flight request. Bissell would be accompanied by CIA boss Allen Dulles and sometimes also General Pierre Cabell, No. 2 at the Agency. Allen's brother Foster Dulles, the Secretary of State, would usually be present, along with the Chairman of the Joint Chiefs of Staff and the Secretary of Defense or his deputy. President Eisenhower would pore over the agency's flight plan map, sometimes suggesting detailed alterations. He was always concerned about the possibility of an aircraft being shot down and the reaction of an increasingly bellicose Soviet leadership under its new master of propaganda, Nikita Khrushchev. After all, what would his own reaction be if a Soviet warplane flew straight across the middle of the US? Eisenhower usually reserved his final decision for a couple of days, often for a further conflab with Foster Dulles, before the verdict was relayed back to Bissell through the Presidential aide, General Andrew Goodpaster. At many other times, the authorization process would not even reach this stage, since Goodpaster would let it be known that the President wasn't entertaining flight requests for the moment. Either the Soviets had been sabre-rattling, or there was some summit meeting or another coming up, and he didn't want to run the risk.

Little of all this was apparent down the line at the operating level. The Agency stuck religiously to the 'need to know' principle; even within the detachments, only a handful of people were supposed to know the full details of an overflight plan. Apart from the old Agency hands, most of the troops in Germany, Turkey and Japan had no idea of how things went back in Washington. The first group of pilots had been introduced to Allen Dulles when he visited The Ranch one time, and some of the second group met Richard Bissell when he visited Incirlik in early 1957. Many others in the programme knew their ultimate boss only by reputation, as the distant and all-powerful 'Mr. B'.

Were the flights now threatened by Soviet countermeasures? It was a moot point. The large, fixed-site SA-1 missile was discounted, but its SA-2 successor was now being fielded and was mobile, although still rather cumbersome. It had a boost motor which the SA-1 apparently lacked, and a different arrangement of control surfaces — possibly sufficient to bestow some degree of control in the thin air above 60,000 feet, if the other missile parameters such as motor and guidance worked properly. ELINT tapes recorded during U-2 overflights now indicated that missiles had been fired against them, although no pilot had yet reported a visual sighting. As for Soviet fighters, these had been observed far below, as they vainly tried to climb high enough to intercept, but neither the MiG-19 nor the Yakovlev Yak-25 seemed capable of more than 40,000 feet, although they might threaten a higher-flying vehicle with a lucky air-to-air missile shot. However, the Soviets were working on mixed rocket/jet power machines; the Ye-50 test vehicle reached Mach 2 and 65,000 feet in just over nine minutes, although this line of development was not ultimately pursued.

In California, Kelly Johnson arranged for his test pilots to fly an F-104 on an attempted interception of a high-flying U-2. The Starfighter didn't come close. The Dragon Lady appeared therefore to remain safe from interference at the higher cruising altitudes, but what if a flame-out forced the aircraft lower for a relight into an area of heavy air defence cover?

Extra precautions were taken. The first piece of defensive gear was mounted on the airframe, in the form of an early radar lock-breaker capable of counteracting enemy fighters approaching from the rear. This was provided by Granger Associates, a small Californian company — the pilots therefore called it a 'granger', but its official designation was System 9. On operational missions, it replaced the drag 'chute in a small compartment at the base of the vertical tail, and was activated by the pilot when the aircraft entered denied territory. If the U-2 was illuminated by an air-to-air missile radar, System 9 would retransmit the received signal with modifications designed to cause the radar to break lock. At the same time, a light in the cockpit would alert the pilot to the potential danger. The System could not spoof the simpler gunsight (range-only) type of airborne radar, but it did provide a warning of them, to enable the U-2 pilot to take evasive action. Overflight routes were now planned so as to avoid known Soviet SAM sites. The points of entry into Soviet territory were varied as much as possible, in the hope that the aircraft could evade radar detection for at least a portion of the flight. This was more easily accomplished by staging missions from airfields other than the main detachment locations, which were presumed to have been identified by Soviet intelligence. Indeed, after the first few overflights 'Det A' had moved from the relatively public location at Wiesbaden to a smaller field at Giebelstadt in the hills south of Wurzburg, but it was soon noticed that a large black limousine was frequently cruising the exterior. This was identified as belonging to one of the Iron Curtain embassies in Bonn. Since Incirlik was a Turkish Air Force base, the Agency had little control over who might enter there, although they did their best to shut off the U-2 area and personnel from outsiders. Security experts at the detachment were constantly concerned about possible Soviet penetration.

Another driving factor behind the search for staging fields was the deep interior location of most of the Soviet test bases. Tyuratam was 1,500 miles from the Turkish base, still within range of a direct out-and-back flight, but not if tactical routing was to be employed, or other targets covered. There were other areas of interest even further east; downrange radars for missile tracking, the nuclear test site at Semipalatinsk, and the nuclear weapons plant near Alma-Ata, for instance. In the far north, there was another nuclear test site on Novaya Zemlya island, and all the military installations on the Kola peninsula. After some delay, a third permanent U-2 base was established at the US-controlled Atsugi airfield near Tokyo in 1957. From here, the aircraft of 'Det C' kept tabs on Soviet military forces around Vladivostok, and further to the north on Sakhalin Island. North Korea and China were also overflown.

If the U-2 was to use additional overseas airfields, however, more countries would have to share the secret, and they would be justifiably nervous of adverse Soviet reaction. Any flight operations from their territory would have to be slick and discreet. This was achieved through a well-organized and tightly-controlled deployment plan. All the necessary support kit could be fitted into two vans, one for PSD and one for maintenance. These would be loaded into one C-130 Hercules transport, along with the twenty-strong ground party and a spare pilot, while fifty-five-gallon drums containing the special U-2 fuel were loaded in another Hercules. If, for instance, an overflight was to be staged from the main base and recovered elsewhere, the two transports would then depart the main base for the designated landing site to await the arrival of the mission aircraft. As soon as it was down, they would spring into action, and once the turn-round was complete, the spare pilot would climb in and fly the U-2 home, while the ground crew packed everything back inside the Hercules transports and followed, leaving virtually no trace of their presence. The whole thing could be accomplished within hours if necessary, although some deployments involved a stay of a week or more, especially if the mission was to be launched from a remote base, and weather aborts ensued.

Bases used in this fashion from 1957 included remote Iranian airstrips near Meshad and Zahedan, Lahore and Peshawar in Pakistan, and Bodo in Norway. Detachment A at Giebelstadt was wound down in 1957, although the German base continued to be used for deployments. All of these countries were, of course, also aiding the US intelligence effort in other ways, for instance by permitting the establishment of NSA ground stations. However, the U-2 flights represented an altogether greater order of potential provocation to the Soviet Bear, and the US promised to declare that it had used the airfields without permission if the programme was ever blown.

Therefore, the Richard Bissell Air Force (as it became known inside HQ) became a worldwide operation. In the fall of 1958, Bissell was promoted within the Agency to Deputy Director of Plans, but he still kept a close eye on U-2 activities. Far-flung trouble spots were now overflown, such as Indonesia during 1958 as it leaned towards Moscow, and Cuba after Castro came to power. Following its early flights over Suez, 'Det B' was often given the task of overflying the Middle East. Syria, Iraq, Jordan and Saudi Arabia were all covered in addition to Israel and Egypt. U-2s were over Lebanon during the factional crisis in 1959, and Frank Powers flew one of the longest-ever missions from Incirlik in 1959 when coverage of civil disorders in the Yemen was requested — a roundtrip distance of 3,700 miles. The following year, the detachment flew a mission over the secret nuclear research facility that France was helping to build for Israel in the Negev desert at Dimona. Israel claimed it was a textile plant, but U-2 photographs proved otherwise, and the Israeli government reluctantly agreed to regular inspections by US scientists in an (ultimately unsuccessful) attempt to ensure that Dimona's true purpose remained peaceful nuclear research.

Flexibility of basing was matched by flexibility of equipment. In September 1958, indications that the Soviets were about to stage nuclear tests at their Arctic test site in the Novaya Zemlya islands were picked up. U-2s from 'Det B' were deployed to Bodo, with the camera pallet removed from the Q-bay and replaced by the special filter and bottle system which was used to collect atomic fall-out samples from the upper atmosphere. A number of sampling missions took place from Bodo to the far north, during which the U-2 pilots flew from night to day to night and day again. On the return trip to Turkey, one of the aircraft was reconfigured with an all-SIGINT package to fly a long ferret mission down the Baltic and over the Iron Curtain countries, recording communications and radar emissions along the way. Fall-out sampling became a major task for 'Det C' in Japan, since airflow in the upper atmosphere tended to bring the particles from Soviet nuclear tests at the Siberian test site near Semipalatinsk in their direction. The Far East detachment also

This CIA U-2A is in a typical late fifties operational configuration for an overflight of denied territory. Slipper tanks are carried on the wings to provide extra fuel, and all national markings have been erased. A range of SIGINT equipment is on board, as evidenced by the L- and X-band dialectric panels at the front of the Q-bay, and the prominent radome covering the P-band antenna protruding from the rear fuselage.

flew frequent SIGINT missions off the Soviet Asian landmass, and there were two or three overflights of the Kamchatka peninsula.[3]

During these flights towards the polar latitudes, variations in the atmosphere detracted from the U-2's altitude performance, sometimes in a dramatic fashion. Ordered to make an overflight of Kamchatka on one occasion, pilot Barry Baker found himself unable to rise above 59,000 feet; he pressed on with the mission, although through the driftsight he could see MiGs trying to reach his cruising level. Baker's plane had encountered a paradox of the upper atmosphere: while the temperature in these cold northern regions was lower than further south at ground level, by the time one reached 50,000 feet, the air was invariably warmer than at the same level over the temperate or equatorial regions.[4] Conditions varied enormously up here, even from day to day, but on this occasion they prevented the J57-powered Dragon Lady from climbing higher. In theory, the ground rules for overflights stated that pilots should not enter 'denied territory' below 65,000 feet; if it looked like this altitude would not be achieved in time, one wide 360-degree turn was allowed in order to gain the extra height. The trouble was, on most flights the pilot was supposed to establish the cruise-climb schedule at 60,000 feet. It could be some time into the mission before any shortfall in ultimate achievable altitude (and also range) became apparent, and a further complication resulted from the fact that the greatest variations in outside air temperature occurred between 60,000 and 65,000 feet.

Altitude performance was also being affected by the increasing amount of equipment now being carried aloft. There was the black paint (eighty pounds), and defensive ECM (another thirty pounds), not to mention the latest SIGINT packages. Following the death of Bob Sieker, and a number of other accidents logged by the Air Force squadron that was now also flying the U-2, an ejection seat had been installed. Various other bits and pieces of safety gear had also been added, so the original zero fuel weight goal had long been exceeded. In 1958, Lockheed came up with a solution to the declining performance: re-engine the Dragon Lady with the more powerful Pratt & Whitney J75. Although this turbojet had essentially the same configuration as the J57 (axial-flow, twin spool, eight burner cans, three turbines), manufacturing and design techniques had advanced sufficiently to permit a significant increase in thrust.

The new engine had entered production the previous year, having been selected for the Air Force's latest fighters, the F-105 and F-106. In the high-altitude J75-P-13A, Pratt & Whitney were offering a 15,800 lb. thrust version, an improvement of 4,600 lb. over the U-2's existing J57. (Ultimately the thrust would be boosted to 17,000 lb. in the definitive J75-P-13B high-altitude version.) Once again, Pratt & Whitney had to adapt the P-13 to use the special U-2 fuel; as well as pay particular attention to the compressor face, where blades in the first two stages were lengthened. The strictest quality control was needed to ensure that extrusions or nicks that might cause compressor stalls in the high regime were eliminated. On the P-13 development, Pratt & Whitney didn't move fast enough for Lockheed's liking, especially when a fuel control problem showed up (the plumbing was causing some cavitation). In typical Skunk Works fashion, they made up the new tubes themselves at Burbank and pressed on with flight tests.

The J75 was a few inches wider and substantially longer than the J57, but fortunately still fitted the Dragon Lady airframe with very little modification, except that the intakes were widened to serve the increased airflow demand of the compressor face.[5] But it weighed some 750 lb. more than its predecessor, and this made for a tricky weight and balance situation on the U-2. The aircraft's centre of gravity (cg) limits narrowed, putting further aerodynamic loads on the horizontal stabilizer, which therefore had to be replaced with a new section of altered camber. Even so, the cg limits of the re-engined aircraft were only one and a half inches apart under most conditions. As a result, the margin of safety between Mach and stall buffet was even tighter than on the A-model (only four knots in some regimes), and a stall buffet could soon result in pitch-up, followed by a flat spin from which recovery was impossible.

In fact, added ballast was now needed to keep the aircraft within the cg limits, even when it was carrying a full operational load in the Q-bay. Lead weights were added at the tail or on the bulkhead rig, depending on the payload, and even the differing weights of individual pilots could demand an adjustment. All this fiddling about was no picnic for the ground crews. In

The substitution of the J75 engine boosted the performance of the U-2, and significantly altered its flying characteristics. Visible in this photograph of the high-altitude J75-P-13B version are the compressor blades, which had to be machined to much finer tolerance than on the standard engine, if compressor stalls at high altitude were to be avoided.

weight and balance, as in many other respects, each airframe exhibited different characteristics; and the whole situation wasn't helped when even the lead weights turned out to be non-standard.

Before the necessity for ballast and the altered camber for the tail was fully appreciated, Bob Schumacher took a re-engined U-2 up with slipper tanks fitted, and found that the cg had got so far forward that he was having to apply back pressure on the yoke to supplement that provided by the autopilot. With the trim having gone out, he reported that if the autopilot were to disconnect, the aircraft might become uncontrollable, unless the fuel in the slipper tanks had been consumed. It was one of many awkward moments in the U-2/J75 flight test programme, which took place in mid-1959. The pilots likened the difference between the old A-model and the re-engined U-2 (designated U-2C) to that between driving a Cadillac and a Porsche. The new version was a sporty aircraft, even more unforgiving of errors than its predecessor.

Tighter margins for the cg and at 'coffin corner' were not the only problems with the C-model. Another concerned the ascent to operating altitude. Partly due to the more rapid rate of climb that it made possible, the new engine proved particularly susceptible to the temperature changes experienced as the plane passed through the tropopause. Above the United States, this is typically encountered at around 35,000 feet, this being the point at which the outside air temperature reaches its coldest value, around minus fifty degrees centigrade. Beyond the tropopause and into the stratosphere, the temperature tends slowly to increase again, so that by 65,000 feet it can be as high as minus thirty-five degrees centigrade. The problem for the U-2 was that the tropopause varied from day to day in height and thickness, as did the temperatures, with some rapid changes likely to be encountered from time to time.

The J75 engine protested loudly when it encountered these atmospheric fluctuations; the usual high-frequency, low-amplitude buzz heard in the cockpit during the climb out would be replaced by coughing, spluttering and even loud banging noises as the aircraft entered the tropopause. Sometimes, the engine would then quit altogether. On one early U-2C test flight, Agency pilot Jim Cherbonneaux experienced six such flame-outs! This hazardous regime became known as 'The Badlands', and an operational procedure was devised to get the aircraft safely through it and on to the higher altitudes. This entailed retarding the throttle at about 35,000 feet so that the recommended takeoff value of 540 degrees centigrade EGT was reduced to 485 degrees at 40,000 feet. The EGT was held at this lower value until 60,000 feet, when it could safely be increased again to 600 degrees. Eventually, EGT would rise to the redline 610 degrees as the aircraft rose further and the air became even thinner, but by this time the pilot would be having to ease back on the throttle to prevent the hot section of the engine from overheating. He would now have reached the point where he could fly no higher.

Maximum altitude continued to depend on weight, outside air temperature and other variables, but was still around 75,000 feet for all practical purposes. However, re-engining the Dragon Lady did achieve its primary purpose: the U-2C climbed like a rocket with all that excess power, even with the heavier payloads, so that the cruise-climb could be established at 67,000 feet. And even at maximum gross takeoff weight, the U-2C could get there in fifty minutes, having covered just 300 nautical miles ground track. With less than full fuel, this aircraft could pass 30,000 feet within four minutes — it could have matched many of its hot-rod fighter contemporaries had Lockheed or the Agency been in the record-setting business.

Zero fuel weights eventually exceeded 14,000 lb. with all the additional gear that was added to the U-2, but the J75 proved equal to the task. With the maximum fuel load weighing just over 10,000 lb., aircraft maximum gross weights rose from 17,000 lb. on the early U-2A to in excess of 24,000 lb. on some mid-1960s configurations, thanks to the re-engining programme.

Although initial U-2 production had ended within two years with fifty airframes completed, an additional five aircraft were subsequently built from spare parts which had been funded under the original contracts. Lockheed was also kept busy with a constant stream of aircraft returning for overhaul, rectification, modification or crash damage repair. No new C-models were built; all the aircraft so designated were conversions of existing aircraft. Lockheed's U-2 depot was transferred from The Ranch to a remote part of Edwards AFB known as North Base in June 1957, although Agency U-2 operations from the secret site continued for three more years, essentially as a sort of headquarters squadron (designated WRSP-4) for the far-flung detachments. By now, the Skunk Works had acquired a formal title — Lockheed Advanced Development Projects (ADP). In 1958, ADP embarked on another top-secret CIA-funded programme to provide a successor to the U-2. This became known as Project Oxcart when formal contracts were issued in 1959, but the first aircraft in what became known as the Blackbird series did not fly until 1962. In the meantime, the revamped U-2 would continue in service.

Article 360 was one of the first aircraft to be converted into a C-model, whereupon it was sent to 'Det C' at Atsugi. It soon created some very unwelcome publicity by crash-landing on a small civil airfield at Fujisawa near Tokyo. According to the story which subsequently circulated within the programme, pilot Tom Cruel had been trying out the high altitude performance of the new hot-rod machine. He managed to set a new altitude record, but left himself short of fuel for the descent and the tanks ran dry just ten miles short of Atsugi. In zero wind conditions, a descent from 75,000 feet in the U-2C would consume thirty-five gallons and take twenty minutes, but the first 10,000 feet could be very slow, and it wasn't simply a case of pointing the nose down. The pilot had to ease the throttle to idle and extend the gear and speed brakes. Even then, the aircraft hardly seemed inclined to leave the heavens, although the C-model was a little easier to coax down than the lighter A-model. There was precious little drag created by the gear and brakes at these altitudes, and the engine thrust was still considerable even on minimum power. To cut it back further would invite a flame-out. At the same time, care had to be taken to ensure that the speed didn't build up so that Mach buffet was encountered. In view of all this, and the possibility of encountering strong headwinds during the descent, the pilot was supposed to leave altitude with at least one hundred gallons remaining.

This is how the Japanese magazine *Aireview* recorded the crash-landing of a Det C aircraft at Fujisawa airstrip near Tokyo on 24 September 1959. The incident brought unwelcome publicity, which the US probably encouraged when guards arrived from Atsugi and ordered the curious local onlookers away at gunpoint.

For whatever reason, Tom Cruel deadsticked onto the unplanned alternate with his gear up, and slithered to a halt in the grass. Not long after, another Agency U-2 ran out of fuel on its way back from a long overflight of China and crash-landed in a rice-paddy in Thailand. That one was hushed up, with the aircraft recovered from the mud and trucked at night through the streets of Bangkok to Don Muang airbase, where a C-124 was waiting to take it back to the States. In Japan, the Agency was not so lucky. In less than a minute, Cruel was surrounded by curious Japanese civilians, some with the inevitable cameras. He remained in the cockpit until help arrived from Atsugi, in the form of an L-20 Beaver lightplane and a security police detachment. The latter cordoned off the area and ordered the onlookers away at gunpoint, which only excited their curiosity further, since Article 360 was flying completely unmarked save for a minute three-digit serial on the tail. The story and pictures reached the local media, where pointed questions were raised as to why a so-called weather research aircraft should cause such a security panic, and be flying in this striking all-black scheme without national insignia or registration marks!

This was by no means the first time that the U-2's true purpose had been hinted at in public. When Frank Powers was shot down the following year, much of the comment from both official and media sources suggested that the secret had been perfectly maintained over five long years by the US. This was nonsense. Even back in 1956, the U-2 had scarcely been deployed to England a month when *Flight* magazine ran two sarcastic editorials scoffing at the cover story and the excessive security surrounding the deployment. 'Very high weather we're having lately!' remarked another British magazine. Soon the aircraft spotters were reporting U-2 sightings to the letters columns of the aeronautical press. By the following year, the NACA cover story was looking shaky when the London *Daily Express* reported on 30 May that 'Lockheed U-2 high-altitude aircraft of the US Air

Force have been flying at 65,000 feet, out of reach of Soviet interceptors, mapping large areas behind the Iron Curtain with revolutionary new aerial cameras'. The true nature of U-2 operations also leaked into the US press, but not into the major newspapers, since top editors were persuaded by the government to suppress the story in the national interest.

The logic behind these requests for self-censorship was interesting. After all, the Soviets knew most of the key details anyway from their radar tracking and other intelligence means. They even published a U-2 article in the air force journal *Sovietskaya Aviatsiya* in May 1958, complaining about the lack of markings on these 'strategic reconnaissance aircraft'. And the CIA knew they knew. The point was, the Soviets had not made any more protests at the overflights after their initial broadside in July 1956, and Khruschev did not raise the subject during his visit to the US in 1959. Maybe this was because the world was fed a constant diet of propaganda about Soviet military prowess, which might lose its effect if the Kremlin's inability to knock these aircraft down became widely known. The US concluded that it would be highly embarrassing for the Soviet leadership if the US trumpeted to all and sundry its ability to soar across their innermost sanctums with shutters clicking. If Operation Overflight was publicized, the Soviets might respond by redoubling their efforts to shoot a U-2 down, or retaliate in some other way. So the US government decided that the less said, the better, even though Eisenhower did briefly consider revealing the U-2 in 1958, when he was under fire for military complacency in the face of the Soviet threat. It was one of the travails of high office that he couldn't rebut his critics by revealing the secret of the black aircraft, and the hard evidence that they were bringing back from Soviet skies.

At the overseas U-2 bases, there had been a number of changes by 1959. Some of the original recruits had been dropped when 'Det A' closed down. Some of the pilots had elected not to renew their initial eighteen-month Agency contracts, while those that did were allowed to bring their families out to join them. At both Incirlik and Atsugi they lived alongside an increasing number of US servicemen, and were therefore able to share the new amenities that sprang up, such as volleyball courts, supermarkets and clubs. Even so, the mantle of security remained tightly wrapped, and those in the detachments kept mainly their own company. When some former colleagues of the U-2 pilots from the F-84 days at Turner showed up at Incirlik in 1959 (flying Air Force RB-57Ds), they were more than a little put out when the Agency flyers chose to ignore them.

One of the best-kept secrets concerned the presence at Incirlik of a British U-2 team, including a navigator, flight surgeon and four pilots. They had been brought into the programme the previous year. The Royal Air Force selected some of its best pilots for the assignment; all were instructors with A2 ratings, in their late twenties and with about eight years behind them in the service. Unfortunately Squadron Leader Chris Walker, one of the original four pilots selected by the RAF, had been killed in a crash less than one month after his first U-2 flight from the training base, which was now located at Laughlin AFB in Texas. The RAF sent out a substitute pilot, and all four moved on to 'Det B' to conduct some 'British' overflights.

This scheme had been devised by Bissell and Dulles at the Agency, when it became clear that the White House was clamping down on permissions for overflights. Since the British were already closely informed about U-2 flights (the CIA had shared much of the intelligence on missiles with them), why not set up a joint operation whereby certain flights could be sanctioned by London instead of Washington? Eisenhower and British Prime Minister Harold Macmillan approved the plan, and an RAF wing commander joined the team at U-2 project HQ. A U-2 was flown to the RAF base at Watton in eastern England, so that the British top brass could familiarize themselves with this new — but strictly unofficial — addition to the RAF inventory. As it turned out, only two Soviet penetrations were made by the RAF flyers, since permission to overfly proved almost as difficult to obtain from Whitehall as from Washington. The RAF *modus operandi* sometimes raised eyebrows among the old hands at Incirlik. On one of the overflights, Flight Lieutenant John MacArthur flew round the target twice, and on the other, Squadron Leader Robert Robinson pressed on through a mission despite leaving a contrail behind him — the Agency pilots had been told that both these deeds were strictly forbidden. The British pilots also took turns for the peripheral SIGINT missions, and their presence at Incirlik swelled the number of U-2 pilots there into double figures.

Even so, the number of overflights continued to dwindle; there were less than a dozen in 1959. At the detachments, months would pass without one, so that when a penetration was finally sanctioned, the tension was all the greater. After nearly four months without an overflight, 'Det B' at Incirlik was scheduled for two in quick succession in April 1960. Colonel William Shelton, who had recently taken over from Stan Beerli as the commanding officer, chose Bob Ericson and Frank Powers to fly them. Ericson had trained with the third pilot group to go through The Ranch, and had been transferred from 'Det C' in Japan together with some other pilots. Powers was the only pilot to have remained at 'Det B' throughout its existence, so he therefore now had more operational missions to his credit than anyone else.

Speculation around the detachment had it that Washington was trying to squeeze in the maximum number of flights before Soviet air defences finally caught up with them. After briefings at Incirlik, the support team for the first flight deployed to Peshawar in Pakistan on a C-130, while a U-2 was ferried along the border route. On 9 April, Bob Ericson soared into the air and across the spectacular Hindu Kush range. His main target was a major test installation near Sary-Shagan in Kazakhstan, where the Soviets were thought to be developing radar and missiles capable of intercepting incoming ballistic missiles. These new high-altitude missiles would also pose a great threat to future U-2 flights (and, perhaps, even to this one). Unlike earlier missions to this target, some good photography was obtained this time, and Ericson flew on to the nuclear test site near Semipalatinsk before turning west and heading for Tyuratam. The flight ended safely with a landing at the Zahedan airstrip in Iran.

Now it was Powers' turn, but his would be a very different mission. For the first time, the flight plan would send him north from Pakistan all the way across the Soviet heartland, and out the other side to a landing at Bodo in Norway. This would be a nine-hour flight covering 3,800 statute miles, of which 2,920 would be over 'denied territory'. While such a flight had been

British participation in the U-2 programme was kept secret for years. These are the two pilots, Squadron Leaders Charles Taylor (left) and Ivan Webster (right), who trained in 1961 and completed a three-year tour on the aircraft in 1964. 'Chunky' Webster subsequently returned to the US to fly U-2s for NASA, and retired in California.

mapped out before, it had never been executed. Why was it necessary now? The overriding reason lay towards the very end of the flight, some hundred miles inland from the White Sea port of Archangel. Here, according to SIGINT from a Norwegian listening post, the Soviets were establishing a new missile base at Plesetsk. Given its northerly location, which offered a shorter flight track to US territory across the North Pole, this might be the first operational base for the SS-6 ICBM, a target that the photo interpreters at Automat had been searching for in vain since 1958. By approaching from a southerly direction, some hitherto unvisited territory deep in the Urals could be covered, such as the military-industrial complex around Sverdlovsk, and the first part of the flight could include another crack at Tyuratam.

If the prior intelligence was correct, the flight would stand an excellent chance of capturing the construction of the new base. Art Lundahl and his team of PIs could learn more from photos of the uncompleted site, before all the buildings and launch pads had been constructed and camouflaged. The Soviets would probably also move in SAMs to protect the site once it was completed, making a direct overflight risky. But as the usual high-level group gathered in the White House to review the results of Ericson's flight, and make a final decision on the planned Powers flight, there was another vital consideration. A summit conference of the great powers was due to open in Paris on 16 May. Following Khrushchev's visit to the US the previous September, there were hopes for a thaw in the Cold War. A partial nuclear test ban treaty was under negotiation, and the US President had been invited to the Soviet Union in June. Could another U-2 flight be risked so close to the summit?

The Agency pleaded a strong case for the flight. In addition to all the other factors, Plesetsk presented a weather problem. At such northern latitudes, the sun angle would only be sufficient for good photography between April and early September. And yet the weather at these times was often bad; there were only a few days in the month when the target would be clear of cloud. What with the summit and the President's forthcoming trip to Moscow, there might not be another chance at Plesetsk this year. But this was an important target; flight tests of the SS-6 had resumed a year earlier at Tyuratam, indicating perhaps that deployment of operational missiles would soon occur. Had the dreaded missile gap become reality? The US had placed its first ICBM on alert on 31 October 1959, and it was essential to gauge the other side's progress.

Could this be accomplished by using the emerging spy satellite technology? Bissell and his CIA group were also in charge of this project, but despite a dozen attempts, there had not yet been a successful orbit. One of the reasons for Eisenhower's unwillingness to approve U-2 overflights throughout 1959 had been his hope that the reconnaissance satellite would soon be in action. What about the aircraft successor to the U-2 that Lockheed was working on? Contracts had only been issued the previous autumn, and the design was so complex that this Mach 3 marvel wouldn't make its first flight for another year or more. Weighing the situation up, Eisenhower approved the U-2 mission with one condition: that it be flown by 25 April.

A fortnight passed, with the detachment at Incirlik at high readiness to deploy, but the weather across much of the route remained bad. Bissell pleaded for more time, and Goodpaster consulted Eisenhower. The deadline was extended to 1 May. The weather appeared to be improving (ironically, in these pre-satellite days the Agency relied on Soviet meteorological broadcasts for some of its weather data). Early on Wednesday 27 April, Shelton, Powers and the rest of the deployment team boarded a C-130 for yet another trip to Peshawar. If all went well, the mission would launch at 06.00 local time the next day.

Shortly after arriving at Peshawar, where the team had to make do with primitive accommodation in a remote hangar alongside the aircraft, Powers and the backup pilot went to bed on folding cots. Woken at 02.00 the next morning, they had barely got washed and dressed before the mission was scrubbed for twenty-four hours due to continuing bad weather. On Friday, Powers had reached the pre-breathing stage when the same thing happened again. Before the afternoon was out, they knew that Saturday would be another no-go. Like many other deployments, this one was turning into a familiar routine. Confined to a small corner of the base, there was little else to do save sit around playing poker, reading novels, and cooking the rations they had brought with them. During the hiatus, the U-2 originally allocated for the mission was flown back to Incirlik, and Bob Ericson flew out on Saturday night with a replacement. It was Article 360, the same aircraft that had dead-sticked into the Japanese glider strip the previous year, and which had since been repaired at Burbank. More than most aircraft types, each of the hand-built U-2s had its own idiosyncrasies, but Article 360 was worse than most, and its latest foible was an intermittent problem with fuel feed. Once again, Powers prepared for an early morning takeoff.

Considering the weather requirements, the delays were hardly surprising. Not only were the chances of sunny skies over Plesetsk low, the odds were not much better further south along the route. In addition, high altitude conditions had to be such that no tell-tale contrail was formed by the cruise-climbing plane, and the weather forecast for late in the day at Bodo had to be acceptable. Now the sun was rising over Central Asia on Sunday 1 May, the last possible day to fly before the Presidential deadline expired. They didn't know of this deadline at Peshawar, of course, and neither were they aware of a breakdown in the secure communication link by which the final signal to go was transmitted from project HQ in Washington. As was subsequently discovered, on the second stage of its three-part journey to Peshawar, the go-message had been relayed from Germany to Turkey on an open telephone line, which was strictly against regulations. Shelton told Powers the delay was caused by the need to get final clearance from the White House. This appears not to have been true; Eisenhower had already given his go-ahead a week earlier, and Allen Dulles later said it was he himself who made the final decision to go at 17.00 on 30 April, Washington time.

This was 03.00 on 1 May in Pakistan, but by the time the go-signal reached Peshawar, Powers had been strapped in the cockpit at the end of the runway for over an hour, and had missed the 06.00 launch time. The sun was already hot, and he had perspired profusely inside the airtight pressure suit. Bob Ericson, who was standing by at the aircraft's side to offer additional assistance, took off his shirt and held it over the canopy to try and shield it from the sunlight. At 06.20 the go-signal finally came through, the canopy was closed and the engine started. At 06.26, Article 360 was airborne and heading for the border. As the whole world was soon to know, it never reached its intended destination.

Frank Powers was shot down near Sverdlovsk after three hours and twenty-seven minutes of flight, and managed to bale out of the falling aircraft at approximately 15,000 feet. He was captured, taken to Moscow, and interrogated by the KGB. Large sections of the aircraft fell to earth in relatively good condition, although scattered over a wide area, and were recovered and put on display. In August, Powers was subjected to a show trial before the Soviet Supreme Court. By that time, the incident had been blown up into a major world crisis by Khruschev, who staged a dramatic walkout from the Paris summit. The Soviet Premier had also cleverly manipulated news of the shootdown and Powers' survival, delaying the release of full details sufficiently long for NASA to issue the pre-rehearsed cover story about a local U-2 weather flight from Incirlik possibly having strayed off course, since the pilot 'had reported difficulties with his oxygen equipment' when last heard from. This was soon exposed as nonsense by Khrushchev, and after a few days of official dithering, President Eisenhower elected to carry the can personally, by admitting that he had authorized Operation Overflight. This was unprecedented — never before had the US formally admitted that it engaged in espionage. The President never made his trip to Moscow, and the Cold War returned with a vengeance. Powers received a ten-year prison sentence, but was released in February 1962 in exchange for Soviet master spy Rudolf Abel.

All this is not in dispute, but much of the detail surrounding the shootdown quickly became surrounded in obfuscation, misunderstanding and scepticism. Every journalist in Washington was on the story, of course, but few were sufficiently expert to evaluate what they were being told, either by Moscow or Washington.[6] Many people in the US government and Congress were guilty of a similar failing. Despite a thorough debriefing of Powers when he was released, followed by a CIA Board of Enquiry, and his own memoirs published eight years later, differences of opinion about what actually happened over Sverdlovsk remain, and will probably never be reconciled.

The crucial point of contention was whether Powers was shot down at high altitude, or only after having been forced to descend for some

"All the News That's Fit to Print"

The New York Times.

LATE CITY EDITION
U. S. Weather Bureau Report (Page 88) forecasts: Cloudy, rain today, tonight. Clearing gradually tomorrow. Temp. range: 64-54; yesterday: 66.6-54.8.

NEWS SUMMARY AND INDEX, PAGE 95

SECTION ONE

VOL. CIX—No. 37,360. © 1960 by The New York Times Company. Times Square, New York 36, N. Y. NEW YORK, SUNDAY, MAY 8, 1960. 15¢ outside New York City, its suburban area and Long Island. 50¢ in 11 Western states, Canada; higher in air delivery cities. TWENTY-FIVE CENTS

U. S. CONCEDES FLIGHT OVER SOVIET, DEFENDS SEARCH FOR INTELLIGENCE; RUSSIANS HOLD DOWNED PILOT AS SPY

JOHNSON ARRIVES IN WEST VIRGINIA AS CLIMAX NEARS

Texan Declines to Choose Between Humphrey and Kennedy in Primary

VOTE DRIVES WINDING UP

City of Clarksburg Invaded by Politicians, High School Bands and Rotarians

By WAYNE PHILLIPS
Special to The New York Times.

CLARKSBURG, W. Va., May 7—Senator Lyndon B. Johnson, an undeclared candidate for President, flew into West Virginia today as the climax approached in the state's Presidential preference primary.

He announced that he would not state a preference between the two Democratic candidates in the primary, Senator Hubert H. Humphrey of Minnesota and Senator John F. Kennedy of Massachusetts.

He described the candidates as "colleagues of mine, both of whom I have the greatest respect and affection for." And he added, "Either one of them would be far superior to anything that the Republican party can possibly offer."

Candidates Everywhere

"I didn't come down here to tell you how to make your decision," he said. "I look forward with confidence to your decision, whatever it may be."

The city of Clarksburg that welcomed him appeared like a rather soggy finale to "The Music Man." For it was also playing host to thirty-three high school bands, a Rotary convention, and an army of hopeful candidates for almost every public office from constable to President of the United States.

No one counted the number of trombones in the bands to see if there were seventy-six. But Clarksburg is about the same size as Mason City, Iowa, which inspired the Broadway musical. And the light rain that fell discouraged none of those who had traveled from miles around to see what was going on.

Humphrey's Sister Stumps

While the crowd was waiting for the bands to parade down Main Street, Mrs. Frances Humphrey Howard, a sister of Senator Humphrey, was passing out coffee and doughnuts on the steps of the County Courthouse.

Near by, one of the many local candidates who was moving through the crowd greeting old friends and trying to make new ones, was Ralph J. Keister, who is running for the state's House of Delegates.

There are more than 200 candidates on the Democratic ballot in this county alone, and nearly that many on the Republican ballot. Together with family and friends, that made a sizable number of people here

Continued on Page 47, Column 2

Nixon Shifts Tactics To Combat Kennedy

By The Associated Press

WASHINGTON, May 7—Vice President Nixon reshaped his campaign plans today on the theory that Senator John F. Kennedy would win the West Virginia primary Tuesday and go on to win the Democratic Presidential nomination.

Herbert G. Klein, Mr. Nixon's press secretary, told the newsmen that if the belief was accurate, Mr. Nixon, as the expected Republican nominee, would be competing against a single opponent instead of half a dozen Democratic Presidential possibilities.

Mr. Klein indicated that if this turned out to be the case, Mr. Nixon, in drafting his speeches during the coming weeks, would concentrate more on Senator Kennedy. Even if Senator Hubert H. Hum-

Continued on Page 46, Column 3

N.A.A.C.P. TO FIGHT CURBS AT BEACHES

Plans 'Wade-In' Campaign at Tax-Maintained Resorts From Jersey to Texas

Special to The New York Times.

ATLANTA, May 7—The National Association for the Advancement of Colored People announced today a "wade-in" campaign against segregation on Southern beaches.

The association's executive secretary, Roy Wilkins, said "hundreds and thousands of miles" of beaches and public parks were maintained with tax funds.

Negroes, he said, pay taxes and "they get hot just like white people do." They like to swim to cool off and with the prospect of warm weather ahead "they intend to do it this summer" from Cape May, N. J., to Brownsville, Tex., he said.

Between Cape May, on the Atlantic, and Brownsville, on the Gulf of Mexico, is a coastline that touches eleven states that practice segregation. North of Cape May, public facilities are generally integrated.

Cites Biloxi Incident

The wade-in drive, Mr. Wilkins told a news conference, was spurred by the recent attempt of Negroes to use a public beach at Biloxi, Miss. They were attacked by whites and for a time authorities feared serious rioting.

Mr. Wilkins met here at the Waluhaje Apartments with state presidents and secretaries of the association from Arkansas, Florida, Georgia, Louisiana, Mississippi, North Carolina, South Carolina, Oklahoma, Tennessee and Virginia.

Besides the beach demonstrations and a continuation of the group's efforts to break down lunch-counter segregation, Mr. Wilkins said the conference had reviewed plans to increase voter registration and N. A. A. C. P. membership and to speed the process of school desegregation.

He pointed out that the civil rights law signed by President Eisenhower this week required establishment of a pattern of discrimination before Federal referees could step in to guarantee Negro voting rights.

Many to Seek Vote

As a result, he said, local N. A. A. C. P. officials will urge Negroes to "apply in numbers."

He also cited recent gains in registration, which he attributed to a strong campaign by the organization and particularly its youth chapters.

A Florida representative reported that 4,000 Negroes had been added to the registration rolls at Tampa since Feb. 1. They are now registered in the only South Carolina county

DERBY CAPTURED BY VENETIAN WAY

Bally Ache Defeated by 3½ Lengths at Louisville

Venetian Way, owned by Isaac Blumberg, won the $158,950 Kentucky Derby yesterday at Churchill Downs in Louisville by three and a half lengths. Bally Ache ran second and Victoria Park third. Tom-

LONDON TROUBLED

Fears That U.S. Stance for Summit Parley Will Be Injured

By DREW MIDDLETON
Special to The New York Times.

LONDON, May 7—The United States position at the summit conference and in the global contest with the Soviet Union may be seriously weakened by the plane incident in Russia, diplomats said today.

The damage to the United States abroad has been heightened by the first hasty denial of spying on the Soviet Union and the subsequent admission forced from the Eisenhower Administration. At first, Allied diplomats thought the Soviet version would be accepted only by the Left-Wing circles, neutralists and pacifists of Europe, Asia and Africa.

The State Department's later acknowledgment of the nature of the plane's mission, even though it was unauthorized by the authorities in Washington, means that distrust of the United States will spread beyond those circles.

'Bad Luck' Evident

The Foreign Office had no official comment on Mr. Khrushchev's charges. This Government has not been informed by the Administration about the incident.

Beneath the diplomatic explanation that it was a United States affair there clearly was concern over what was considered to be rather offhand treatment of a stanch supporter of the Administration at a critical moment.

The British are clearly worried about what professionals call "bad luck" and by the effect of the incident on the diplomatic position of the United States, Britain and France at the summit meeting.

The immediate effect in this field, it was said, is that Mr. Khrushchev's report will raise doubts all over the world about the true willingness of the United States to work for a reduction of tensions.

The Image of Eisenhower

The British and their European allies seldom read the fine print in Presidential and Senatorial speeches. Consequently, they have perhaps been too impressed by what they considered American readiness to work for such a relaxation.

The European image of General Eisenhower, supported by his speeches abroad last September and December, is that of a man of peace and goodwill. The latest incident is said to

Continued on Page 24, Column 5

'CONFESSION' CITED

Khrushchev Charges Jet Was 1,200 Miles From the Border

Excerpts from the Khrushchev speech are on Page 24.

By OSGOOD CARUTHERS
Special to The New York Times.

MOSCOW, May 7—Premier Khrushchev jubilantly reported today the capture of the pilot of a United States plane that he said had been shot down on May Day. He said the American had admitted attempting to carry out a photo-reconnaissance mission all the way across the Soviet Union from Pakistan to Norway.

Mr. Khrushchev said the American was being held and probably would be tried, presumably for espionage, in Moscow.

The Premier said the plane had been shot down by a Soviet rocket near Sverdlovsk, 1,200 miles from the Afghan-Soviet border.

To wildly cheering Deputies of the Supreme Soviet, who had been called into a three-day session to pass on internal legislation, Mr. Khrushchev cried:

"We have parts of the plane and we also have the pilot, who is quite alive and kicking. The pilot is in Moscow and so are the parts of the plane."

Provocation Implied

He implied once again that he felt that the United States military had sent the plane across the Soviet Union as a provocation aimed at sabotaging the summit conference, which is scheduled to open a week from Monday in Paris. But he indicated that he still intended to meet with the Western leaders and play host to President Eisenhower. Mr. Khrushchev said, however, that the incident was "bad preparation" for the East-West talks.

He displayed a handful of large photographs of what he said was part of the "espionage equipment" taken from the plane's wreckage and from the pilot.

This was Mr. Khrushchev's reply to the State Department's contention that the plane may have been a weather reconnaissance aircraft that has been missing on a high-altitude flight over northeastern Turkey after the pilot, Francis Gary Powers of Jenkins, Ky., had reported trouble with his oxygen equipment.

The Soviet leader said the plane had taken a cluster of photographs of industrial cen-

Continued on Page 25, Column 1

ACCUSES PILOT: Premier Khrushchev displaying before the Supreme Soviet in Moscow one of the views of Soviet territory he said had been obtained by United States flier. Associated Press Radiophoto

VOROSHILOV QUITS AS CHIEF OF STATE

Brezhnev, Khrushchev Aide, Rising in Party Councils, Succeeds Marshal, 79

By MAX FRANKEL
Special to The New York Times.

MOSCOW, May 7—Leonid I. Brezhnev, a 54-year-old Communist party functionary, was chosen today to succeed 79-year-old Marshal Kliment Y. Voroshilov as titular chief of state of the Soviet Union.

Marshal Voroshilov, a slight and wizened figure in the otherwise robust ranks of the Soviet hierarchy, resigned because of health in an emotional scene at the close of today's session of the Supreme Soviet, this country's version of a parliament. The move was not unexpected.

The retiring chief of state was praised, decorated and finally kissed on both cheeks by Premier Khrushchev and by his successor. Minutes later the Premier, who remains unt disputed leader of the Soviet Government and the Communist party, nominated Mr. Brezhnev.

The nature of Marshal Voroshilov's illness is not known.

Long a Revolutionary Symbol

Marshal Voroshilov was a Bolshevik revolutionary for a decade before his faction seized power in 1917. To nearly all Russians he has been a popular symbol of stability and national dignity. He was associated with Lenin and the revolution, with the growth of the Red Army and the civil war, with Stalin, the defense of Leningrad and the defeat of Germany in World War II.

Mr. Brezhnev, on the other hand, represents a new generation of Soviet figures, men who were born in this century and are now coming into their own. They were weaned on the machines, in industry, agriculture and politics.

A personable and respected figure in party circles and a particularly steadfast ally of Mr. Khrushchev in recent years, Mr. Brezhnev will continue to

Underground Atom Blasts Set by U.S. to Aid Detection

By E. W. KENWORTHY
Special to The New York Times.

WASHINGTON, May 7—President Eisenhower announced today a six-fold expansion of the United States program to improve the detection of underground nuclear explosions. The President said that the expanded program, known as Project Vela, would involve a series of underground nuclear explosions of various sizes in different kinds of geological formations.

The announcement was made this morning at Gettysburg, Pa., by James C. Hagerty, the President's press secretary.

The President explained that the explosions would be limited

Text of White House statement is printed on Page 34.

to those "essential to a full understanding of both the capabilities of the presently proposed detection system and the potential for improvements in this system."

All Underground Blasts

He emphasized that all the explosions would be conducted underground "under fully contained conditions and would produce no radioactive fall-out."

Officials here also stressed that the planned explosions would not be used in any way for the development of nuclear weapons. There was no intention, they insisted, to end the present moratorium on weapons testing, which is now being renewed on a month-to-month basis.

The officials emphasized also that the decision to conduct such tests was not a pre-summit maneuver and that the timing of the announcement had not been determined by the disclosure of Premier Khrushchev

Continued on Page 35, Column 1

INTELLIGENCE ACTS ADMITTED BY U. S.

Both Soviet and American Efforts in Field Cited in Statement on Plane

By JACK RAYMOND
Special to The New York Times.

WASHINGTON, May 7—The United States statement today on the plane downed in the Soviet Union contained the first official Government disclosure that this country was engaged in aerial intelligence efforts.

Heretofore such activities have been only hinted at through announcements of Soviet activities and military strength.

For example, it was the United States that first announced Soviet atomic and hydrogen bomb explosions. The United States has consistently made public, officially or indirectly, statistical estimates of Soviet military strength, including missiles.

The State Department's announcement called attention to both the United States' own intelligence-gathering efforts and those of the Soviet Union, saying that these are "certainly no secret."

Planes Active in Turkey

It has been known for many years that in Turkey the United States has sent aircraft high into the skies, equipped with radar and various electronic instruments, to seek to learn secrets of the Soviet nuclear explosions, missile-launching bases and other military materials.

The United States has constructed giant radar antennas

ACTION EXPLAINED

Officials Say Danger of Surprise Attack Forces Watch

Text of the State Department statement is on Page 29.

By JAMES RESTON
Special to The New York Times.

WASHINGTON, May 7—The United States admitted tonight that one of this country's planes equipped for intelligence purposes had "probably" flown over Soviet territory.

An official statement stressed, however, that "there was no authorization for any such flight" from authorities in Washington.

As to who might have authorized the flight, officials refused to comment. If this particular flight of the U-2 was not authorized here, it could only be assumed that someone in the chain of command in the Middle East or Europe had given the order.

President Clears Statement

"It appears," said the statement, "that in endeavoring to obtain information now concealed behind the Iron Curtain, a flight over Soviet territory was probably undertaken by an unarmed civilian U-2 plane."

The statement was issued by the State Department after clearance by President Eisenhower. It came at the end of a day of uneasy silence. All through the day the highest officials of the Government had worked on an answer to Premier Khrushchev's charges that the United States had been caught red-handed in an aerial-intelligence operation behind the Soviet borders.

The statement contained what was probably the first official admission that extensive intelligence activities were being conducted along the Soviet frontiers. It gave no assurance that these activities would be curbed in the future.

Soviet Activity Cited

But it justified this intelligence work on several grounds. "The Soviet Union," it pointed out, "has not been lagging behind in this field." Furthermore, it said, the excessive secrecy practiced by the Russians and their refusal to accept a United States plan for mutual protection against surprise attack obliged the free world to take every precaution.

"It is in relation to the danger of surprise attack that planes of the type of the unarmed civilian U-2 aircraft have made flights along the frontiers of the

Continued on Page 28, Column 4

John Reed Kilpatrick, 70, Dies; Headed Madison Square Garden

Arena President From 1933 to 1955—Was Soldier, Builder and Athlete

John Reed Kilpatrick, honorary chairman of Madison Square Garden and its president from 1933 to 1955, died yesterday of cancer in Roosevelt Hospital. He was 70 years old.

Cuba Is Exchanging Envoys With Soviet

Today's Sections

Index to Subjects

The shootdown of Powers' aircraft on 1 May 1960 led to a major confrontation between the superpowers, as the Soviet Premier strove to obtain maximum propaganda value from the incident. This is the front page of the New York times for 8 May 1960, *after* the US had been forced to admit that the flight had taken place, but *before* President Eisenhower revealed that he had authorized it.

U-2 flies over the XB-70 bomber at Edwards AFB. After pilot Powers was downed, stories circulated in Washington that he had been forced to descend to a lower altitude before being shot down. This version of events suited the manned bomber lobby, who were fighting hard to prevent the cancellation of the then-new XB-70. If the Soviets had missiles capable of reaching the U-2's cruising altitude, the high-flying XB-70 would be redundant. They did, and it was.

reason to medium altitude. Upon his return to the US, Powers said that the U-2 was disabled by an explosion somewhere behind him, probably external to the airplane. He said that he had managed to convince his interrogators that this event had occurred at 68,000 feet. Furthermore, he said he had also managed to convince them that this was the maximum altitude at which the U-2 could fly. Powers eventually stated (in his autobiography) that he had not actually been flying at 68,000 feet. His strategy during the interrogations had been to maintain the two fictions (about the altitude from which he had been shot down, and the maximum altitude of the U-2) in order to protect the pilots of any future overflights. Perhaps the Soviet height-finding radar had not actually pinpointed the aircraft's true altitude for the missiles, he reasoned. Fortunately, according to Powers, such intercept data that the Soviets had at their disposal inclined them to agree with the 68,000 feet figure that he gave them.

The pilot's memoirs give the impression that he deceived the Soviets on his actual altitude by many thousands of feet, but given the type of mission being flown that day, it is very unlikely that Powers could have been much above 70,000 feet as he approached Sverdlovsk. This will be further explained later. Also, for their determination of the aircraft's altitude, the Soviets were partially reliant on technical analysis of the camera and film recovered from the wreckage, making their deductions from the focal length and the size of known objects as they appeared on the film.[7] There is also a body of opinion which holds that the Soviets were anxious that the West believed their missiles capable of striking at the higher altitudes. This being so, it would be in their interest for the story of a shootdown from 68,000 feet to gain currency, even if they themselves had some doubts about its veracity. Of course, Powers did indeed manage to prevent them learning the aircraft's true maximum altitude, a variable figure as we have already seen, but around 75,000 feet under most conditions.

Unknown to Powers, who was kept incommunicado in Lubianka Prison for the first two months of his captivity, the analysts at the National Security Agency (NSA) had come up with SIGINT interceptions of Soviet air defence communications which indicated that the aircraft had wandered off course in an erratic fashion, and descended. Interceptors had been launched and missiles had been fired — a large number, indeed — but Powers' plane had spent a full half an hour in an unexplained downward excursion before apparently being hit. The first strands of this intelligence reached U-2 project HQ in Washington within twelve hours of the incident, and was immediately taken at face value there, so it was hardly surprising that much of the officialdom in Washington also latched on to it over the next few days. Goodpaster told Eisenhower a very similar story, and it soon reached the public via the media. (Journalists were told in only the vaguest terms that the US had certain 'sources' for this version of events; this was 1960, and the NSA's very existence had not yet been officially acknowledged.) In secret testimony before the Senate Foreign Relations Committee exactly a month after the downing, Allen Dulles stated, 'We believe that (the aircraft) was initially forced down to a much lower altitude by some as yet undetermined malfunction.' A flame-out was, of course, the most likely possibility. Eisenhower repeated the flame-out story in his memoirs, published in 1965.

Powers later contended that the story of a descent before the shootdown had been deliberately advanced by those for whom it would be very inconvenient if the Soviets were proved to have a SAM capable of reaching the U-2 at the highest altitudes. He was referring to the top brass in the Air Force, who were anxious to protect their shiny new XB-70 high-altitude bomber from criticism that it had now become a sitting duck for missiles. Powers also accused the CIA of touting the story in the belief that it would get them off the hook; it would look better for them if pilot error, or some other mischance, had caused the plane to fall to within reach of Soviet air defences.

From this distant perspective, it is impossible to evaluate whether or not the manned aircraft lobby played a part in disseminating the not-shot-down-at-high-altitude story. Those in the US who favoured the development of missiles rather than aircraft certainly suspected this was the case, since the B-70's future was being debated at the time in Congress. (It was subsequently cancelled.) It is equally difficult to determine whether the Agency attempted to deceive, although they could certainly be faulted for failing to correct the impression that neither pilot nor aircraft would likely survive a Soviet downing, except in very small pieces.

Even the President was led to believe this. Eisenhower later maintained he had been told that the destruct device which was always carried on the U-2 during overflights would ensure that the aircraft was blown to bits. He also related how Bissell and Dulles had given him the impression that the pilot would not survive if knocked down. Neither assertion could be advanced with any great confidence.

The destruct mechanism was a two and a half pound package of cyclonite attached to the aft bulkhead of the Q-bay. (It had an inbuilt seventy-second time delay to allow the pilot to set it and then make his escape from the cockpit.) The charge was powerful enough to make a mess of the camera and other systems in the Q-bay. However, it was not of itself sufficient to completely destroy the aircraft. There was already evidence enough from earlier high-altitude U-2 upsets to prove that even when abandoned by the pilot, the aircraft often remained in one or two pieces all the way down. As for the pilot's chances of survival, high-altitude escapes from disabled U-2s had already been made, and in acknowledgement of this, the CIA pilots were provided with an augmented survival kit designed to help them escape from enemy territory if the worst should happen. It contained 7,500 Soviet roubles and twenty-four gold coins, together with messages in fourteen languages to the effect that those rendering assistance to the downed pilot would be rewarded. Powers also carried a concealed poison pin which gave him the option of committing suicide should he be captured and tortured. As the U-2 story unfolded after the shoot-down, there was much lurid and uninformed speculation in the press that, unbeknown to him, the destruct device had been rigged to blow as soon as the pilot activated it, that he was under orders to use the poison pin whatever the circumstances of capture; and so on.

This was all rubbish, but the serious questions which remain unanswered to this day concern whether Frank Powers had any reason to mislead his employers on the circumstances of his shoot-down, and whether the SIGINT evidence —

which suggested that he had done just that — could be trusted.

Regarding the SIGINT evidence, there is a big question mark. Technically, it was by now possible for ground-based monitors to pluck SIGINT on the Soviet air defence system out of the air from great distances, but such data could only give a very partial picture. The 'spooks' were able to monitor long-range HF communications between different air defence sectors, but not short-range VHF or UHF links between a sector controller and his fighter interceptors or SAM operators. As for direct radar tracking, by now the big GE job at Samsun in Turkey had the range to cover the Sverdlovsk area, but only at extreme altitude because of the earth's curvature, and with performance degradation. Despite subsequent reports that Powers' U-2 had been 'tracked' all the way to Sverdlovsk by the US, the Samsun radar cannot have played a key role in monitoring this incident. It is also worth noting that, because of the intense security surrounding U-2 overflights, the radar and listening posts were apparently not pre-alerted when one was about to take place.

So, the SIGINT sources were imperfect. Furthermore, there is evidence to suggest that the Soviets were in some state of confusion as they attempted to down the intruder; in 1961-62 Colonel Oleg Penkovsky of Soviet Military Intelligence passed information to the West indicating that the Soviets had shot down one of their own MiG-19 fighters and fired fourteen SAMs during the interception. (Incidentally, his account tallied with Powers' conclusion that his aircraft had been disabled either by the shock wave or by debris from a SAM salvo as it exploded, rather than by a direct hit.) Now if the Soviets were confused, how could the NSA draw firm conclusions from (at best) a partial glimpse of their deficient reactions to the intruding aircraft? The danger of jumping to conclusions based on partial SIGINT from the Soviet air defence system was well illustrated in 1983 when the Korean Airlines Boeing 747 airliner was brought down. The US administration went public within hours with a version of events which proved to be seriously flawed.[8] The CIA Board of Enquiry which sat in judgement on Powers' performance after his return to the US eventually discounted the SIGINT evidence.[9]

Did Powers have any reason to mislead? It is important to appreciate that the essential elements of his story were established early in his interrogation by the KGB, before he became aware that conflicting evidence had emerged back home. He says that his strategy was to tell his captors the truth where they might have a reasonable chance of cross-checking what he told them, but to pretend ignorance or inexperience in other areas. For instance, he told them this was his first overflight, and that he knew virtually nothing about the camera and SIGINT gear on board. In this way, he could limit the

The comprehensive display of U-2 wreckage mounted by the Soviets even included these shattered components of the SIGINT system. The helical spiral antenna on the left is from System 6, while unwound spools of tape from the recorder are visible on the far right.

A Soviet SA-2 surface-to-air missile. Developed in the mid-fifties, it was a significant improvement on the SA-1, which posed little threat to the U-2. As the new missile was deployed around key military bases in the USSR, U-2 overflights were routed around the danger zone, but the SA-2 had a degree of mobility. On 1 May 1960 the U-2 became the first aircraft ever to be shot down by a surface-to-air missile.

areas of questioning, and thereby protect as many secrets about the U-2 programme as possible. Only a poor dumb pilot! They were inclined to believe him — after all, that was the way they ran their own air force.

This being so, why should he risk lying to them about his altitude? If he had descended to medium altitude, he would figure that they surely would know this through their ground radar tracking or reports from fighter interceptors. If they found him out on this important point, it would be very awkward. Even if he had succeeded in such a lie, why should he maintain it once safely back in the US? Well, there might be the traditional pilot's ego, which makes him very jealous to protect his reputation. Others might screw up, but never me! Most U-2 pilot veterans insisted that it really wasn't all that difficult an aircraft to fly if you were competent. Then there was the more practical consideration of whether Powers would get his back pay — $52,000 was at stake, and there were plenty of suggestions that he didn't deserve it. (These stemmed from all sorts of misconceptions about his orders and duties; that he disobeyed orders by failing to activate the destruct device after the plane was hit — an impossible task, since he couldn't reach the activating switches located in an awkward position inside the wildly spinning cockpit[10]; that he was under instructions to kill himself with the poison pin if captured — untrue; that he had collaborated with the enemy by telling them all about the U-2 operation — unfair, to say the least, given the amount of information he managed to protect.)

To the CIA, to the public, and to most of his colleagues in the U-2 programme, Powers consistently maintained his story about being downed from cruising altitude by 'some upset from behind', which had accompanied 'a kind of hollow explosion and an orange flash'. Kelly Johnson was asked to examine the evidence, in particular some detailed photographs of the display of U-2 wreckage which had been staged in Moscow. He came to the conclusion that an SA-2 had hit the right horizontal tail. Given the nature of the aircraft in the high-altitude regime, this was enough to pitch it forward and onto its back, so that the wings broke off in down-bending. Powers laboured his point repeatedly: 'I had encountered neither engine trouble nor flame-out. During training I had experienced the latter, on several occasions, and there was no comparison. Nor had I descended to 30,000 feet. Whatever happened to my plane had occurred at the assigned altitude.'

And yet, and yet . . . there are at least two former U-2 pilot colleagues who maintain that subsequent to his return, Powers admitted to them that he *had* flamed out and descended to relight before being disabled. One of them was a close friend, Sammy Snyder, who went through U-2 pilot training with him and was also posted to 'Det B'. Shortly before his death, Snyder told author Michael Beschloss that Powers admitted as much during a fishing trip that the two took together shortly after he was released.[11] Another CIA pilot told the present author of a meeting during which Powers described to him a descent to 45,000 feet and an attempted relight there (unsuccessful because he was still too high), followed by a further descent and relight attempt at around 40,000 feet. Kelly Johnson had backed up Powers' story of being shot down at cruising altitude, but it was not inconceivable, of course, that the damage to the plane which Johnson had analysed could still have been caused by a SAM, but at a lower level. This was less likely, however. Both of the former U-2 pilots were recalling meetings with Powers which had occurred twenty-five years earlier, but it seems an important and memorable point on which to be mistaken. However, another pilot interviewed for this book who was even closer to the action on May Day 1960 than the other two, has never doubted that Powers' original story was true. Similarly, Air Force U-2 project manager Brigadier General Leo Geary, who met Powers on the Glienicker Bridge in Berlin upon his release and was intimately involved in his subsequent debriefing, remains convinced that he was telling the truth.

Were the circumstances of this flight such that an upset was more likely than on previous missions? By his own account, Powers was being asked to fly for eight and a half hours on a south-north routeing never before attempted, after spending four days in a hot, noisy hangar and being woken at 02.00 on three of those days. By the time he reached Sverdlovsk, he had already been awake for eight hours. He must have been in less than perfect shape by then.

Then there were the maintenance faults on Article 360, both of which increased the chances of a flame-out or loss of control at high altitude. Occasionally, the fuel feed from one of the tanks had malfunctioned on previous flights. In the flight manual for the U-2C, pilots are warned that 'Mach tuck and buzz characteristics, and buffet, can be aggravated by high gross weights . . . or by wing heaviness due to uneven fuel load'. Powers did not relate any such malfunction on this flight, but an altogether more serious problem cropped up when he was already three hours into the flight. The autopilot began misbehaving, causing the aircraft to pitch nose-up. Powers disengaged it, retrimmed, and flew manually for a few minutes, before re-engaging it. It worked fine for a short while, but then the problem recurred. He repeated the same sequence, with the same end result, and so decided he would have to fly the aircraft manually for the rest of the flight.

It was a brave decision; had this problem cropped up an hour earlier, he would have turned back. By this time, he had twice observed fighters below him through the driftsight; they were obviously being vectored towards him by ground radar, so he couldn't afford a flame-out which would drop him down to their level. Yet as we have seen, merely keeping the plane stable called for complete concentration at high altitude. Powers was also having to operate the camera and SIGINT systems, make written notes on the progress of the flight, and navigate. And the latter task had been proving particularly difficult. Because of the delayed take-off, all the precomputations of sun elevation that had been made for the sextant were wrong, thus rendering it virtually useless for determining position (although it could still provide heading checks). Powers was having to rely instead on basic dead-reckoning, supplemented by bearings from the radio compass. What about the driftsight? This, too, was useless for most of the first three hours of the flight, since there was a solid overcast below. Through occasional gaps in the cloud, and by taking a bearing from a radio station at Dushambe, Powers had twice managed to determine that he was drifting off-course, and make the necessary corrections. The weather was much worse than he had been led to expect; luckily, the cloud had also cleared briefly as he approached Tyuratam, enabling the 'B' camera to achieve some photo intelligence, and then the cloud had thinned and dispersed almost completely.

As if all these problems weren't enough to contemplate, Frank Powers had the additional mental discomfort of knowing that this flight would stretch the U-2's endurance to the very limit. At Peshawar, he had left behind at least two colleagues who believed that the mission could not be completed as briefed. He was being asked to fly too high for too long. Certainly, the Dragon Lady had flown for nine or more hours before now, but in order to achieve this endurance the U-2C model had to be set into the cruise-climb portion of the flight at only 57,000 feet. Flown thereafter according to a strict airspeed schedule, the aircraft would slowly rise as fuel was burnt off. This mission profile gave the very best miles per gallon by providing the optimum relationship between gross weight and altitude throughout the flight — but even if the schedule was flown precisely, the U-2 would reach only 67,000 feet at the end of nine hours flying. Powers had flown such a mission before — the one from Incirlik which routed all the way down the Red Sea to overhead Yemen and back — but on that occasion there wasn't any need to get as high as possible to avoid Soviet SAMs.

If the pilot needed to get to maximum altitude in a U-2C, the steep initial climb had to be continued until at least 66,000 feet, whereupon the pilot set up a different sort of cruise climb which used maximum power. On a standard day, the aircraft would rise to about 74,000 feet at the end of eight hours flying — but by that stage there would only be eighty-five gallons of fuel left, and all the pilots reckoned that you needed to have a hundred gallons remaining at the point where the descent for landing was started. Another significant point to watch throughout this maximum altitude profile was that the pilot had to adhere strictly to the speed specified for the particular altitude reached. It didn't vary much — 394 knots TAS at 67,000 feet and 392 knots at 70,000 feet, for instance — but if the pilot exceeded it even by small increments, there would be a further penalty to pay in fuel consumption.

It *was* possible to combine the techniques of flying for maximum endurance and maximum range, to satisfy the requirements of particular missions. One combination was to fly the maximum altitude profile until reaching the desired height, and then to level off by reducing power, taking care to maintain the schedule airspeed. Another was to descend to 60,000 feet as soon as it was prudent to do so towards the end of a long flight, this being the altitude which offered the best fuel burn to the now lightly-loaded U-2. Both of these techniques granted

Map of the ill-fated U-2 flight. Never before had a complete crossing of the Soviet mainland been attempted. Mission planners at project Headquarters in Washington apparently exceeded the capability of the aircraft by calling for this 3,800-mile trip. Pilot Powers would probably have had to depart from the specified route in order to reach his destination. As it transpired, he never had to make that decision!

the pilot some extra range, and one or other was required during the Powers flight on May Day 1960.

Whatever fine calculations and permutations were worked out at project HQ in Washington before the flight plan was transmitted down the line, the fact remained that Powers was being asked to fly for 3,300 nautical miles (3,800 statute miles) of which three-quarters was over denied territory and therefore required a sacrifice of range for height. His aircraft had full internal fuel of 1,345 gallons, augmented by the wing slipper tanks which each provided a further hundred gallons. The planned route was to head north-west from Peshawar, so that Afghanistan was crossed during the initial climb phase. The Soviet border would be penetrated south of Dushambe (formerly Stalinabad) on a direct line towards Tyuratam. Then he would turn north towards Chelyabinsk, turning to the right there, followed shortly thereafter by a left turn to take him over the southern side of Sverdlovsk. The next turns would be made before and just after Kirov, the latter lining him up for the main target at Plesetsk. Overhead Archangel he would make a left turn, so that he would fly directly over the ballistic missile submarine yard at Zhdanov. There would be a further five turns within a relatively short time as the plane was positioned for various targets over the Kola peninsula before heading north to 'coast-out' over Murmansk. He was then supposed to follow the Norwegian coast for 600 statute miles until reaching Bodo.

In order to complete the flight as planned, Powers would have to have flown with great precision for a very long time. If he were to encounter any problems en route — unfavourable atmospheric conditions, an autopilot malfunction, an inadvertent straying from the planned course, to name but three possibilities — he would be struggling to reach his planned destination. In the hangar at Peshawar, the mission navigator had wrestled with all the parameters, but failed to reconcile his calculations with those of project HQ. He conferred with Powers and Colonel Shelton. They decided to give it a go, and the navigator at Peshawar drew up an 'escape route'. If fuel was running out, Powers could cut the overflight short at Kandalaksha by turning west and heading straight for Bodo across northern Finland and Sweden. He also pointed out the airfield at Sodankyla in Sweden to which the pilot could divert *in extremis*. Whether this was of much comfort to Powers has gone unrecorded; he chose not to discuss this aspect of the flight in his memoirs. He may have contemplated abandoning the planned course after Plesetsk, especially since he-had already flown into Bodo and knew it was not an easy field to approach because of the local terrain — he would really be pushing his luck to start down for there with much less than a hundred gallons of fuel remaining.

There again, by even daring to proceed with this May Day enterprise, the Agency was pushing its luck to the limits. Back in project HQ, more than one mission planner thought that the flight should be scrubbed because of the secure communications breakdown, and because it was known that the Soviets had moved some more SAMs to the areas being overflown. The fact that Sverdlovsk was defended by SA-2 missiles had been determined by no less a person than Vice-President Richard Nixon, who had reported seeing the characteristic 'Star of David' installation when the Soviets flew him there during his visit to the USSR in July 1959. Since then, however, Marty Knutson had flown a U-2 as far as Sverdlovsk to photograph this significant Soviet military-industrial complex, and returned safely. Even so, the odds were shortening that a flight over denied territory would soon encounter some sort of serious trouble. During the four years that the Agency had been conducting operations, thirteen Dragon Ladies had been destroyed or seriously damaged in accidents and eleven pilots killed. There had also been countless flame-outs, and a good proportion of these had ended in deadstick landings after the pilots had failed to get a relight.

Admittedly, the majority of these incidents had occurred on training flights over the US, which were the responsibilty of the Air Force U-2 unit. The CIA reckoned that its pilots, procedures and maintenance were better, but how much longer could their good fortune last? When Operation Overflight had begun, no-one had expected it to last much more than eighteen months, before Soviet countermeasures forced a halt. It was now nearly four years later, and the Agency had experienced a few close calls of its own during operational missions. On a peripheral SIGINT flight out of Incirlik one night, an aircraft had suffered complete instrument failure, but the pilot brought it safely back without straying over the border. On an overflight of the Chinese mainland, Sammy Snyder had experienced a flame-out caused by the freezing engine bearings, but Taiwan was fortunately within gliding distance, and he made it down safely to a deadstick landing at an airbase there. Then there were the two forced landings in Japan and Thailand, and another caused by complete electrical failure. But as Allen Dulles pointed out: 'To stop any enterprise of this nature because there are risks would be, in this field, to accomplish very little.'

Poor Frank Powers! Whatever the exact circumstances over Sverdlovsk — and we shall probably never know — a great deal was demanded of him that day. If he had died in the wreckage, or killed himself with the poison pin, he would surely have been enshrined as a national hero. Instead he survived, conducted himself admirably during his confinement, and came home to a less than rapturous welcome from misinformed detractors. Not only that, he became embroiled in a messy divorce suit within a year of his return. (His marriage to Barbara had been under strain since long before May Day 1960.) He subsequently remarried successfully, but his new colleagues at Lockheed noticed the emotional scars left by his unfortunate experiences. 'He remained kind of bitter about the whole thing,' said one. As a much-needed vote of confidence, Kelly Johnson had offered him a job at the Skunk Works, flying the U-2 as an engineering test pilot. Powers clocked up a further 1,400 hours on the type over the next eight years, until a decline in U-2 flight test work in 1970 meant that his services were no longer required.[12]

After a couple of years in limbo, he started work with a Los Angeles radio station, reporting on traffic conditions from the air in a Cessna light plane. In late 1976 he was hired by the local TV station KNBC and sent on a helicopter flying course. He would be piloting the station's heavily-equipped 'telecopter' to provide live relayed coverage of fires, police chases and other newsworthy events. On 1 August 1977 Powers and his cameraman were returning to Burbank after covering a brushfire near Santa Barbara when the engine quit. The 'telecopter' crashed onto a sports field in Encino just three miles from the Skunk Works, killing both Powers and his passenger. The chopper carried a TV camera and data-link antennae on unwieldy-looking external mounts, which may have contributed to Powers' failure to control the machine's unpowered descent. Some suggested that Powers did not have enough helicopter flying experience to cope with such an emergency. In any case, it seemed an inappropriate end for the man who had cheated death over the Soviet Union, but it was more than that. Having spent over 2,000 hours of his life in the U-2, where fuel management was a cardinal skill, the subsequent official verdict on the helicopter accident represented a huge irony. Powers had allowed the fuel tank to run dry![13]

'Frank tried so damned hard — he was the kind of guy that couldn't say no,' an associate was reported as saying. 'He was just out to please the newsroom and do the best job he could, even if it meant stretching it further than he should. Which is what happened.'

1 This most famous of all U-2 pilots was a twenty-seven year-old First Lieutenant in the 508th SFW at Turner when the call to join the U-2 programme came. A small-town boy from the Appalachian mountain area along the Virginia-Kentucky border, he met Barbara Gay Moore at Turner in 1953 and married her nineteen months later. It was she who insisted that he be known by his middle name, Gary. As far as his flying buddies were concerned, he was always called Frank. When he was shot down over Sverdlovsk on 1 May 1960, he was one of the most experienced of all the CIA pilots, with some 500 hours on type.

2 In vast unpopulated areas of the Soviet Union, railways provided the main artery of transport and communication. It was therefore a safe bet that military installations would be situated alongside, especially if the movement of heavy equipment such as long-range missiles was involved. Sections of a number of U-2 overflights therefore followed railroad lines, especially the Trans-Siberian Railway, in an attempt to locate IRBM and ICBM launch sites.

3 'Det C' also made a significant contribution to the 'legitimate' U-2 role of weather research, since it flew widely in an area that was frequently subjected to typhoons. Much was learnt about this destructive atmospheric phenomenon during these flights, and on one occasion a U-2 became the first aircraft to fly over the eye of a typhoon.

4 The variation in temperature was mainly due to the different height of the troposphere in the polar regions, where it could be encountered as low as 25,000 feet, compared with 55,000 feet over the equator. A U-2 climbing out would encounter rising air temperatures once it was through the tropopause. When heading out on a northern mission, it would therefore have passed through some 35,000 feet of rising temperature by the time it reached 60,000 feet. What with all the atmospheric variations, the importance of accurate meteorological predictions to the planning of U-2 flights becomes apparent, for flight performance reasons as well as to ensure cloud-free conditions over the target. U-2 operations often relied upon advance data from the upper atmosphere, supplied to meteorological stations by radiosonde balloons.

5 In fact, this was not to be the definitive U-2 intake. In the mid-sixties, when development of the scaled-up U-2R model powered by the same J75 was initiated, Lockheed engineers re-examined the airflow requirements and came up with a 'coke-bottle' type of flared intake design. This was included on the new model, and also retrofitted to the earlier J75-powered aircraft.

6 Even the technical press made some inexcusable errors and assumptions. For instance, *Aviation Week* published photos of the wrecked U-2 on display in Moscow, and commented that 'condition of the blades at the bottom of the engine was an indication that pilot Francis G Powers made an emergency belly landing', (issue of 30 May 1960). It also uncritically accepted a government suggestion that Powers had radioed back to base from over Sverdlovsk that he was in trouble, despite having also published a list of U-2 equipment which (correctly) included only one piece of communications gear — the AN/ARC-34 UHF radio which had a range of less than 200 miles.

7 It is not clear whether Soviet analysts were able to examine the film that was actually being exposed at the moment the shootdown occurred; it is quite possible that they made their estimate according to earlier exposures along the route, which would not have revealed the precise altitude at which the plane was disabled.

8 For an excellent analysis of why this is so, see *The Target is Destroyed* by Seymour Hersh, Random House, (New York, 1986).

9 A summary of the Board's conclusions was made public, and the relevant section made only oblique reference to the NSA evidence. It reads: 'Some information from confidential sources was available. Some of it corroborated Powers and some of it was inconsistent in parts with Powers' story, but that which was inconsistent was in part contradictory and subject to various interpretations. Some of this information was the basis for . . . stories . . . that Powers' plane had descended gradually from its extreme altitude and had been shot down by a Russian fighter at medium altitude. On careful analysis, it appears that the information on which these stories were based was erroneous or was susceptible of varying interpretations. The Board came to the conclusion that it could not accept a doubtful interpretation in this regard which was inconsistent with all the other known facts . . .'

The CIA certainly went to some lengths in evaluating the NSA version of events. There was some indication that Article 360 had wandered up and down, as well as off course, before being brought down. Could this be due to some engine problem — an overtemp for instance? Out in California, Lockheed test pilot Bob Schumacher was asked to try and recreate such a problem at altitude in a U-2, but no firm conclusions were drawn.

The Board's final verdict was borne out by later experience when U-2s were flown over communist China. Listening posts on Taiwan monitored the reaction to these incursions by the mainland's air defence system (which mainly comprised Soviet equipment). Nearly every time, mainland radars underestimated the U-2's cruising altitude by some 3,000 feet, and frequently reported fluctuations of 5,000 feet or more, when none had occurred. The communists also plotted the aircraft's track imperfectly, suggesting a zig-zag course when, in reality, the U-2 pilot had maintained a straight-line course.

10 Powers' failure to operate the destruct device did allow the Soviets to examine the U-2 SIGINT and ECM systems in a relatively undamaged state. Opinions differ as to whether the USSR thereby scored a scientific intelligence coup. They may have learnt something new about travelling wave tubes from those in System 9, or in the X-band amplifier of System 6, and System 3 contained some recent advances in swept-frequency receivers. However, they might well have recovered significant parts of the systems intact, even if the destructor in the Q-bay had gone off, since there were components scattered all over the aircraft. And, of course, they may already have obtained undamaged examples of this equipment through technical espionage.

11 In his book *May-Day: Eisenhower, Khrushchev and the U-2 Affair,* Beschloss advanced a tortuous thesis worthy of the best traditions of conspiracy theory. He suggested that it suited the Soviets to have everyone believe their missiles capable of reaching 68,000 feet, even if it were not the case. This also suited Powers, who might stand accused of inattention or incompetence if he had allowed the plane to descend. During the subsequent negotiations for the pilot's release, according to this thesis, the US quietly pledged not to refute the Soviet version of his downing, (so that the Soviets would not be embarrassed) and Powers subsequently concurred in this 'out of patriotism and gratitude for his release'. There is no evidence to support this hypothesis, and this author agrees with Richard Bissell, who found it 'too complicated to be very appealing'.

12 Since this event coincided with the publication of his memoirs, some have suggested that Lockheed fired Powers (perhaps at the CIA's behest) for re-opening old wounds. The truth appears to be that test work on the U-2 was indeed declining, now that the new U-2R model had successfully entered service. He might have been considered for transfer to other Lockheed flight test programmes, but of all the test pilots then employed at Burbank, Powers was by far the least qualified. He didn't even have a twin rating, and seemed disinclined to make the effort to get one.

13 During the accident investigation, it was revealed that the helicopter had developed a faulty fuel gauge some days earlier. Powers had written up the fault. There was some evidence to suggest that a mechanic had fixed it without signing the repair off. His failure to do so might have led Powers to believe that he had a few more gallons remaining than was actually the case.

Chapter Four
Crowflight

The U-2 programme was probably one of the greatest bargains ever realised by the American taxpayer. Of course, some people were unhappy with the set up, but as time went on they began to understand that what we really had was the best of both worlds — CIA and Air Force.
Brigadier General Leo Geary, 1987.

Unless you were a thorough-going country boy, it was not one of the more desirable locations for an airbase. Driving west from the bright lights of San Antonio, Texas, you were soon the only car on the road as the rolling farmland stretched all around, interspersed with only an occasional settlement. West of Uvalde, even the farmland gave way to a desert scrubland of cactus and sagebrush, and yet it was still sixty miles before you reached the gates of Laughlin AFB, just outside the small Texas border town of Del Rio. The town itself had grown up around the San Felipe Springs, which provided a bountiful supply of good drinking water for travellers and horses following the stagecoach trail to San Diego. Just beyond Del Rio lay the muddy waters of the Rio Grande, crossed here by a bridge leading straight into the ramshackle main street of Ciudad Acuña on the Mexican side.

As the airmen and families of SAC's 4080th Strategic Reconnaissance Wing (SRW) made their way to this isolated spot in the early months of 1957, the area was suffering from an unprecedented dry spell. Rainfall had been way below average for the past three years, and the heat and dust of the long summer would prove almost intolerable for many of them. They were used to the green fields back east, for the 4080th had largely been created at Turner AFB, Georgia, out of the personnel and assets of the two SAC F-84 fighter wings which were disbanded there.

SAC had activated the 4080th on 1 May 1956 to operate the the Martin RB-57D variant of the British Canberra which had been secretly built by the Martin Aircraft Company at Baltimore. The squadron which was to fly them was designated 4025th Strategic Reconnaissance

U-2A from the 4080th SRW captured in flight above the cloud line. This is how the SAC aircraft looked in the late fifties and early-sixties, in highly-polished natural metal finish. Clearly visible as it protrudes from the cockpit is the pilot's rear-view mirror, which he used to check whether the aircraft was leaving a condensation trail.

Following the arrival of the first Air Force U-2A at Laughlin AFB, two pictures of the aircraft in military markings were released. This is one of them, showing the aircraft in typical climb-out attitude. The bulge under the right wing is the engine oil cooler air scoop.

Squadron (SRS), and they called themselves the 'Black Knights' after the codename which SAC had issued for the project. Unlike Bell's X-16, the Martin plane had survived the advent of the U-2, and the first of an eventual twenty of the big-wing aircraft had arrived at Turner on the last day of May 1956. They were delivered in four distinct configurations, each designed for different tasks. Some carried cameras, while others had only ELINT sensors. One even had an early side-looking reconnaissance radar. Some could be refuelled inflight, others not. The six ELINT versions carried two crew; the rest were single-seaters.

Despite the rebuff that had been dealt to the Air Force when President Eisenhower assigned control of the U-2 to the CIA, SAC was determined to stake its place in the high-altitude reconnaissance business, and a detachment of photo RB-57Ds had been sent off to Yokota AB near Tokyo in September 1956. From here, they had penetrated the Soviet landmass around Vladivistok and flown quite some way into the interior, drawing a stiff protest from Moscow.

While the Black Knights were still in the Far East, SAC decided to move the unit's home base to Laughlin. The wide open spaces of west Texas were deemed more suitable for training and operating, especially since the 4080th was now also slated to become parent unit for the Dragon Lady training outfit which the Air Force was running at The Ranch. Once the CIA operation was under way, SAC had therefore selected some more pilots for U-2 training at The Ranch. In similar fashion to the CIA recruitment, these pilots were drawn exclusively from the ranks of SAC's disbanding fighter wings. They were joined at The Ranch by groundcrew from the 4080th, who received their first instruction on the new super-secret aircraft from Lockheed at Oildale.

Colonel Jack Nole was designated the first commander of the Air Force U-2 squadron, the 4028th SRS, and on 13 November 1956 he was the first of its pilots to be checked out on the U-2A. Another twenty Air Force pilots made their first flights at The Ranch, while Laughlin was being made ready to receive the Dragon Lady. This was no easy task; the base was short on most things — it didn't even have a perimeter fence when the troops from Turner began moving in. Among other deficiencies, there was no married accommodation, since the previous unit had been a flying school, using T-33s to train young — and mostly single — pilots. The 4080th people had to find accommodation downtown in Del Rio, or travel fifty miles each day to and from billets at a run-down Army post in Bracketville.

On 11 June 1957 Jack Nole led the first of two three-ship U-2 formations from The Ranch to Laughlin. Their arrival went virtually unnoticed by the local population. As far as SAC was concerned, things could stay that way, since the less attention given to the 4080th's activities, the better. All the wing's personnel had received strict instructions to keep their mouths shut while off-base; many married airmen didn't even tell their wives a thing about the U-2 until three years later, when it truly became public property in 1960, after Frank Powers was shot down.

However, SAC's policy of absolute secrecy was shattered after just eighteen days. The first U-2 pilot to be checked out at Laughlin decided to take a trip downtown — with his aircraft! Lieutenant Ford Lowcock had set up house in the Hunter subdivision of Del Rio, on the outskirts of town near to the small civil airport. Just before nine in the morning on Friday 28 June, only three dayş after his first U-2 ride, Lowcock flew low across town and over his little house on Avenue P. It looked like he was staging a private airshow for the folks back home — a wife and two small sons. He made some spiralling left hand turns, but somehow failed to pull out as he neared the ground. The aircraft struck a ceniza-covered hill next to the airport, left wing low, and slid along the ground for 150 feet before being flipped upside-down. There was no fire, but the twenty-eight-year-old Lowcock was knocked unconscious, and died from asphyxiation.

Only two hours later, First Lieutenant Leo Smith took off from Laughlin for a high-altitude training flight and he, too, did not return. The plane crashed about forty-five minutes later some thirty miles north of Abilene, Texas. The accident investigation determined that the pilot had allowed fuel to move from the right wing to the left. The autopilot was holding this wing up, and when Smith lifted the control to make a scheduled left turn over the Abilene VOR beacon, the left wing dropped and put the aircraft into a spin, from which recovery was impossible. Fuel imbalance had possibly been Lowcock's downfall as well.

The base was obliged to put out short statements about this double disaster to the local papers, although the type of aircraft involved was carefully not specified. A fortnight later, however, 4080th SRW commanding officer Colonel Hubert Zemke was cleared by SAC HQ to belatedly announce the arrival of the U-2 at Laughlin. His statement consisted of the now-familiar rubric about high altitude weather research, in support of NACA and Air Force objectives.

In fact, the U-2s of the 4080th *were* about to embark on a high-altitude research programme, but the object of attention wasn't so much the rain clouds as atomic clouds. The SAC squadron would be taking air samples in the lower stratosphere to determine the exact composition of fall-out debris from nuclear weapons tests, and the pattern of its dispersal around the globe. The amount of information about a nuclear weapon that could be deduced from a study of the fall-out was quite substantial. Indeed,

Nuclear tests in the atmosphere were mounted by all the nuclear powers until 1963. Although the Soviet tests were conducted in remote regions, a great deal of intelligence could be gleaned from analysis of the fall-out particles brought back from high altitude by the U-2s of Operation Crowflight.

the very first Soviet nuclear test had only been detected when a USAF RB-29 on a sampling mission between Japan and Alaska had encountered high amounts of radioactivity. This was on 3 September 1949, five days after the event, and a long time before any announcement was made by the Soviets. Since then, the esoteric science of 'weapons diagnostics' had advanced, so that careful analysis of airborne samples could reveal precise details about a bomb's yield, construction and composition, the nature of the fusion and/or fission reaction, and whether it had been detonated at ground level, underwater or in the air.[1]

By now, the US had conducted over eighty nuclear tests in the atmosphere, and twenty Soviet tests had been announced (sometimes by Moscow, other times only by courtesy of the US). There had also been nine British tests. So there was now a good data base, especially since American sampling and diagnostic techniques had been proved by reference to the country's own nuclear test programme. Of course, the patterns of fall-out dispersal from nuclear tests varied considerably according to the weather and the winds, and this could complicate the process of collecting samples from the other side's test shots. There was also a need to sample the fall-out clouds as soon as possible, since some of the fall-out particles in which the analysts were interested had only a short 'half-life'.

Ironically, the advent of thermonuclear weapons ('H-bombs') had made the collection process a little more predictable, provided that high-altitude aircraft were available. This was because the fireball from a weapon of megaton yield, being larger and much hotter than that produced by an A-bomb, would inevitably rise into the stratosphere, where winds were negligible and airflow more stable. So the U-2 and the RB-57D were prime candidates for the sampling mission.

As far as the Dragon Lady squadron was concerned, sampling operations fell into two categories, one highly classified and the other which suffered from fewer security constraints. One category concerned the long-term study of fall-out behaviour in the stratosphere, a project which was officially entitled the High Altitude Sampling Program (HASP). On these flights, the 4080th's prime 'customer' was the Defence Atomic Support Agency (DASA), which eventually published the results of the U-2 sampling. DASA also supplied the findings to a United Nations committee which had been set up in 1955 to monitor the effects of nuclear radiation in response to growing worldwide concern. But DASA did not publish results from the second category of sampling flights, which were flown for weapons diagnostics purposes, mainly from Alaska whenever the Soviets were conducting a series of nuclear test shots. The samples gathered on these flights were analysed by defence scientists at the Livermore and Cambridge laboratories, where they learnt much about the Soviet state-of-the-art in nuclear weapons.

The DASA surveys were of great interest, especially to those nations which were beginning to get very agitated about atmospheric nuclear tests. More and more megaton-yield shots were going off, causing long-range radioactive fallout to spread around the globe. India and Japan led the protests, but there was also concern in Western scientific communities. Two reports from British and American scientists in 1956 attempted to soothe public concern, but even they were forced to admit 'the inadequacy of our present knowledge'. Quite apart from the increasing amounts of plutonium and harmful uranium isotopes being flung into the air — not all of which was confined to the immediate vicinity of the test site — the presence in long range fallout of the strontium-90 isotope was causing great concern. This fission product remained radioactive for years, and was known to concentrate in the bone structure once it had penetrated the human body. A heavy dose could theoretically cause bone marrow cancer.

So the Dragon Lady pilots were sent out to look for strontium-90, cesium-137 and all the other potentially harmful particles which were known to be flowing around the stratosphere, and eventually returning from there to contaminate mother earth. The idea was to determine how much nuclear debris was up there, and how long it was staying put before 'falling out' to the ground. The brains at DASA drew some lines of longitude on a map of the Americas, and told the 4080th to fly up and down them twice a week at altitudes varying from 50,000 to 70,000 feet. The lines covered thousands of miles; many of the flights lasted seven hours or more. Like scientists everywhere, their appetite for mountains of data seemed insatiable, and they wanted the same sorties flown over and over

General Thomas White, US Air Force chief of staff, gets a briefing on the SAC U-2 operation during a visit to Ramey AFB, Puerto Rico, in late 1957. This is a 'hard-nose' U-2A dedicated to the sampling mission. The nose duct through which the air entered is open, and two access panels have been removed to show the workings of the sampling system, in which filter papers were rotated into the airflow to capture particles of debris from nuclear tests.

again. They called it 'minimizing the sampling bias'. For variety, they threw in a few orbiting missions and climbing flights as well, but mostly they required the U-2 pilots to cruise along prescribed 'sampling corridors'. The scientists had determined that over a sufficiently long period, the earth's zonal circulation could be expected to carry all stratospheric air through such a corridor. What with the rigid twice a week schedule that was set up, and the repetitive nature of the flight tracks, the 4080th felt at times as if it was running a regular scheduled airline rather than a go-anywhere, anytime military outfit.

During the summer of 1957, Lockheed delivered six late-production U-2A models which had been specially modified for HASP, or Operation 'Crowflight', as it soon became known to the men of the 4080th. These aircraft had a small intake door at the tip of the nose, which was opened and closed by a cockpit control. The door led into a duct, which gradually widened (thus slowing the airflow down) until the point was reached where a filter paper was placed across the airflow in a ring holder. The filter papers consisted of cotton fibres with a gauze backing which were impregnated with an oily substance designed to retain minute particles. Four filters could be exposed on the same flight; the ring holders were mounted in a circular rack which pivoted at its centre above the duct. By means of an electric actuator operated from the cockpit, the papers could be rotated in and out of the duct in sequence. Lacking the fibreglass panels used for SIGINT on other U-2s, the six HASP U-2s were known as the 'hard nose' aircraft.

Training flights with them began at Laughlin on 22 August, and the HASP began in earnest in late October 1957, when three aircraft each were despatched to Ramey AFB in Puerto Rico ('Det 3') and Plattsburgh AFB, New York state ('Det 4'). It was the beginning of what turned out to be a seven-year effort during which some 45,000 flying hours would be clocked up at more than a dozen operating locations worldwide. A typical Crowflight detachment consisted of four or five pilots and fifty groundcrew from Laughlin. This show got on the road courtesy of at least two C-124 transports, plus a KC-97 or KC-135 which accompanied the U-2s during their ferry flight. The wing adopted the black crow mascot from the label of a well-known brand of bourbon. This bird appeared somewhat dishevelled and definitely the worse for wear. After a typical forty-five-day tour of duty at an operating location, the same could often be said for the U-2 pilots and groundcrew!

From each deployment base, the U-2s flew north and south along the predetermined tracks, usually launching two aircraft at a time to head in opposite directions. From Plattsburgh, for instance, they ventured along the seventy-one degrees west meridian to beyond the Arctic Circle (sixty-seven degrees north) and the Tropic of Cancer (twenty-one degrees north). This entailed flying across some quite inhospitable terrain for hours on end; the frozen wastes of northern Canada or the wide open spaces of the western Atlantic Ocean. Three hours before the U-2 was due to take off, therefore, an Air Force search-and-rescue (SAR) C-54 would depart from the Crowflight base and proceed along the flight path. Since the U-2 flew about twice as fast, it would eventually overhaul the lumbering transport, but this procedure gave the maximum SAR coverage of the route, since both aircraft would reverse course at about the same time. The C-54 carried a para-medical team to assist the pilot if he went down. Their services were to prove necessary more than once over the next few years.

Back at Laughlin, the training effort gathered pace, with another twenty-six pilots checking out on the Dragon Lady by year's end. Once the six HASP aircraft had deployed, there were only some half-dozen U-2s remaining at home base. The maintenance troops worked round the clock to produce serviceable aircraft to meet the heavy flight schedule. It wasn't easy; the peculiar circumstances in which the U-2 had entered the Air Force inventory meant that the normal military paperwork routine had been side-stepped. Technical manuals were hard to come by and had non-standard layouts. Sometimes the groundcrew worked from handwritten notes they had made in the classroom during U-2 ground school at The Ranch. If they wanted a particular part, it came direct from Lockheed rather than through the usual logistics channels. And there was the constant problem of every plane being different — it was axiomatic that, for instance, a canopy or a balance arm from one U-2 wouldn't fit another one! So the 'Article' number had to be specified each time a U-2 part was ordered. A contraption consisting of lead bars suspended from an A-frame in line (hence its nickname — 'the glockenspiel'), had to be carried in the Q-bay for training flights when the mission payloads were not carried. Even these ballast weights were non-standard!

The maintenance shops at Laughlin did their best to cope with these problems, and with a different set posed by the peculiar construction of the RB-57Ds, but it was an uphill struggle. Then a series of accidents focused the attention of higher headquarters on the Laughlin wing. The 4080th suffered eight crashes in five months. Following the twin U-2 disasters of 28 June, there had been two non-fatal incidents involving the Dragon Lady in late September. Jim Qualls had deadsticked one aircraft into a ranch, and squadron boss Jack Nole had made a remarkable escape by baling out of an aircraft which had broken up at high altitude after the flaps had somehow extended. It was said to be the highest parachute escape in history.

This was not all; the wing's pilots flew T-33 trainers for proficiency, and two were lost while in the traffic pattern, one at Laughlin and one at Offutt AFB. In the latter accident, which took place right under the noses of SAC's top brass at command headquarters, the two crew had been killed. About the same time, two 4025th SRS pilots had been killed in a B-57 accident at Andrews AFB. General McConnell of SAC's 2nd Air Force (to which the 4080th reported) decided to pay a visit to Laughlin. On 5 November, he flew over from Barksdale in his VIP C-54, but as he entered the traffic pattern at Laughlin, a Black Knights B-57C trainer making a touch-and-go landed with the wheels up, and the runway was blocked. The fuming general had to divert back to Barksdale!

Within a week, McConnell had fired the 4080th wing commander, Colonel Hubert Zemke. In flying circles, they didn't come more distinguished than this forty-four-year-old wartime ace. His gung-ho leadership of two fighter groups in England in 1943-44 had won him lasting fame and a bunch of medals. Zemke had only taken charge of the 4080th when it moved to Texas, but he was already a familiar figure to many of its personnel, since he was a former wing and division commander at Turner AFB. But he was ill at ease in the tightly-controlled world of SAC in the late fifties, when standardization and accountability were the watchwords, and a commander had to work with large staff elements at wing level and higher headquarters. So Zemke was shuffled off into a desk job at North American Air Defense Command (NORAD), and McConnell drafted in Brigadier General Austin J. Russell to replace him. Russell had commanded a SAC bomb wing and was a dour disciplinarian in the LeMay tradition. He ran the show at Laughlin for the next year before handing the command back to an officer of Colonel rank, Andrew Bratton. Russell had only been in charge for ten days when there was yet another U-2 accident. The weather in this isolated corner of Texas was generally good, but on the evening of Friday 22 November, there were thunderstorms brewing. Operations went ahead and launched Captain Benedict Lacombe on a long night training sortie. Shortly after 11 p.m., after five hours in the air, Lacombe crashed thirteen miles south-east of the base in open country. He had apparently lost control in IFR conditions; a malfunctioning gyro indicator could have been responsible. By the time that Lacombe had baled out of the aircraft he was too low for the parachute to deploy.

The pace of new pilot training slackened during Russell's tenure, but Crowflight continued at Ramey and Plattsburgh until June, and the Dragon Lady squadron made the first of many deployments to Eielson AFB, Alaska in February 1958. They were sent there for 'hot' sampling of the fall-out from a new series of Soviet nuclear shots, which were mainly conducted at their Arctic test site on the Novaya Zemlya chain of islands. As the fireball from the Soviet shots topped out in the stratosphere, the debris started moving due east, and within twenty-four hours it would come within range of the U-2s flying north and west from Eielson across the frozen wastes of Alaska. Once again, a rescue C-54 was in attendance for these missions across inhospitable terrain, which were codenamed North Flight and frequently took place in the constant darkness of the polar winter. The squadron found a couple of nooks and crannies in the U-2 fuselage into which were stuffed a sleeping blanket or two, and the survival gear in the pilot's seat pack was augmented. The upper surface of the aircraft's ailerons and elevators, together with both sides of the rudder, was painted bright red, in the hope that the rescue aircraft would be able to spot a downed U-2 against an icy background, but the pilots were under few illusions about their prospects of survival, should the worst happen.

As for the sampling gear, a new Q-bay hatch called the F-2 was introduced, with a large streamlined duct protruding from its port side. This new particle sampler could accommodate six filter papers, two more than the hard-nose version, and yet still leave enough room in the Q-bay for another form of collection, that of gaseous samples. In the P-3 Platform System, these were collected by a bleed from the engine

Close-up view shows the F-2/P-3 sampling system, which fitted in the aircraft's Q-bay and superseded the earlier nose-mounted arrangement. In addition to the filter papers, gaseous samples could also be taken in six bottles. Back on the ground, scientists could determine a surprising amount of information about Soviet nuclear weapons from a careful analysis of the samples.

compressor, with the gas samples being stored in six spherical shatterproof bottles. The pilot monitored the pressure build-up in these bottles. After about fifty minutes, upon reaching 3,000 pounds per square inch, a shut-off valve was automatically activated and the pilot selected the next bottle on his P-3 control panel. There was one other panel which he would be well advised to keep an eye on during these sampling operations. The radiation count of the air entering the sampling duct was measured by a dosimeter and presented to the pilot on the right hand system panel. There was a warning light which could be set to come on if the count got too high.

Most of the U-2 pilots reckoned there was no great personal risk, especially considering how insulated they were in the partial pressure suit, immersion suit and all the other gear. In any case, much of the most penetrating radiation (that from gamma rays) was known to decay within the first twenty-four hours, so it really wasn't a worry, at least not on Crowflight. Nevertheless, if the dosimeter had registered a 'hot' count during a sampling sortie, the returning aircraft was washed down with soap and water before the groundcrew were allowed to crawl over it. The people that really needed to worry about radiation were their colleagues in the 4025th; in March 1958, RB-57Ds were despatched to the Marshall Islands to monitor Operation Hardtack (Phase 1), a series of thirty-one US nuclear tests at the Eniwetok Proving Grounds which lasted until August. There they got much closer to the fall-out clouds, and more quickly. In the light of modern research, such exposure would be cause for concern, but things were different then. In any case, exposure to radiation hazard was not a new experience for many of the Dragon Lady pilots and groundcrew. While training at The Ranch, which was just a few miles from the AEC's Nevada test site, they had all been required to wear film badges, and on one occasion the secret base had been evacuated for two weeks when a test shot was scheduled. None of them had apparently come to any harm.

After a six-month pause, Soviet nuclear tests resumed in late September 1958. Once again, the 4080th U-2s were in attendance at Eielson, although the aircraft had only just been cleared to fly again at altitude after a two-month grounding. This had been imposed by General Russell after a traumatic twenty-four hours at Laughlin on 8/9 July when two pilots had been lost in mysterious circumstances. The first of these was Chris Walker, one of the four British pilots now being trained for the Agency. He went down in the Texas panhandle near Amarillo on the Tuesday afternoon. By midnight, the wing had despatched an investigation team led by operations chief, Colonel Howard Shidal, to the scene. They had only been at the scene a few hours when the news came through that Captain Alfred Chapin had been killed in another crash near Tucumcari in eastern New Mexico, less than a hundred miles away.

Both pilots were out on high-altitude training sorties, but neither had reported any difficulties. Although Walker was a rookie on the Dragon Lady, he had lots of experience in RAF Canberras, and had been progressing smoothly through the training programme. Chapin had been flying the U-2 for over a year, and was considered one of the wing's best pilots. He had just received the Distinguished Flying Cross after bringing back a U-2 for a deadstick landing the previous August following a power failure. He had been chosen as General Russell's instructor pilot when the wing boss checked out on the Dragon Lady in March 1958.

All sorts of rumours spread round the base. The possibility of sabotage was seriously considered. Almost the entire wing was corralled into the base theatre at Laughlin and lectured about loose talk, security precautions, Soviet spies, and so on. The exact cause of the crashes was never pinned down, but oxygen problems were suspected. It would have been easier to determine what went wrong if the pilots had survived. The fact that they hadn't served to confirm a decision taken a few months earlier, that the U-2 would have to be equipped with an ejection seat. Lockheed was now working on this, but the seat didn't reach the operating units until the turn of the year.

In the first week of August 1958, Russell partially lifted the grounding imposed a month earlier to allow low-altitude training flights. Within four days, another pilot was dead. Paul Haughland, a twenty-seven-year-old First Lieutenant making his first U-2 flight, crashed on final approach to the base. Although the rescue services were quickly on the scene, there was nothing they could do for Haughland since the aircraft had caught fire. Well, everyone knew what a difficult thing it was to land a U-2 if you weren't real sharp, so training flights were not suspended this time.

Operation Crowflight now entered a new phase. Radiochemical analysis of the filter papers from the early flights had shown that fall-out was not mixing across the stratosphere as much as had been thought. It was flowing round and round the earth as expected, but was staying at roughly the latitude at which it had been injected. Even though the HASP sampling corridors were half a world away from some of the test sites, debris from individual shots could nevertheless be identified almost every time the U-2s flew along them, and sometimes within a fortnight of the test shot. And for the first few weeks, the debris flow was still sufficiently concentrated to merit the description 'hot cloud' by the scientists. Another, somewhat disturbing, conclusion was that strontium-90 and other nuclear debris was falling out of the stratosphere more rapidly than had been anticipated. Since HASP had already shown that the debris wasn't dispersing very well up there, it became clear that higher concentrations of fall-out than anticipated were descending to the ground in certain regions. Now that the US and Britain were staging megaton-range nuclear tests in the Southern Pacific, a further extension of the sampling corridor into the southern hemisphere was required.

CROWFLIGHT

The Crowflight mascot of the SAC U-2 squadron.

Argentina gave permission for HASP flights to be staged from its territory, although the communist opposition protested. On 11 September 1958 'Det 4' relocated from Plattsburgh to Ezeiza airport near Buenos Aires, where it would remain for the next eleven months. From here the U-2 pilots flew up and down the sixty-four degrees west meridian, going as far as Trabajares, Brazil (nine degrees south) in one direction, and the Falkland Islands (fifty-seven degrees south) in the other. The sampling gear soon picked up debris from the new American and British tests, which were codenamed Operations 'Hardtack' and 'Grapple' respectively. Meanwhile, 'Det 3' continued at Ramey, from where monthly deployments were made to Plattsburgh to cover the northern latitudes. The six hard-nosed aircraft were outfitted with the F-2 hatch in June 1959, and from then on a total of ten filter papers could be exposed on any one HASP sortie.

The atomic scientists now called for an intensive sampling effort in the northern latitudes, so in September 1959 the Crowflight deployment pattern changed again. Ezeiza was temporarily closed down. Three aircraft operated out of Laughlin, and made a monthly deployment to Ramey. The other three HASP-dedicated aircraft were sent to Minot AFB in North Dakota. Flights as far as seventy-one degrees north were staged from here; DASA wanted to take samples even further north, but the scientists were told that the compass on the U-2 would not function so close to the magnetic pole. There was also a problem in determining in advance the winds that the flights would encounter in the upper atmosphere. In the lower latitudes, this was accomplished by taking data from radiosonde balloons sent aloft by weather stations. There were none of these stations in the polar regions. DASA eventually arranged three B-52 flights to the North Pole instead. However, a technique known as grid navigation, in which the compass was unslaved from the gyro, did enable U-2 flights from Eielson to venture as far as seventy-three degrees north by now, and later the 4080th flew right up to the Pole. This capability was apparently not relayed to the DASA people.

The far-flung Crowflight detachments were

fine for the single men, but not so good for those men of the 4080th that had family commitments. Periods of TDY were usually forty-five days, and there were plenty of them, leaving little time for the wife and kids back home. The Air Force hadn't even got round to providing decent married accommodation on base yet. The wives resigned themselves to long periods on their own. Their ranks were swelled when 'Det 4' returned from Argentina in August 1959; nine Americans married local girls during the deployment. The intensive Spanish language training that SAC had organised for those going on the Argentine TDY had produced an unexpected bonus for some!

The sister squadron of RB-57Ds was also still picking up regular overseas deployments, so there were constant comings and goings around the little Texas border town. The Black Knights flew from the UK, West Germany, Turkey, and Japan on reconnaissance sorties along the borders of the communist world. Unfortunately, the big wing on the Martin RB-57D was causing problems. In October 1958, Major Bob Schueler had made an emergency landing at Laughlin due to an unsafe gear indication. As he touched down, the entire left wing broke off at the engine mount. As the aircraft slewed off the runway trailing pieces of wing and engine behind, it caught fire, and Schueler was lucky to escape. A subsequent inspection of the other RB-57Ds revealed several cracked wing spars. A similar incident — this time to the right wing — occurred to a test aircraft at Kirtland AFB. Modifications were made, but safe operation of the RB-57D fleet on a long-term basis could not be guaranteed. Following their return from a European deployment in April 1960, most of the remaining RB-57Ds were placed in storage and the 4025th SRS was deactivated. Many of its people were absorbed into the Dragon Lady operation, including pilots.

On 6 May 1960, the news that Frank Powers had been shot down over Sverdlovsk began reverberating around the world. Attention soon focused on the remote Texas airbase where the U-2 pilots were trained. While the Agency operation at The Ranch and Edwards North Base remained under close wraps, people in the 4080th wing were pointed out and questioned wherever they went.[2] To make things worse, the news broke just as a new Crowflight deployment to Argentina was being mounted. It went ahead, with three aircraft, but as political fall-out from the Powers flight increased, the Argentine government withdrew permission for the SAC aircraft to remain at Ezeiza, and they were flown out to Ramey on 8 June.

It was a time for lying low. Premier Khrushchev blustered and threatened the countries that had allowed U-2 operations to be launched from their territory. Following anti-American demonstrations in Japan, the Agency U-2 detachment at Atsugi was withdrawn in early July. Within a month, the operation at Incirlik was closed, and the aircraft and personnel returned to the US. To ease the tension which had developed, President Eisenhower declared that no further overflights of Soviet territory would be approved by his administration.[3] Despite this, Bissell and his colleagues pleaded for a retention of the Agency's U-2 capability. They pointed out how successful the aircraft had been when sent out to gather information on other countries in which the US had an interest . . . the Middle East, Cuba, China and so on. Even if Powers had been shot down from high altitude by a missile, many of these countries did not have such weapons.

There was, in any case, no reliable alternative to the U-2. Reconnaissance satellites (also sponsored by the CIA) were in their infancy, and Lockheed's Mach 3 spyplane was still a year or more away from its first flight. After some debate within the administration, the CIA was allowed to continue with a scaled-down U-2 operation. The closure of 'Det B' in Turkey and 'Det C' in Japan was confirmed, but the headquarters squadron in the US was retained. It was moved down from The Ranch to Edwards North Base in June, displacing the Lockheed U-2 flight test operation, which moved in turn to Burbank. From the new, but still secret, location at the far north end of the giant dry lakebed, the Agency continued training flights and despatched the occasional aircraft on low-profile detach ents. WRSP-4 also geared up to support a permanent joint U-2 operation on Taiwan, to be run in conjunction with the nationalist Chinese government, and using pilots from both countries.

On 24 October 1960, in a hangar at Laughlin AFB, President Eisenhower finally came face to face with the spyplane that had caused him so much trouble in the previous few months. He had flown into the base en route for a meeting with the Mexican President, which took place

Major Pat Halloran, detachment commander, conducts a briefing for U-2 missions codenamed Toy Soldier at Eielson AFB in the early-sixties. The three SAC pilots scheduled to fly are all in attendance, along with the operations officer, an unidentified corporal, and a civilian tech rep from Lockheed. Halloran went on to complete 1,600 hours in the U-2, later flew the SR-71, and became 9th SRW commander at Beale AFB.

on the bridge linking Del Rio with Ciudad Acuña. There they signed the initial agreements authorizing construction of a huge dam across the Rio Grande, which would provide irrigation control in the lower river and create a 67,000-acre lake in the canyons to the north of the twin border towns. Returning from the bridge to Laughlin by helicopter, the President and his senior aides were ushered through the 4080th facilities by Colonel William Wilcox, who had taken over as wing commander three months earlier. The accompanying party of pressmen was left outside on the deserted flightline — all the U-2s had been hidden from view.

While the CIA U-2 operation was now reduced to some ten aircraft and seven pilots, the fleet at Laughlin was over twice this size, and the 4080th continued to clock up the hours on training and Crowflight missions. But Argentina withdrew permission for the U-2 flights, so new arrangements had to be made to fly a sampling corridor in the southern hemisphere. The Australian government obliged, and in mid-October 1960 five pilots and three hard-nose aircraft were despatched to East Sale airbase in the state of Victoria for a two-month stay. They flew there by way of Hickam AFB, Hawaii and Fiji. Also deployed were four JB-57s from Kirtland AFB for sampling at lower altitudes (up to 40,000 feet), and two C-54s for search and rescue. DASA also required a continuation of HASP flights in the tropical latitudes, which were now mounted out of Hickam and, from mid-1962, Andersen AFB on Guam and Albrook AFB in the Panama Canal Zone.

Laverton was an enjoyable TDY, with friendly hospitality provided by the Royal Australian Air Force. Further deployments to the Australian base were made the following April, and again in October 1961. To reciprocate the hospitality, the 4080th put their highly-polished aircraft on display at the base open house on 14 May 1961 — one was renamed 'City of Sale' in honour of the occasion. Six days later, a base open day was held back home at Laughlin, with a 4080th U-2 forming part of the static display for the very first time.[4]

At first, the new series of Crowflight missions were sampling nothing but two- or three-year-old fallout particles. Neither the US, UK or USSR had conducted any nuclear tests in the atmosphere since November 1958, when an international moratorium agreed by world leaders had taken effect. This was intended to allow negotiations on a permanent test ban treaty to proceed smoothly, but the talks at Geneva soon bogged down, and the Powers incident of May 1960 effectively scuppered the prospect of further progress. As the Iron Curtain descended again, the Soviets walked out of Geneva and laid plans for a new series of nuclear tests in the atmosphere. These commenced on 1 September 1961 at the Siberian site, but soon extended to the Arctic site on Novaya Zemlya. There, on 28 October 1961, they set off the mightiest nuclear blast ever recorded, a fifty-eight-megaton blockbuster which turned out to be a version of their hundred-megaton bomb in which the yield had been kept low for test purposes by replacing the bomb's uranium

This 1964 view of SAC U-2A serial 66952 shows the HF aerial tuner housing and wire antenna added to the aircraft in the early-sixties. The vertical tail is decorated with the 4080th SRW badge and the ribbon which signifies that the wing has received a distinguished unit citation. The aircraft has been painted light grey — another recent innovation.

casing with other material. The three U-2s of the 4080th at Eielson collected samples of the new Soviet tests for diagnostic purposes, in an operation which was now codenamed 'Toy Soldier'.

Alaskan U-2 operations were now made easier by the first of a number of further improvements that were made to the U-2 in the early 1960s. Although the SAC planes were already equipped with an HF radio so that the pilots could communicate during their long, lonely flights across remote wastelands with no UHF or VHF coverage, the system was a rudimentary, fixed-channel affair. This radio had not been fitted to the CIA aircraft at all, for fear that any HF transmission from an overflying U-2 would give away its position to the enemy on the ground below. Frank Powers and the others had therefore been sent out with only a UHF set which had line-of-sight range onboard. When things started going wrong over a thousand miles from base, Powers had no way of communicating the problem. With very little idea of what had happened to him, the Agency issued the pre-rehearsed cover story, thus setting up the US for Khrushchev's carefully-staged propaganda coup.

A new burst transmission device known as 'Birdwatcher' was now developed for the U-2. Birdwatcher monitored a wide range of aircraft performance parameters, such as altitude, G-force, airspeed, engine RPM and EGT, fuel pressure and oil pressure. It also monitored whether the DC generator was online, what the AC output was, and whether ECM equipment had been activated. It automatically sent this data back to base every so often, in short-duration HF pulses so that the opposition would not have time to pinpoint the aircraft's position with direction-finding gear. There was also a cockpit control so that the pilot could initiate Birdwatcher transmissions manually. He used the 'A' switch at preselected points in a mission to indicate that everything was going well. The 'B' switch was used to indicate that the mission was being aborted. If some disaster occurred, Birdwatcher would also report whether the canopy had been jettisoned, whether the ejection seat had been fired, and whether the tail assembly had separated. In these circumstances, it would continue to transmit as long as electrical power was available, and the antenna was still intact.

A Collins Mk 618T single sideband radio formed part of the new HF suite. It was a 400 watt system which also allowed voice contact at ranges of 3,000 miles or more. There was simply no room in the U-2 airframe to make a fully-internal installation of this radio. The antenna wire was strung from the leading edge of the vertical tail to the top of the fuselage aft of the Q-bay, where the tuner was housed in a small, pressurized fairing. The rest of the equipment was put in the nose.

Other improvements were made as a result of the Powers incident. Lockheed ran a competition for a new autopilot, in which Honeywell challenged the incumbent, Lear Siegler. Lear reduced the price of their Model 201 AFCS and stayed on board. To provide the pilot with timely warning of an autopilot disconnect, an aural tone was fed to the headset. Dual engine ignition was another safety feature incorporated at this time.

Next, an inflight refuelling capability was added. In the wake of the Powers incident, it was clearly not going to be easy to gain foreign country approval for a U-2 detachment, or even for a one-stop call to refuel. If the aircraft's range could be extended, some of these political problems might be avoided. There would also be an operational bonus. For instance, a U-2 could take off and fly along the periphery of denied territory for up to 1,500 miles at low level, thus avoiding detection by the opposition's early warning radar net. It could then be refuelled before climbing rapidly to cruise-climb altitude, and making its penetration from an unexpected direction. A full eight-hour overflight was now possible and hopefully would now go at least part of the way without alerting unfriendly air defences. In fact, a refuelled U-2 could cover 7,000 nautical miles and fly for fourteen hours — provided that the pilot had the endurance to match. Indeed, given sufficient engine oil, the U-2 could theoretically go on even longer, but fourteen hours was about the limit for both pilot and his life support system, although a fourth oxygen cylinder was now installed.

Lockheed designed a neat, retractable receptacle for refuelling, which was installed in front of the recently-added HF antenna tuner fairing. The receptacle was illuminated for night refuelling operations, and the boomer in the tanker was given added visual reference for these operations by the placing of white reflective strips on the leading edge of the vertical and horizontal tail surfaces of the Agency's all-black aircraft. A standard military AN/APN-135 rendezvous beacon was installed aft of the HF tuner fairing, so the U-2 had now sprouted a spine which extended a third of the way along the fuselage. Using its X-band radar to interrogate the beacon, the tanker could determine range and bearing to the Dragon Lady while maintaining radio silence. When the original J57-engined models were equipped with the refuelling gear, they were redesignated U-2E. The U-2C models, when similarly modified, were redesignated U-2F.

Test hook-ups with the refuelling tankers were conducted over the Edwards range. The Air Force provided KC-97s, an old, propeller-driven type, and jet-powered KC-135s, which were its replacement. Lockheed and Agency pilots soon worked out a *modus operandi*. With the KC-97, refuelling would take place at 20,000 feet and 170 knots. If a KC-135 was to be used, the operation would take place at 35,000 feet and between 200 and 220 knots. In either case, the Dragon Lady pilot would have to select the gust control, to dampen the effects of turbulent airflow on the airframe. Even so, bringing the two aircraft together called for careful manoeuvring. With the U-2 stabilized at the right altitude and speed, the tanker was supposed to overtake it about a quarter of a mile to the right, and at the same altitude. The tanker

Photographs of the U-2 being refuelled inflight are few and far between. This one from the mid-sixties shows a U-2F taking on more gas from a KC-135. Missions of over thirteen hours became possible, and pilot fatigue became the limiting factor.

would then reduce speed, and the U-2 would slide in towards it, while descending about 500 feet. Upon reaching a position directly below and in line with the tanker, the U-2 would rise slowly to the refuelling position. In this fashion, its delicate airframe would not be exposed to jetwash or propwash from the tanker's engines. But if the U-2 pilot were to stray even slightly off centre, the aircraft would begin to roll as it encountered the tanker's downwash. This was to happen in an early training hook-up, with tragic consequences.[5]

With some 900 gallons to transfer, at a maximum rate of 250 gallons per minute, the actual refuelling operation would be completed in five minutes. The U-2 pilot had a panel of lights on the upper right instrument console with which to monitor progress. Through a series of motor and solenoid-operated valves, the ARS system was supposed to operate automatically, but a manual override was provided via a switch on the left console. It was vital that the pilot kept an eye on the valve operations; this aircraft was considered sufficiently fragile that if the tanks were filled too quickly by the incoming fuel, the resulting overpressure from displaced air might lead to structural failure of the wing. After refuelling, the pilot had to ensure that the U-2's rudimentary fuel counter was reset, there being no other indication of the quantity on board, save for the low fuel warning light.

The procedure for separating tanker from receiver was for the U-2 to decrease power and descend about one hundred feet below the tanker, before sliding away left or right. Once he was a quarter of a mile abeam the tanker, the U-2 pilot could climb away on a parallel course until he was well clear of any downwash effects. When refuelling took place on an operational mission, radio silence could be maintained throughout the process. The tanker had a system of coloured indicator lights so that the U-2 pilot could manoeuvre more precisely during the approach to hook-up, and the boomer was provided with an illuminated chalk board with which to communicate gallons received, together with latitude and longitude information, to the U-2 pilot.

During the refuelling flight tests, the ever-resourceful Skunk Works came up with a couple of supplementary ideas. Since Kelly Johnson was now fully occupied with development of the A-12 Blackbird (which made its first flight on 26 April 1962), the U-2 team at Lockheed ADP was now led by Fred Cavanaugh as programme manager. Cavanaugh suggested that the aircraft's range could be increased even further if the tanker could actually tow the U-2 along behind it for a distance! This was a variation on a Skunk Works idea from way back in the U-2 design phase, when it was thought that the aircraft might be towed aloft and released at medium altitude by a B-47. Why not turn the tanker into a glider tug instead? Some preliminary flight tests of this towing arrangement were actually conducted by Lockheed pilot Bob Schumacher. Once hooked up, the idea was for the fuel coupling to be locked, whereupon the U-2 pilot would slowly ease off the power. The scheme didn't work out in practice. Another idea which never left the drawing board was for one Dragon Lady to refuel another inflight, by means of a wing-mounted drogue, with the U-2 tanker pilot monitoring operations from the cockpit by means of mirrors.

At Edwards North Base, the CIA's pilots were checked out in the refuelling procedures. The unit had a few Lockheed T-33s for proficiency training, and pilots under instruction would first fly in one of these alongside a refuelling operation, to see how it was done. Then the long, uncomfortable flights in the refuelled U-2 began. One was flown all the way across the Pacific from Edwards to Taiwan, refuelling en route and making only one landing at Wake Island. It was a good demonstration of the possibilities, but the exercise wasn't repeated. Future transpacific ferry flights made two stops, one at Hickam and one at Guam.

Rivalry between the Air Force and CIA U-2 outfits was growing. At Laughlin, the Agency's highly-paid pilots were held in disdain by some of the 4080th pilots, who dubbed them 'the mercenaries'. At higher levels of command, the Air Force argued that now the U-2 programme was common knowledge, the military should take over the entire show. The government disagreed. So the Air Force lobbied within the intelligence community for an expansion of the SAC U-2 unit's mission. After all, much of the aircraft's potential for intelligence gathering had direct relevance to the SAC war-fighting role, such as ferreting out the other side's electronic order of battle, so that the bombers could get through if ever the balloon went up. SAC devoted considerable resources to the ELINT mission, with a fleet of converted RB-47 bombers constantly on patrol around the Soviet periphery. (Only two months after the downing of Frank Powers' U-2, one of these RB-47s had been shot down by Soviet fighters over the Barents Sea.) The RB-47s could only climb to about 45,000 feet, and a higher-flying platform was needed to eavesdrop on transmissions emanating from further inside denied territory. The RB-57Ds of the 4080th Wing had offered such a capability, but they had been consigned to the boneyard with fatigue problems.

The argument over U-2 missions was but part of a wider battle for turf between the Agency and the Air Force, as they jockeyed for control of the National Reconnaissance Office (NRO) and the associated Committee on Overhead Reconnaissance (COMOR) and National Photographic Interpretation Center (NPIC). The NRO was supposed to be a joint Agency/military co-ordinating body to run the growing spy satellite programme. NPIC was a similar joint operation which had been set up in 1961 as a successor to Automat, Art Lundahl's original U-2 imagery analysis shop. It was rehoused in a much larger building at the Washington Navy Yard, where CIA and military analysts worked alongside each other (and sometimes continued to draw different conclusions from the same imagery). COMOR was the successor to Richard Bissell's Ad Hoc Requirements Committee, which had sorted and prioritized requests from the CIA, NSA and military services for U-2 reconnaissance missions. Along with requests for satellite coverage, sensitive U-2 missions were now approved here.

As far as U-2 operations were concerned, a compromise was reached. The Air Force was tasked with an increasing number of SIGINT missions, especially out of Eielson. Some of the improvements that the CIA had recently funded — such as the J75 engine, Birdwatcher, and new ECM systems — would remain exclusive to its small fleet of aircraft. But in order to accomplish its newly-assigned role, some of SAC's U-2A models would receive the inflight refuelling modification, along with a version of the System 3 COMINT gear. In the meantime, some military pilots would receive training on the latest Agency models, prior to these being 'borrowed' by the 4080th for some peripheral SIGINT missions. In February 1962, Majors Rudolph Anderson, John Campbell and Richard Heyser were despatched from Laughlin to Edwards, where they began flying the higher-powered and refuellable aircraft. These three were among the most experienced of the SAC pilots, each having amassed 1,000 flying hours in the U-2 over a five-year period.

Shortly after 21.00 on 1 March, disaster struck. As he moved in for a night refuelling behind a KC-135 at 35,000 feet, Campbell somehow entered the tanker's slipstream and was flipped over. The U-2 fell away out of control, with the SAC pilot desperately trying to assert his authority over the tumbling aircraft. Failing to do so, he attempted to eject, but the seat did not leave the plane. He was apparently still trying to extricate himself from the cockpit when the plane hit the ground; his body was found in the wreckage, halfway out of the cockpit, the canopy having been manually opened.

Although the accident had taken place virtually overhead Edwards AFB, it took thirty-seven hours to find the wreckage. The search was started at 02.00 Friday, and occupied twenty-five pilots who flew fifty-seven sorties throughout that day and the following morning. They used T-birds, Cessnas, T-28s, helicopters and a C-123, which ranged far and wide over Southern California, Nevada, Arizona and Utah. In the dark, there had been no way for the tanker crew to determine what had happened to Campbell, so they figured he could be anywhere. The crash site was eventually located by Bob Schumacher at 11.00 on Saturday morning, while he was flying the faithful Beech Bonanza owned by the Lockheed Flight Test Department. The wreckage was lying in the Kramers Hills, a section of the Mojave Desert just a few miles east of Edwards.

As accident investigators pored over the

remains, Campbell's failure to escape caused much concern. Only two months earlier, Captain Charles Stratton had enjoyed an amazing escape from a SAC U-2A during a night training flight from Laughlin. He recalled pulling the ejection ring, and he did indeed depart the airplane and float gently down to the ground. But when they found the plane, the seat was still in it! It seemed that the U-2 ejection seat, installed three years earlier after a number of fatal crashes, was not functioning properly.

Escape system failures were not the only cause of concern back at Laughlin. There had been an unforgivable mix-up in the arrangements for notifying next-of-kin. Word of the accident had reached the base within a couple of hours, but somehow, Rudy Anderson had been listed as the missing pilot. As usual in these circumstances, the base commander and chaplain were despatched to the Anderson family home with the dread news. But in time-honoured fashion, pilots who were also flying or scheduled to fly at the time of an accident would rush to the nearest phone and call home to reassure their family that they were not involved. Rudy Anderson had attempted to do just this, putting in a call to his wife from Edwards, but since the U-2C/F training programme was highly classified, he had felt obliged to not mention that an accident had occurred. Instead, he contented himself with some mundane pleasantries, and rang off secure in the knowledge that if the news broke, she would know from the phone call that he was still alive.

When Jane Anderson answered a knock at the door in the middle of the night, and found the chaplain standing there and about to embark on the usual litany, it was therefore a particularly confusing, as well as traumatic moment for her. She insisted that her husband could not be missing. What had she heard on the telephone a short while before — a ghost? The whole distressing experience was to leave a mental scar, for which she never forgave the powers-that-be, especially since the dread deputation was back on her doorstep eight months later, and this time the bad news was for real.

1 Moreover, it had become possible to identify the Krypton 85 isotope which was produced and vented into the atmosphere from Soviet reactors which were producing plutonium. By this means, US intelligence could estimate the other side's rate of nuclear weapons production.

2 To satisfy the media, a U-2 photo-call was arranged at Edwards, but not at the secret North Base site. It so happened that, in the same week as Frank Powers was shot down, the Aviation and Space Writers Association were meeting in Los Angeles. NASA had arranged a bus to take the journalists out to their hangar at Edwards, for a first-hand look at the Bell X-15 rocket aircraft, which was then some ten months into its record-breaking flight test programme. The tour was scheduled for Friday 6 May, just twenty-four hours after NASA had issued the Agency-prepared cover story about one of 'their' U-2s being missing on a weather reconnaissance flight from Incirlik. Someone in the Agency evidently hit upon the bright idea of backing up the cover story by showing the journalists a suitably-marked U-2 during their visit to the NASA facility. For this to happen, they would have to move quickly out in California!

Early on the morning of the planned tour, an unmarked, all-black Agency U-2A model was towed the three miles along the lakebed from Lockheed's North Base facility to the NASA hangar at the main Edwards base. Here, a bemused NASA mechanic was told to find a stepladder, a paintbrush and a NASA decal in double-quick time, and to paint the NASA identity and a tail number on the black aircraft. He chose the Air Force serial number of a target drone which had been shot down in operations at Edwards some days earlier, and so it was that when the U-2 was revealed to the pressmen a short while later, it carried a neat yellow tailband containing the letters 'NASA' and the numbers 55741 below. What with all the rush, the mechanic's main concern was that the press or someone else would climb up the tail and smudge the paint, which wasn't yet dry! As for the visitors, they climbed off the bus and rushed over to the unfamiliar black U-2, almost completely ignoring the X-15, which was standing some twenty yards away with a deputation of famous test pilots surrounding it. There was no-one around the U-2 to answer questions, but the press got its pictures of a 'NASA' U-2, which were duly published alongside the cover story. The Agency's effort was in vain, of course, since within a week President Eisenhower had gone on record repudiating the cover story and taking full responsibility for the U-2 operation.

3 When he assumed office in January 1961, President Kennedy renewed this pledge.

4 Many thought that these were the first occasions at which the public had been allowed to get close to the Dragon Lady. This was not quite true. In the spring of 1958, during the very first Crowflight deployment to Ramey, a U-2 had been rolled out and flown in front of an open day crowd. In March 1959, the Argentinian press were allowed to crawl all over the Crowflight aircraft at Buenos Aires. Six months after the Powers shootdown, on 2 November 1960, the first U-2 to be displayed to the public on the American mainland had gone on show at Patrick AFB, Florida. It was one of the special test aircraft operated by a unit of the Air Research and Development Command from Edwards.

5 An alternative method of bringing tanker and U-2 to a rendezvous was sometimes employed, when the U-2 was already at operating altitude, perhaps having already performed part of the mission. The two aircraft approached each other from opposite directions. When the U-2 was fifty miles from the tanker, its pilot would start a descent. Upon reaching 35,000 feet he would turn through 180 degrees, which would put him in trail with the KC-135.

Chapter Five
The Missiles of October

I must say, gentlemen, that you take excellent pictures, and I have seen a good many of them . . . I think that you can take every satisfaction in what you are doing, what you have done, and in what you will do. We are very much indebted to you all.
President John F. Kennedy, 26 November 1962.

There was no doubt about it. The pictures showed SA-2 surface-to-air missile (SAM) sites at the western end of Cuba. Thanks to the 'B' camera's work during Operation Overflight, and the growing volume of satellite photography since, the photo-interpreters at NPIC were by 1962 quite familiar with the pattern of six launchers surrounding a radar which betrayed one of these sites. The new evidence about Cuba had been obtained by a CIA U-2 on 28 August 1962, but it only added to a growing body of evidence that the Soviets had stepped up their military aid to the Castro regime. This had been obvious since the spring, and the Agency had been scheduling U-2 flights over Cuba twice a month since then.

Two SAM sites could be positively identified, and there were indications of another six in preparation. Analysts at CIA headquarters (which had moved by now from Washington to a purpose-built complex at Langley, Virginia) concluded that the Cuban air defence system was being refurbished — and nothing more. At least one man at Langley disagreed — the Director of Central Intelligence, John McCone. He had replaced Allen Dulles the previous year after the abortive CIA-sponsored invasion of Cuba known as the 'Bay of Pigs' affair. (Richard Bissell had also been heavily involved in this fiasco, although he hung on as CIA Deputy Director of Plans until February 1963, before being replaced by Richard Helms.) McCone now warned President Kennedy of his theory that the Soviets had decided to install offensive, as well as defensive, missiles in Cuba. Senior Congressional leaders voiced similar fears in public, and on 4 September the Soviet Ambassador in Washington was summoned to the White House, where he flatly denied that this was the case. Nevertheless, the Agency stepped up its schedule of U-2 flights. The next sortie on 5 September covered the central and western end of the island, discovering more

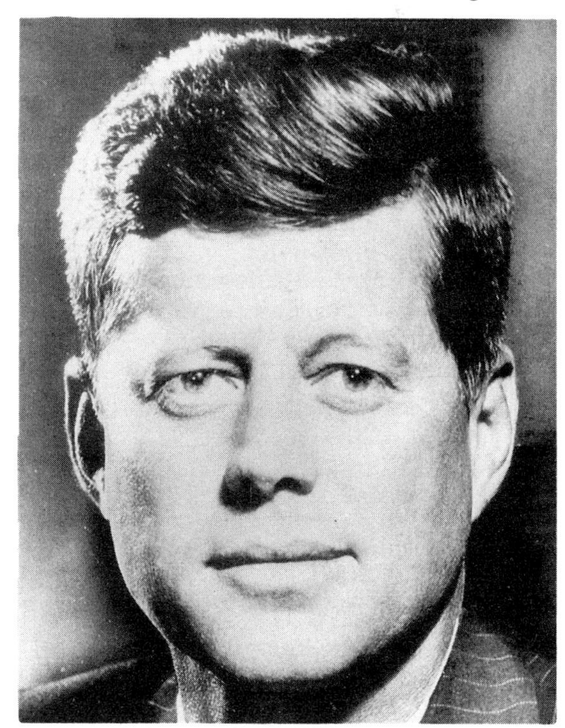

When President Kennedy came to power in January 1961, he extended his predecessor's ban on further U-2 overflights of Soviet territory. Little did he realize that the aircraft would play an invaluable role in a new crisis less than two years later, when the young President successfully faced down the challenge of offensive missiles deployed on America's doorstep.

SAM sites as well as a MiG-21 interceptor on the ground at the newly-constructed Santa Clara airbase. Alongside were four crates presumed to contain further MiG-21s. Another flight on 10 September was less successful, due to cloud cover.

In Washington, a meeting of the high-level Committee on Overhead Reconnaissance (COMOR) was called for 11 September to assess the situation in Cuba, and other developments. There was bad news from Taiwan, where the Agency's joint U-2 operation with Nationalist China had been running for over a year. The Chinese communists had just downed one of the aircraft over Nanchang, apparently with an SA-2 SAM. Nearly two and a half years after the Powers shootdown, there could now be little doubt that this missile posed a lethal threat to continuing U-2 operations. The operation on Taiwan had been kept secret up till now, but Peking had boasted of its success in downing the aircraft, and a statement would now have to be issued. Not only this, but another U-2 had created an international incident with the Soviets on 30 August. An Air Force U-2 on a peripheral SIGINT flight in the Far East had strayed over the southern part of Sakhalin Island for nine minutes in the middle of the night of 30-31 August, prompting a Soviet protest.

COMOR decided on a cautious approach. The surveillance of Cuba by U-2 would be continued, but flights would be routed clear of known SAM sites, in particular by flying abeam the coast wherever possible. The geography of Cuba contributed to this decision; at no point was the 730-mile-long island more than a hundred miles wide, so the flights might still obtain some inland coverage, as well as footage of the ports where Soviet military shipments were unloaded. By the same token, geography apparently worked against any possibility that the new reconnaissance satellites might make a contribution. Since their south-to-north orbital trajectories and control facilities were designed for maximum coverage of the Soviet landmass, they would cross the east-west lying island very quickly, and not be capable of returning helpful imagery.

On 13 September President Kennedy attempted to calm an increasingly nervous situation with a televised statement: 'Soviet influence in Cuba has increased . . . there have been new shipments . . . the island is under our most careful surveillance . . . but there is no threat to any other part of this hemisphere.' As it subsequently turned out, the opposite was the case — it later became known that the first offensive missiles had arrived on the island five days before the President made this statement. The trouble was, COMOR's new restrictions on U-2 flightpaths had the effect of eliminating the area west of Havana from further scrutiny, and it was here that the Soviets were siting those first offensive missiles. Five further U-2 missions were flown by Agency pilots (on 17, 26, and 29 September, and 5 and 7 October), but although further SAM and fighter activity was detected, that was all.

CIA Director McCone left Washington on a three-week honeymoon in mid-September, still proclaiming his belief that offensive missiles were on the way. But his analysts at Langley differed, and a formal national estimate which discounted this possibility was approved and sent to the White House on 19 September. The ink on this estimate was hardly dry before reports from reliable CIA agents about new missile shipments began to flood in, and an alert photo-interpreter in the Defence Intelligence Agency noticed something interesting about the U-2 pictures of western Cuba taken almost a month earlier. Having studied U-2 and satellite photographs of strategic missile deployments inside the Soviet Union, Colonel John Wright realized that the particular pattern in which the SA-2 sites had been laid out in Cuba matched the way these SAMs were deployed to protect offensive ballistic missile sites in the USSR itself. The tell-tale trapezoidal pattern of the SAMs in Cuba indicated that an offensive missile site would soon be established in the San Cristobal area. The evidence that the Soviets were mounting a big power-play was becoming persuasive.

When, therefore, COMOR met again on 4 October it was faced with a renewed request for overhead reconnaissance of western Cuba. There was some hesitation, and an urgent examination of the alternative options to the U-2. The obvious one was to send in low-level tactical reconnaissance jets — Air Force RF-101 Voodoos and Navy/Marine Corps RF-8 Crusaders — but this would alert the Cubans and Russians to the fact that they had been discovered. There was also some consideration given to the use of Ryan target drones which had been converted to take pictures. The drones, however, were still in

This photograph is representative of the U-2C configuration at the time of the Cuba missile crisis. The intake has been widened from the original model to accommodate the increased airflow requirement of the J75 engine. The sextant bubble is prominent in front of the cockpit, as is the HF antenna housing to the rear. The aircraft's UHF antenna is visible beneath the forward fuselage.

an operational test phase, with only two pre-production models available. This was also a highly classified project; it was eventually decided not to risk compromising them. It would have to be the U-2 again, but now there was a further complication. The Air Force, supported by Secretary of Defense Robert McNamara, insisted that it fly all future U-2 missions over the island, since a potential military conflict was developing. The CIA resisted. Unable to resolve the dispute within the NRO, the number two men at the Pentagon and Langley took it to the White House, where a senior national security official told them they had better stop squabbling before the President found out about it. They did. The Air Force won the argument, and on Wednesday 10 October renewed U-2 flights over western Cuba were sanctioned by the President.

The word reached Laughlin when Major General Compton from SAC headquarters arrived and called 4080th wing commander Colonel John DesPortes and his senior staff into a classified meeting. At a higher level, it had been decided that the CIA's better-equipped U-2C models would be used by the Air Force pilots, whenever possible. These aircraft were equipped with the System 9 intercept warner and jammer, and had also now been fitted with System 12, a receiver which warned the pilot when SAM radars were trained on his aircraft. Having already been checked out in the C-model, Majors Anderson and Heyser were despatched to Edwards North Base, where they prepared themselves for the vital missions. The meeting also decided to recover the first flights at McCoy AFB in Florida, and to stage all future overflights of Cuba from there. Being much nearer to the island than Laughlin or Edwards, the U-2s would be able to take off with a lighter fuel load — 660 gallons — and hence cross the island at 68,000 feet or higher, in an attempt to counter the threat posed by the SAMs.

A flurry of activity ensued, as all the necessary personnel to support an operating detachment were recalled to Laughlin and embarked upon a SAC VC-97 transport. The arrangements were made in great secrecy — it was still ten days before the world would know that a major crisis was at hand. Late on Saturday night, 13 October, the VC-97 took off and headed for McCoy. Meanwhile, the 4080th's own U-2A models were also made ready — that is, those that were actually available. For the crisis could not have hit at a worse time; the 4080th fleet was dispersed all over the globe on the sampling mission. There were four Dragon Ladies in Alaska, three on Guam, and three more in Australia. There were even three U-2s at Upper

Major Steve Heyser, pilot of the first Air Force U-2 mission over Cuba, which confirmed the presence of offensive Soviet missiles there.

Heyford airbase in Britain, where the first-ever European HASP deployment had been made in mid-August. In the end, the wing was able to muster ten aircraft and eleven pilots for the first few weeks of US Air Force U-2 flights over Cuba.

As the C-97 from Laughlin landed at McCoy in the small hours of Sunday morning, Heyser took off from Edwards. Meeting the sunrise across the Gulf of Mexico, his U-2C flew past the south-western tip of Cuba before turning north to cross the Isle of Pines and the Gulf of Batabano. Within six minutes, he had passed over the mainland and was clear on the northern side, heading for the Florida Keys. There was no opposition. At 10 a.m., Heyser landed the black plane at McCoy. The film from Mission Golf 3101 Victor was rushed to NPIC at Washington by a waiting Air Force jet. Early next morning, Anderson launched out of Edwards to fly a second U-2C across the island. By the time he had landed at McCoy, the photo-interpreters at NPIC studying the film from Heyser's flight the previous day had found the evidence they were looking for. They called in Art Lundahl to confirm their conclusions: they were looking at SS-4 MRBMs (codenamed 'Sandal' by NATO) being set up among the trees at San Cristobal. Making use of the scale and clarity provided by the long shadows of the early-morning sun, they identified seven missile trailers and four erector-launchers. The missiles themselves were probably hidden beneath the two large field tents next to the trailers. There were more tents in rows nearby — accommodation for the rocket troops — and plenty of military trucks. A few miles away, another missile site was being prepared. The whole set-up was identical to those that had been photographed in the USSR.

At 5.30 p.m. Lundahl informed CIA headquarters, and was told to go back over the film and double-check his conclusions. But there was no mistake, although a layman would have been hard put to identify anything sinister in the pictures. As the evening wore on, news of the find was spread through the appropriate channels. At Laughlin, plans to launch three U-2As on further overflights went ahead despite deteriorating weather, and three pilots prepared themselves for the mission. At 4 a.m. on Tuesday 16 October, they taxied out and took off in a blinding rainstorm. Visibility was so poor that they only found the end of the runway thanks to the lights on the mobile van which led them out!

The three aircraft were still airborne when the news that offensive missiles had been confirmed in Cuba reached President Kennedy at his breakfast table in Washington. By lunchtime, Art Lundahl was in the White House with blow-ups, maps and charts to brief the President and the Executive Committee (EXCOM) of the National Security Council.

Offensive missiles in Cuba! The SS-4 had a known range of 1,100 nautical miles, enough to reach well into the American heartland, not to mention Washington. There were also now intelligence indications that SS-5 'Skean' IRBMs with a 2,200 nautical mile range were being deployed as well, and these would threaten the ICBM bases in the northern United States. By Wednesday, U-2 photographs had revealed preparations to install IRBMs at Guanajay and Remedios, as well as two more MRBM sites at Sagua La Grande. The pilots of the 4080th also discovered crates containing Ilyushin Il-28 bombers at San Julian airfield. On the Wednesday, a flight over Santa Clara airfield photographed no fewer than thirty-nine MiG-21s in

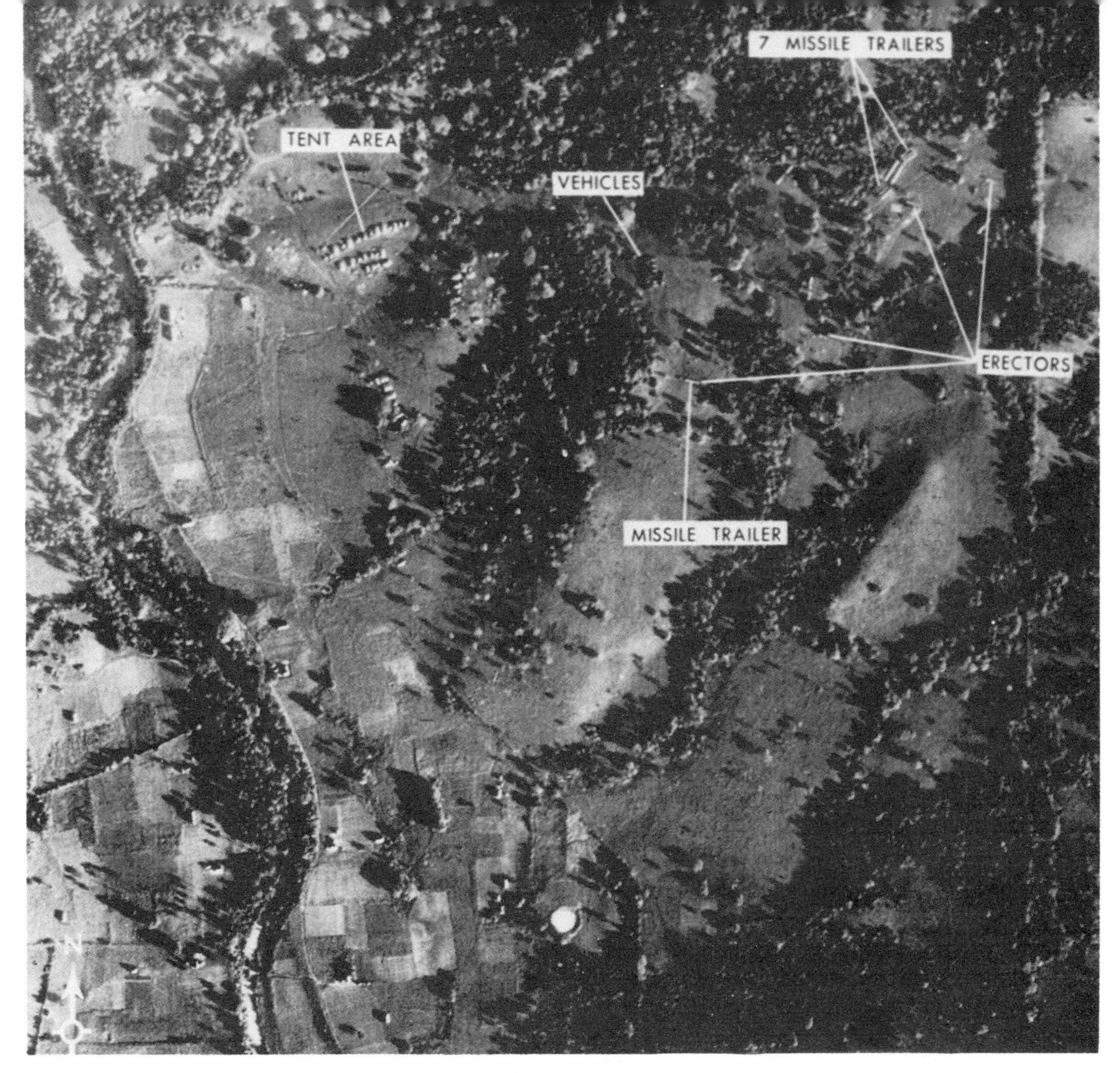

One of the vital photographs of the Soviet missile launch site at San Cristobal taken on 14 October 1962 during U-2 mission G 3101V. It shows eight SS-4 missiles on trailers, as well as missile erectors, support vehicles, and temporary accommodation facilities clustered among the trees. This vertical shot from the Type B camera was released to the public ten days later, during the United Nations debate on the Cuba crisis.

various stages of assembly. The next day, one of these was actually captured taking-off on film by a U-2 flying directly overhead.

Meeting with his senior advisors in EXCOM once or twice a day from now on, Kennedy ordered the U-2 flights stepped up. The Agency's best instructor pilot, Jim Barnes, was drafted in to check out more of the 4080th pilots in the C-model. Between the Sunday when Heyser and Anderson had flown the first Air Force missions, and Monday 22 October, when the US went public with news of the missile build-up, 4080th pilots flew some twenty times over Cuba. It was an increasingly dangerous enterprise; no fewer than twenty-four SAM sites had now been established the length and breadth of Cuba. Given that the SA-2 was thought to have a lethal range of twenty-five miles, this meant that virtually the entire island was covered.

The 4080th mission planners at McCoy did their best to counter this formidable SAM threat. Flight tracks across the island were drawn so that no U-2 would fly straight and level for longer than thirty seconds before making a turn of at least sixty degrees. The pilot needed those thirty seconds to acquire the photo-target through the driftsight and line up correctly, but the missile operators on the ground were thought

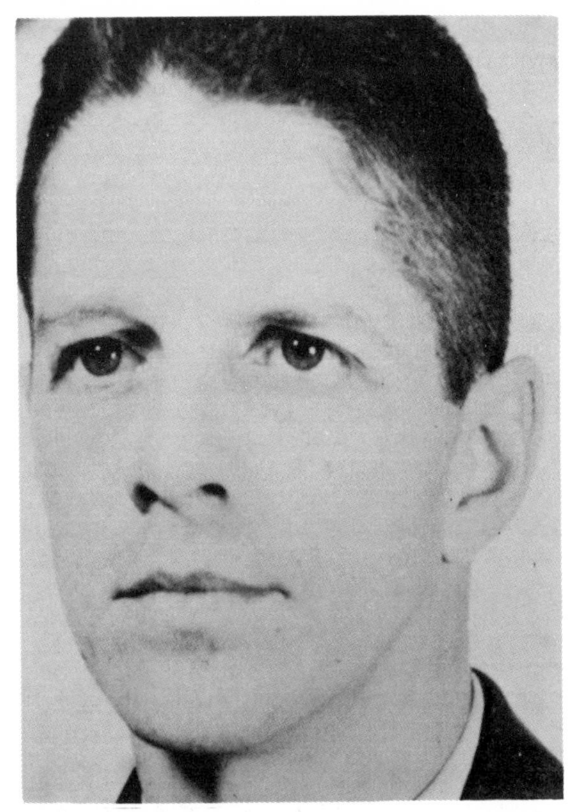

Major Rudolph Anderson, killed when his U-2 was shot down over eastern Cuba on 27 October 1962, at the height of the missile crisis. Like Frank Powers two years earlier, it seems that an SA-2 missile exploded at close range, but unlike Powers, Rudy Anderson had the misfortune to take a hit from flying shrapnel which penetrated the cockpit.

to need at least as long to track, acquire and fire their SAMs at the overflying aircraft. Where possible, missions were flown offshore paralleling the coast, in an attempt to avoid the SAM rings.

Over the weekend of 20/21 October, the US administration considered its options. Some senior military men were in favour of immediate attacks on the missile sites before they could become fully operational. This was ruled out for the time being when the Air Force confirmed that it could not guarantee total success on a first strike. Since intelligence believed that four of the MRBM sites were already operational, with four launchers on each site, the chance of one launcher escaping an attack and then firing a missile against an American city in retaliation was not worth taking.[1] Another option was an invasion to take out the sites (and maybe topple Fidel Castro at the same time), but the Bay of Pigs still rankled in the government's collective memory. The President and his advisors chose the option of a naval blockade of Cuba, to be imposed from 10.00 a.m. on Wednesday, and this was announced to the world on Monday 22 October by the young President: 'Let no-one doubt that this is a difficult and dangerous effort on which we have set out. No-one can foresee precisely what course it will take . . . but the greatest danger of all will be to do nothing.'

Now that the crisis was out in the open, low-level photo sorties were approved, and these commenced on Tuesday 23 October with RF-101s bringing back further evidence of the startling pace at which the Soviets were progressing. From now on, the overhead reconnaissance effort would be shared between the 4080th and the 363rd TRW's Voodoos, along with RF-8 Crusaders flown by Navy and Marine pilots. The next day, more low-level flights revealed the presence of four 1,200-man Soviet Army battle groups complete with tanks and battlefield missiles. As tension mounted, the Soviet Ambassador to the United Nations was challenged on the floor of the UN debating chamber to deny the existence of the offensive missiles. When he demurred, US Ambassador Adlai Stevenson produced the U-2 photographs for all to see.

The first sign of a break in the crisis came on Wednesday 24 October, when Soviet freighter ships en route to Cuba were seen to stop in mid-Atlantic. Late in the week, Moscow began to indicate a compromise; it seemed that they might agree to withdraw the missiles if the US would dismantle its Jupiter IRBMs in Turkey. For its part, the US pledged that it would not invade Cuba if the missiles were removed.

Early on Saturday morning, 27 October, Rudy Anderson took off from McCoy AFB in Article 343 for yet another U-2 overflight. He did not return. His route was designed to remain clear of the known SAM sites by flying along the northern coast of Cuba. But although he was flying one of the Agency's C-models equipped with the SAM warner, he was apparently taken by surprise by a salvo of SA-2s which were fired at him from the Banes naval base at the eastern end of the island. One exploded above and behind his aircraft, and fragments from the blast penetrated the cockpit and Anderson's pressure suit at shoulder level. He was presumed to have died as the cockpit depressurised and his suit failed to inflate.

Old Agency U-2 hands who had stayed on at McCoy to observe the Air Force operation shook their heads; they figured this was bound to happen, so long as the military continued to insist on sending the U-2s up day after day in the same pattern. They were scheduling them across like airliners down there! It was a far cry from their own concept of operations, in which single flights would be mounted sparingly, approaching from as many different directions as possible, to keep the other side's air defence system guessing.

Air Force U-2 people took the greatest exception to these comments. The Agency's operating techniques would not have provided a fraction of the quantity and coverage that was now being demanded by the nation's leaders. Colonel Tony Martinez and his colleagues in 4080th operations at McCoy had taken every precaution possible under the circumstances. They received the suggested routes from Washington, and never accepted one that even remotely resembled a previously-flown route. They employed the tactical routeing already described, and other variations. The Agency criticism was out of order, they protested.

This Saturday was not a good day for the 4080th. At almost the same time as Anderson was shot down, one of the wing's U-2A models got into trouble at the other end of the North American continent. Major Charles Maultsby made a navigational error during a sampling flight from Eielson AFB and drifted across the eastern edge of Siberia, alerting the Soviet air defences. Two months earlier, the Alaskan detachment had extended their sampling flights to the North Pole. Captain Donald Webster had made the first such flight on 25 August 1962, an eight hour forty minute effort during which he covered 3,121 nautical miles before landing back at Eielson.

As back-up pilot on this Saturday in October, Don Webster was sitting in the command post at Eielson while monitoring the progress of Chuck Maultsby, a former Thunderbirds formation team pilot who had been flying U-2s since January 1961. Maultsby reached the pole in darkness and turned back towards Alaska, but apparently didn't roll out on the correct course. He was using the grid navigation technique whereby the gyro is unslaved from magnetic north, because of the wild fluctuations which it would otherwise encounter in the polar regions. But this technique required some very careful flying and constant plotting, as well as cross-checks with other methods of navigation. In the minimally-equipped U-2, this amounted to taking star shots with the sextant. At any rate, Maultsby progressed steadily southwards, without detecting his error, until the pilot of the rescue C-54 came on frequency to report that he could see the sunrise beginning. From the command post, Webster called Maultsby on HF to check whether he was seeing the same thing. The answer was no!

Maultsby now took a star shot, but the results didn't match his celestial chart for the planned route. He was clearly off course, and was told to turn left ninety degrees immediately. Meanwhile, a radar post at the western extremity of Alaska detected Soviet fighters climbing to intercept an intruder over the Chukotsky Peninsula! An anxious time followed as the U-2 slowly made its way back east, with the MiGs in pursuit below. US Air Force F-102 interceptors were scrambled to meet the wandering U-2. Maultsby eventually reported that he could now see the sunrise. He

Captain Don Webster, pictured shortly after he became the first man to fly a U-2 to the North Pole and back on 25 August 1962. Webster was back-up pilot for Major Charles Maultsby's ill-fated flight two months later, when a navigation error placed him overhead Soviet territory during the Cuba missile crisis.

reached the Alaskan landmass safely, but his problems were not yet over, since he was fast running out of fuel. He couldn't make it back to Eielson. Maultsby elected to shut down the engine to gain a hundred miles or so in the glide, intending to make a relight before landing on a remote airstrip in the tundra. As it turned out, he was unable to relight, and had to deadstick into the small field with his faceplate frozen over. It was a remarkable recovery, and Maultsby was exhausted; he had been airborne for ten hours twenty-five minutes, a record for an unrefuelled U-2 at this time.

It was the height of the missile crisis; and the Soviet air defence system had gone into top gear. They had every reason to link the errant U-2 with US preparations to attack the motherland. Maybe this was a final reconnaissance sortie before the bombers were sent in! In his message to President Kennedy the next day, Premier Khrushchev made specific reference to Maultsby's flight, which he termed 'a provocation . . . at such an anxious time . . . when everything has been put into combat readiness'. President Kennedy made a prompt reply, explaining the circumstances: ' . . . a serious navigational error . . . I regret this incident . . . the aircraft was engaged in an air sampling in connection with your nuclear tests'. In private, the President was less circumspect, as he shook his head and made the now-famous remark: 'There's always some sonofabitch who doesn't get the message.'

Less than twenty-four hours after Anderson was shot down, Premier Nikita Khrushchev announced on Moscow Radio that the Cuban missiles would be withdrawn. The previous evening, the US had called up reservists and delivered a note to the Soviet Ambassador in Washington, which threatened to 'remove the bases'. The U-2 pilot proved to be the only man lost to enemy action during the crisis, although there were other deaths during the massive US military manoeuvres that accompanied the blockade; SAC's RB-47 reconnaissance wing lost two aircraft and crews in takeoff accidents.

There now followed a strange period when the US remained unconvinced of true Soviet intentions. There were no formal agreements, merely 'assurances' and 'understandings' passing back and forth between the superpowers, mainly through the Soviet Ambassador in Washington, Anatoly Dobrynin. Of most immediate concern to the 4080th crews in Florida was the question of whether further U-2 sorties over Cuba would meet with opposition from the SA-2 missiles or MiG-21 interceptors. The day after Khrushchev's announcement, a two-ship RF-101 sortie encountered anti-aircraft fire at low-level over the San Cristobal MRBM sites. Neither plane was hit. Despite the hostile reception, film from the two aircraft revealed that the Soviets had begun to dismantle the missiles. But the air defence situation seemed 'hot', and no further U-2 missions were scheduled for the next six days.

More low-level flights detected some other sites being dismantled, and on Friday 2 November the White House announced that the missiles appeared to be on their way out, but wide-area photographic coverage — which only the U-2s could provide — was still needed to settle the question of whether the Soviet withdrawal was complete.

General Thomas Power, SAC Commander-in-Chief, was present at McCoy for the morning briefing of the 4080th on Sunday 4 November, the day that U-2 flights were resumed. 'I wish I could give you all some encouragement,' he said, referring to the SAM threat, 'but I can't!' The pilots were instructed to abort their flight as soon as the SAM missile radars went active. If the RB-47 ELINT aircraft flying offshore detected any such activity, an RC-121 radio relay plane would alert the U-2 pilots. This was their cue to take immediate evasive action and clear the area. On this Sunday, no fewer than five U-2s would attempt to make a photographic clean sweep of the island, flying a five-minute separation and ten nautical miles apart on east-to-west parallel tracks. But the SAM radars came up hot soon after coast-in. The RC-121 transmitted the abort codeword 'Green Arrow' and the U-2s scattered in different directions before setting course for McCoy.

The next day's U-2 sorties fared rather better, with one aircraft photographing two freighters taking on missile transporters and erectors at the port of Mariel. Other U-2 flights over Casilda and La Isabella confirmed that missiles were also being embarked at these ports. The abort rules were relaxed as time went on, but the SAMs remained in place, and were not included in the understandings reached with the Soviets on the withdrawal of their other forces.

There was also still a fighter-intercept threat. On one mission a couple of days later, Chuck

The officers and men of SAC's U-2 unit became some of the most-decorated in Air Force history, and medal award ceremonies became regular events. Here, 4080th SRW commander Colonel John Des Portes hands out the latest batch to (left to right) Major Harold Melbratten, Captains Tony Bevacqua, Dick Bouchard and Bob Spencer.

Kern overflew José Martin airbase near Havana. Through the driftsight, he observed two of the newly-arrived MiG-21s take off in pursuit. Seconds later, the autopilot surged and his engine flamed out. Kern reported his predicament to the US airborne controller orbiting at a safe distance in the RC-121, turned left to avoid the SAM belt over Havana, and tried to head south-west and out of danger. However, as he was descending, the two MiGs were climbing as fast as they could towards him! Kern's canopy frosted over, but through the driftsight he saw the fighters smoking heavily as they passed through the contrail level. The U-2 pilot considered making for Mexico, but changed his mind and, skirting the western end of the mainland, turned north towards Florida. With the aircraft's UHF radio now losing strength, another U-2 which had been flying ahead of him relayed his anxious calls to the RC-121 for US fighter support. Kern now had to forget about the MiGs, as he was below 40,000 feet and had to concentrate on trying to get a relight. He was successful, and climbed out of danger just as every US fighter that was airborne in that part of the world converged on the area to try and splash the MiGs!

Despite such alarms, the crisis gradually wound down, but there was no sign that the Il-28 bombers were being withdrawn. The US continued to insist that this be done, and the USSR finally gave way on 20 November. Five days later, U-2 imagery confirmed that the bombers at San Julian airfield were being dismantled, and on 27 November a run over Holguin revealed that all the crated bombers there had been removed. These ones had never even been assembled. By now, the 4080th's tasking had been reduced to one flight per day. Even so, in the first week of December the wing notched up its one hundredth Cuban sortie.

It was time for the commendations and the medals. Richard Heyser was summoned to Washington, where a formal meeting with the President had been arranged, in company with Colonel Doug Steakly from the NRO, Lieutenant Colonel Joe O'Grady from the 363rd TRW, and General Curtis LeMay, who was by now the Air Force Chief of Staff. As they stood by to enter the Oval Office, LeMay advised the two pilots to keep quiet while he did all the talking! On 26 November, President Kennedy flew to Homestead AFB, where he presented the 4080th SRW and the 363rd TRW with outstanding unit awards.[2] Addressing the men of the U-2 and RF-101 squadrons drawn up in front of him on the apron, he said they had 'contributed as much to the security of the United States as any unit in our history, and any group of men in our history'. Later, there was another ceremony in General Power's office at SAC headquarters, Offutt AFB, when the ten 4080th pilots who had participated in the first phase of Cuban overflights were each awarded the Distinguished Flying Cross. In addition to Richard Heyser, those decorated were Majors

Pilots scramble to their MiG-21F interceptors, the first Soviet aircraft to challenge the U-2 at high altitude. The MiGs reached the upper atmosphere in a zoom climb, but were unable to manoeuvre or maintain height, and therefore posed little threat to an alert Dragon Lady pilot.

Buddy Brown, Ed Emerling, Gerald McIlmoyle, Robert Primrose, and Jim Qualls; and Captains George Bull, Roger Herman, Charles Kern and Dan Schmarr.

There was, of course, one notable absentee that day. Rudy Anderson's body had been returned by the Cubans, and was buried at his home town of Greenville, South Carolina, on 6 November. General Power led the eulogies. 'It is because of men like Major Anderson that this country has been able to act with determination during these fateful days,' he told Jane Anderson as he handed over her husband's Distinguished Service Medal — the nation's highest peacetime decoration — which he had been awarded posthumously.

There were other U-2 pilots who could reasonably have laid claim to participating in the Offutt ceremony. First there were the Agency's pilots, who had flown all the missions prior to 14 October. One of them, Bob Ericson, had discovered the SAM sites which alerted the intelligence community in the first place. But it was part of the whole 'sheep-dipping' deal that they should remain unsung heroes, at least as far as the public was concerned.[3] There were also some half-dozen 4080th pilots who were left out, since the DFC had only been awarded to those pilots who flew across Cuba prior to Khrushchev's climb-down on 28 October. The others started flying after that date, and the decision to deny them a medal implied that the skies over Cuba were safe thereafter.

No-one in the 4080th could understand this. After all, General Power himself had made it clear that the SAM threat still existed when he spoke at the wing's mission briefing on 4 November. Over the next few weeks, as more of the U-2s were equipped with the System 12 radar warning device, many a U-2 pilot would sweat out an overflight with the red light blinking and the headset screeching to indicate that a SAM radar was tracking him and shifting into firing mode.

Had Khrushchev and Kennedy reached some secret agreement that the reconnaissance flyers would not be shot at? This seemed possible at the time, but when the recorded deliberations of EXCOM were released twenty years later, a different picture emerged. On 3 November, a week after the Soviet agreement to withdraw, the President and his senior advisors were told by the State Department that the Soviets 'reportedly could not guarantee that Cuba would not try to shoot down the overflights — they were a sovereign country'. The CIA believed that the

Soviets were in control of the SAM sites, but there was clearly no formal agreement about this or any other aspect of the Soviet withdrawal. Three days later, President Kennedy asked what would happen if another U-2 was shot down. He was told by General Maxwell Taylor that all further missions would be halted while an attempt was made to find out whether the Soviets had ordered the action. 'If so, we should make an air attack on the SAM site,' he was advised. Robert Macnamara chipped in to say that the Russians had been advised that they would be held responsible if any US aircraft was shot down. Responding to these concerns, Under-Secretary of the Air Force Dr Joe Charyk reversed the earlier decision not to use the new reconnaissance drones over Cuba. Their launch aircraft was almost ready for take-off when Charyk's move was countermanded by General Curtis LeMay.

It is therefore clear that the nation's top leaders were not at all sure that further U-2 flights would go unopposed. In mid-November, Fidel Castro wrote to the UN Secretary-General pointing out that US military aircraft were still 'invading our airspace on . . . illegal and aggressive acts . . . Cuba possesses a legitimate right to defend its territory against such violations'. All in all, it seemed hard on those half-dozen U-2 pilots that didn't receive a medal. But at least they could join the others in the satisfaction of knowing that the job had been well done. Between them, they had laid to rest that old joke in the intelligence community, about how the only overflights that the SAC U-2 unit was good for were those that went over Mexico during the final approach to Laughlin!

Those Cuban MiGs had given everybody a fright. This was the first time that the top-of-the-line Soviet interceptor had challenged the Dragon Lady; the first production MiG-21F models had only just begun to reach the Soviet Air Force in 1960, when U-2 overflights were called off. Now they were reaching the U-2 at altitudes beyond 60,000 feet, but hadn't managed to fire off a shot. One 4080th pilot over Cuba even reported a MiG-21 passing right over the top of him, but his aircraft had not been hit by missiles or guns.

It was soon realized why. The MiG simply could not manoeuvre in the thin upper atmosphere — its control surfaces were too small. It couldn't even get up there unless it snapped into a zoom-climb while going at maximum speed. Once they had a U-2 on radar, the Cuban air defence controllers would instruct the MiG pilot to climb to some 42,000 feet and accelerate, while placing him on an opposing track some ten miles to the side of the U-2. Upon passing abeam his target, the MiG pilot would be told to make a 180-degree turn to place him directly behind, and in position for the snap-up. Then he would zoom up to the intruder's altitude, but in so doing he would lose effective control as his interceptor described a ballistic arc in the stratosphere, before tumbling back to earth. As he did so, the MiG pilot's engine would most likely flame out, and he would have an interesting time on the way down trying to regain control and relight the engine, with possibly very little fuel remaining! The U-2 pilot could usually watch this entire performance; he would be alerted to the initial radar-tracking by System 12, whereupon he would scan the driftsight and probably spot the fighter below him giving off a large contrail. He would follow it through the driftsight until the contrail ceased — this indicated that the MiG was zoom-climbing through the tropopause. This was the signal for the U-2 pilot to make a simple turn, which the flailing fighter could not follow.

Under these circumstances, a successful fighter-intercept seemed unlikely. But to guard against a lucky air-to-air missile shot, Lockheed fabricated an eighteen-inch extension to the U-2 tailpipe which effectively shielded the hottest part of the engine from infra-red seekers. It was known as the 'sugar scoop', and the original extension covering only the bottom quarter of the tailpipe was extended a year later to cover 140 degrees of the circular rear orifice. To validate the sugar scoop, and confirm that U-2 pilots need have no fear of enemy fighters, a series of intercept tests were flown over Edwards. Air Force F-104C Starfighters — the type which most closely matched the MiG-21 in design philosophy and performance — were flown against Agency U-2s. They had the same problem as the MiGs in the thin upper atmosphere, and very few F-104 pilots got within Sidewinder missile range of the U-2. Even then, thanks to the sugar scoop, they needed to be at the same level or above before the missile's seeker could acquire the target.

But what about radar-guided missiles? In a related test, Air Force F-106 interceptors equipped with Hughes radar and Falcon missiles were flown against the U-2s, to test the effectiveness of the System 9 warner and jammer. The F-106 pilots reported that it was working very well. The nearest they got to threatening a U-2 was when one fighter flamed out at the top of its zoom-climb and nearly collided with its quarry! Some of the F-106s which flamed out failed to get a relight, and had to make a deadstick landing on the dry lakebed below.

In view of all this, ideas about changing the U-2 paint scheme were dropped for the time being. While the fighter threat was being taken seriously, it was reasoned that the current midnight blue scheme would stand out against the clouds and haze below, if a MiG pilot managed to climb above the U-2. Various alternatives were flight-tested later on U-2s at Edwards, such as polka dot patterns and zebra stripes, but none proved particularly effective. In any case, a dark matt paint would always be the best camouflage against an aircraft approaching from underneath, because of the black sky all around at high-altitude. In 1965, a very flat black paint known as 'Black Velvet' was developed by Lockheed and introduced on the U-2. Until this time Agency U-2s remained midnight blue (which under certain light conditions, and in monochrome photographs, appeared as a black sheen). Air Force U-2s swapped their polished metal finish for a gloss grey coat of paint in 1961-62, (Colonel DesPortes named it 'confederate grey'), but did not adopt a black paint scheme until 1964. The SAC planes eventually received the Black Velvet scheme in 1966-67.

In 1963, overflights of Cuba continued, settling into a regular routine. The 4080th detach ent moved from Florida to Barksdale AFB, Louisiana, from where a sortie would be mounted almost every morning. In tandem with the U-2 sortie, ELINT-gathering RB-47s from SAC's 55th SRW would fly circuits around the island. The routine was interrupted with a jolt on 20 November 1963, when Captain Joe Hyde crashed into the Gulf of Mexico on his way back from Cuba. His autopilot had failed just prior to coast-out, so he was hand-flying the aircraft back to base. He lost control and entered a flat spin from which the aircraft never recovered. Hyde attempted to eject, but the seat failed again. Although the wreckage was recovered from shallow water off the Florida coast, there was no sign of Hyde. He was presumed to have attempted a bale-out — unsuccessfully. Hyde was flying Article 350, one of the black C-models; accident investigators wondered if he had lost control in the first place because he was not so familiar with the critical cg characteristics of the re-engined aircraft.

The SAC U-2 wing remained active all over the globe. Detachments were now known instead as Operating Locations — OLs. Barksdale was OL-19. The wing received three refuellable U-2E aircraft, which were also equipped with TRW's System 192 for SIGINT. They sported a peculiar pair of large scimitar antennae on the

The 4080th SRW moved from Laughlin AFB, Texas, to Davis-Monthan AFB, Arizona, in 1963. Here is one of the wing's U-2A models on display at an open day at the new base in February 1965. In addition to the pogos, the wings are supported by a metal stand.

upper rear fuselage, which were nick-named 'ram's horns' or 'elephant's ears' by the crews. SAC's codename for the refuelled mission was 'Muscle Magic', and the wing started making thirteen-hour transpacific deployments to the Philippines and Okinawa. Out of Kadena (OL-8), they would fly nighttime SIGINT sorties up to the Sea of Japan, where a KC-135 tanker would be waiting to prolong the mission. Meanwhile, sampling operations continued out of Eielson (OL-5), Albrook or Howard AFBs in the Panama Canal Zone (OL-18), Andersen AFB on Guam, and Laverton RAAF base near Melbourne in Australia (OL-11).

Those Crowflight missions seemed to keep on coming with monotonous regularity. Always on the same days each week, and the same routes each flight. They were now flown at constant altitudes between 50,000 and 70,000 feet. If the one hour pre-breathing was included, a typical sampling flight would require the pilot to remain wrapped up in the pressure suit for nine hours, as he trundled out and back along the allotted track. There wasn't a lot to do up there, except amuse oneself with an occasional sextant shot, maintain the flight log, press a switch to rotate the filter papers once in a while, and check two-way communications with the rescue C-54 or a ground station.

From the time that the Soviets broke the moratorium on nuclear tests in September 1961, both sides had embarked upon the most intensive series of test shots yet, and a large amount of new fallout had reached the stratosphere. The Soviets were responsible for the higher proportion of this; their series of tests in the second half of 1962 — duly monitored and sampled by the U-2s in Alaska — had injected sixty megatons of fission yield, compared with eleven tons from US explosions the same year. Although they didn't set off another fifty-eight-megaton burst that year, a couple of thirty-megaton shots were detected. In January 1963, about eighty per cent of the strontium-90 detected in the atmosphere by the U-2 and other sampling flights at lower altitudes had been put there by tests conducted in 1962.

Negotiators in Geneva inched towards a new test ban agreement, as both the superpowers rushed to complete their current series of shots. Agreement between the UK, US and USSR to ban permanently atmospheric tests was reached in July 1963, and the Partial Test Ban Treaty came into effect three months later. Operation Crowflight continued for another year and a half, and was finally discontinued in February 1965. By this time, the 4080th had gone to war in South-east Asia, and there were plenty of more pressing calls on U-2 flight time.

After six months of planning, the 4080th Wing moved in July 1963 from Laughlin to Davis-Monthan AFB, at Tucson, Arizona. It was goodbye to the sleepy little Texas border town, and to its more rowdy Mexican equivalent on the other side of the river. Ciudad Acuña had become known as 'Boy's Town' to the airmen at Laughlin, on account of the wide variety of entertainment available there, some of which

The SAC U-2 unit was one of the first to go to war in south-east Asia, in 1964, and one of the last to depart, in 1975. OL-20 was based at Bien Hoa AB near Saigon, and this group photo dates from the early days of the operation. The aircraft is a J57-engined aircraft modified for inflight refuelling, and carries the strange-looking antennae for COMINT System 192.

was most definitely not suitable for a family audience! Laughlin was reclaimed by the Air Training Command, and the whine of circuit-bashing T-birds replaced the sudden roar as a U-2 applied power and took off into the wild blue yonder.

At the Arizona desert base, the U-2 aircraft had a peculiar new stablemate. The small cadre of Air Force people who had been working up the new photo-reconnaissance drones at Holloman and Eglin AFBs moved in alongside the 4028th SRS, to which they were now officially assigned. The highly-classified programme was codenamed 'Lightning Bug'. The converted Ryan Firebee target drones were launched from specially-adapted C-130 Hercules transports, and recovered in mid-air by a snare trailed below a CH-3 helicopter. These drones gradually increased in size and sophistication over the next few years, and eventually came to usurp much of the U-2 mission, especially overflights of denied territory. From the start of Lightning Bug, therefore, there was the potential for a strained relationship between the two teams of erstwhile rivals, gathered together within one wing to operate SAC's only overhead reconnaissance assets at the time. In 1965, SAC revived the old RB-57D squadron number and allocated it to the drone operation, which became the 4025th Reconnaissance Squadron.

On the last day of 1963, President Johnson gave his personal approval for U-2 aircraft to be sent to South Vietnam. SAC codenamed the deployment 'Lucky Dragon', and the chosen base — Bien Hoa near Saigon — became OL-20. On 14 February 1964 the 4080th deployed four U-2s to OL-20 by way of Hawaii and Guam. These were still early days in the America's long involvement in the south-east Asia conflict, and there were not yet any US fighters or bombers deployed. Although four RF-101 Voodoos had been stationed at Saigon airport since October 1961 for low-level reconnaissance flights over South Vietnam, and the CIA had flown several U-2 missions over the North out of Taiwan in 1963, there had been no systematic effort to photograph the wider area of interest, especially the infiltration routes through North Vietnam, Laos and Cambodia that were being used to resupply the communist guerillas in the south. This the U-2s of OL-20 proceeded to do, averaging one or two sorties each day. Although the flights were only about two-and-a-half hours in duration, they produced so much film that the rudimentary processing laboratories already set up at Bien Hoa were unable to cope with the load.

As the year wore on, the U-2s ranged far and wide over North and South Vietnam, Laos and Cambodia. The US was dragged further into the conflict, and a watershed was reached in early August when the USS *Maddox* was attacked by North Vietnamese patrol boats. Congress passed the Tonkin Gulf resolution, and B-57 bombers were moved from the Philippines to join the U-2s at Bien Hoa, with fighters deployed further north at Da Nang. Within a couple of days, the North Vietnamese Air Force received its first fighters, quickly flown in from southern China. A U-2 photographed thirty-four MiG-15/17s at the newly-built Phuc Yen airfield. Hitherto, some of OL-20's missions over North Vietnam had been flown at medium levels, due to the lack of potential opposition, but now the U-2 flights were forced upwards, and the drones were called in. Agency U-2s were also in the area. Aircraft and Chinese pilots were deployed from the base on Taiwan to Takhli in Thailand, from where missions were flown over southern China to scan the overland routes by which Peking kept Hanoi supplied with military hardware.

Following the Tonkin Gulf incident, the Lightning Bugs had been quickly deployed to Kadena AB on Okinawa, from where the C-130s flew south-east to launch the drones into that same area of southern China. They were recovered on a paratroop training area close to the Agency's U-2 base at Taoyuan AB on Taiwan, but there was no co-ordination between the blue-suiters and the Black Cat squadron. In early October, the Lightning Bug team moved to Vietnam, settling in alongside the Air Force U-2 boys at Bien Hoa, although the drone recovery crews were based at Da Nang. If any one at OL-20 had any illusions about this conflict, they were rudely shattered on the night of 1 November 1964, when the Viet Cong crept close to Bien Hoa and shelled it with eighty-one millimetre mortars. Five B-57s were destroyed and fifteen others damaged, but the OL-20 area escaped serious damage.

Aerial bombardment of North Vietnam by US fighter-bombers started with carrier-based strikes in the southernmost part of the country on 7 February 1965, and really got under way when the Air Force launched the 'Rolling

Thunder' campaign on 2 March. Given the Soviet and communist Chinese commitment to support the North, the response was inevitable. On 5 April a U-2 sortie revealed that SA-2 SAM sites were being prepared around Hanoi and Haiphong. By early July, five sites had been identified, and the first was going operational. The US Air Force lost its first aircraft to a SAM on 24 July — an F-4C Phantom flying at 20,000 feet.

Before long, they were also shooting at the Dragon Lady. Captain Ed Perdue flew a mission to overhead Haiphong harbour, where he was due to make a 110-degree turn and roll out again to head for home. Observing heavy cloud cover to the west, Perdue elected to turn right and go the long way round before rejoining the south-bound leg. This was an alternative option allowed by his flight plan. Upon rolling into the right hand turn, Perdue was startled to see an SA-2 streak past his left wing. Had he turned the other way, he would surely have been shot down.

Fifty-six SAM sites were located by the end of 1965, extending outwards from Hanoi and Haiphong to cover the railroads leading north towards China. In December, the first MiG-21s appeared. Things were clearly getting hot up there for the U-2s, especially since OL-20 was still operating those old A-models, with minimal protection against enemy air defences. All they had was a two-light SAM warner (and even this hadn't worked on Perdue's aircraft the day he was nearly wiped out). There was no scope to indicate the direction of the threat, and no ECM to throw the missile off course, once it was launched.

So the U-2 photo flights were routed around known SAM sites, and the approach routes to the heavily-defended target areas in the North were varied as much as possible, to keep the air defences guessing. There were also an increasing number of SIGINT missions which did not penetrate the North at all, since they were flown offshore over the Gulf of Tonkin. But even when flying here, there was still a threat from enemy fighters, and some U-2 missions were flown with a pair of F-4 Phantoms as escort. These would cruise many thousands of feet below the Dragon Lady, hoping to intercept any MiGs which might rise in challenge.

The drones began to take over the photo mission up North, and low altitude versions were deployed in 1966 to augment the higher-flying versions. A big effort was mounted to get the proximity fusing signals for the SA-2, so that missile countermeasures could be developed. Some drones were equipped with an ELINT relay system and a travelling wave tube to augment the radar return, making them appear like the U-2 on the North's radars. SAC RB-47s accompanied the drone at a safe distance to record the signals relayed as it penetrated and lured the North Vietnamese into firing a missile. After a number of failures, the effort finally met with success on 13 February 1966, when a 55th SRW RB-47H captured the drone's destruction on tape.

The 4080th gained an unprecedented third outstanding unit award for its pioneering re-connaissance work in south-east Asia during 1964-65. The ten-year-old outfit now laid claim to be the most highly decorated US military unit in peacetime. But U-2 numbers were dwindling fast, thanks to accidents and the Chinese communists, whose air defences were taking their toll of the Taiwan-based aircraft. By the end of 1965, the 4080th Wing was reduced to eleven U-2 aircraft, and a number of pilots and other personnel (including the wing commander, Colonel John DesPortes) had been transferred into the SR-71 Blackbird programme, which was just getting started at Beale AFB, California. The Air Force wanted to keep the U-2 wing intact, but this was getting hard to justify with the diminishing fleet numbers. At SAC head-quarters, they made plans to deactivate the U-2 operation completely in 1968.

1 The CIA was never entirely certain that nuclear warheads for the missiles had reached Cuba. It also appears that the SS-5 IRBMs never reached their destination, although three sites (two at Guanajay and one at Remedios) were in an advanced state of construction, with the necessary excavations and concrete structures being prepared.

2 This was the second such award for the 4080th wing. The first had been made in March 1960, for the unit's work on Operation Crowflight over the previous three years.

3 In fact, most of the CIA U-2 pilots had already been awarded the Distinguished Flying Cross, which they could proudly wear on their uniform once they had returned to the military. In a secret ceremony the following April, many of them would also be awarded the Intelligence Star, one of the CIA's highest decorations, which most certainly was *not* allowed to be worn in public!

Chapter Six
Parting the Bamboo Curtain

A unit of the People's Liberation Army has shot down another U-2 spyplane belonging to the Chiang Kai-Shek gang, which intruded over Shantung province. Let there be no doubt of the people's resolve to resist these aggressive acts! We can shoot them down by day, and also by night . . .
Radio Peking, 11 January 1965.

It was a quiet night in early August 1959, and the security guard was dropping off to sleep at his post in a hangar at the Montezuma County Airport just outside Cortez, Colorado. Midnight was approaching, and all the Cessna and Piper light plane owners and mechanics had gone home hours ago. Suddenly, there was an urgent knocking at the door. The sleepy guard stirred and went to investigate. He found what seemed like a creature from outer space standing there!

Upon closer investigation, the creature appeared to be a pilot, but he was dressed in all sorts of strange gear. Underneath a drab coverall with various zips and toggles attached, he was wearing something which severely restricted his movements. He was very agitated, and he most definitely wasn't an American! He seemed to be an Oriental type, and as he hopped up and down in discomfort he pointed out onto the darkened airstrip, and kept repeating the words 'Come quick! Bring gun! Maximum Security!' at the confounded guard. He didn't appear to know any other English.

The pilot was Major Mike Hua from the nationalist Chinese Air Force (CAF) in Taiwan, and the garb which was causing him so much difficulty was his tight-fitting pressure suit. The US Air Force was teaching him to fly the U-2 at Laughlin AFB, prior to the start of an ambitious programme organised by the CIA during which the high-flying spyplane would be sent deep into the communist mainland to gather vital information on Peking's military progress. In common with five other colleagues now being checked out by Air Force instructors at the Texas base, Hua did have a basic command of English. But in the excitement of the occasion — an emergency deadstick landing following a flame-out at high altitude — Hua forgot nearly all that

The first group of pilots from nationalist China to undergo U-2 flight training in the US. From left to right, Hua Hsi Chun, 'Tiger' Wang, 'Gimo' Yang, Chih Yao Hua, and Chen Huai Sheng. Yang would later become commander of the squadron on Taiwan, while Chen was fêted as a national hero after he was shot down over the mainland and killed in September 1962.

he had learnt of the foreign language.

Word of the strange happening at Montezuma County Airport inevitably leaked out, but the guard and other locals who came into contact with the Chinese pilot that night were told to keep their mouths shut by the posse of serious-looking security men who descended upon the place. In Texas, the 4080th wing commander, Colonel Andrew Bratton, was authorized to issue a brief statement describing the incident, but it was hoped that the secret of the proposed U-2 flights from Taiwan could be maintained.

All six Chinese pilots had arrived in the US in early 1959. Led by Colonel Yang Shih Chu,[1] the group had all made their first flights in the U-2 during May, and were now learning the finer points of high-altitude reconnaissance flying. Major Hua was on a five-hour night flight to test his navigation skills, such as use of the sextant, when the flame-out occurred. He tried four times without success to get a relight, before abandoning the attempt and searching for a suitable landing site. He saw lights from cars on a nearby road, and prepared to deadstick the aircraft there until he realised that a small airport lay nearby. It was nearly too late to change his mind! He pulled the landing gear back up in an attempt to gain greater gliding distance. In the event, he did make it to the field, but was unable to extend the gear again as the U-2 floated down to land. He slid across the runway and came to a halt some fifty feet beyond. There was some minor damage to the fuselage, enough to necessitate dismantling the plane and trucking it back to Texas.

By all accounts, these first six Chinese U-2 pilots were just about the best that Taiwan had to offer. Many more would follow over the next fifteen years, but in the opinion of their American instructors, not all of them displayed the same aptitude as the first cadre. The language barrier, of course, was a major impediment. During flight training, a Chinese interpreter would stand by on the mission frequency, but misunderstandings occurred nevertheless. Air Traffic Control was warned whenever one of the Chinese was about to fly — 'Limited English Pilots on frequency,' they were told. The boys in ATC soon took to calling them the LEPs. Returning from a particularly awkward training flight one day, one of the 4080th instructor pilots remarked in exasperation: 'The trouble with these people is, there's not enough generations between them and the water buffalo! They just can't hack it!'

This was unfair. In fact, all the Chinese pilots sent for U-2 training had plenty of previous military flying experience, mainly on fighter jets. The Dragon Lady was a difficult aircraft for the best of pilots to master, let alone those trying to struggle with a training regime in a foreign language, where very little of the written material was translated. Some Americans in the U-2 training programme thought that the pilot's manual at least should be rendered into Mandarin, but this was never officially done. In the classrooms and on the flightline, instructors would ask the Chinese if they understood the point being made. The answer was always 'yes'. It took time for the Americans to realize that there was a cultural difference here: the Chinese didn't like to say no, because it might represent a loss of face. It didn't help when the 4080th assigned an instructor pilot to a Chinese student, the two would become familiar with each other's ways, and then a second instructor would have to take over halfway through the course because the first had been sent off on an overseas deployment.

Upon completion of the training programme, the Chinese pilots returned to Taiwan, to Taoyuan airfield some thirty miles south of Taipei. This was a major nationalist airbase, from where RF-101 and RF-104 camera-equipped fighters regularly took off on fast, low-level forays across the mainland opposite. Such flights had been taking place ever since the nationalists had been expelled by the communists from the mainland in 1948. Chinese Air Force flyers inserted and recovered intelligence agents, dropped propaganda leaflets, and flew reconnaissance missions. The two sides were still technically at war, and the nationalists kept up their vow to reclaim the lost territory. There had been a major outbreak of fighting (including aerial combat) during the Taiwan Straights Crisis in 1958. During that episode, the Black Knight squadron of the 4080th had become involved in the situation, when the US decided to allow nationalist pilots to fly the RB-57D across the mainland on photo-reconnaissance sorties. Initially, three CAF pilots had been trained on the big-winged machine, and three of the 4025th's aircraft were sent to Taoyuan 'on loan'. Two of them were shot down over the mainland in February 1958 and October 1959.

The operation ended when the RB-57D was withdrawn from SAC in 1960.

From now on, two of the black Lockheed planes would be permanently stationed there instead. Officially, they were assigned to the 35th Reconnaissance Squadron of the Chinese Air Force. This matched the prepared cover story, to the effect that Taiwan had purchased two U-2s from Lockheed in July 1960. In fact, from this date the U-2 operation was a joint American-Chinese affair, with the CIA providing the aircraft, the maintenance, and other support such as PSD, while the nationalist government provided pilots, base facilities, and additional support people. Most of the American personnel were Lockheed employees working under contract to the Agency, but there were also technical representatives from Pratt & Whitney, Lear Siegler (the autopilot manufacturer), Hycon and Itek (the camera people), and from the contractors whose SIGINT and ECM gear was carried on the black aircraft. The American contingent was headed by a US Air Force Colonel, but since there was not supposed to be any official US government involvement, he was known simply as 'the manager' and always dressed as a civilian. The same was true for the dozen or so other airmen and officers assigned by the US military to supervise operations, maintenance, PSD and photo-interpretation. As in Operation Overflight, the Agency provided the security, administrative and communications staff, and some payload specialists.

On the Chinese side, there was a squadron commander and his executive officer (who were supposed to be the only two who knew of direct US involvement through the CIA), and the pilots, flight surgeon, weather forecaster and others. One or two of the Agency's American pilots were always present, not only to conduct further U-2 flight training and give advice, but also to fly some of the missions which for various reasons could not be flown by the Chinese. This hybrid outfit became known as the 'Black Cat' squadron, from its emblem of a black feline head on a red background. Colonel Lu, who had been in charge of the CAF RB-57D operation, became its first commander, but it was Colonel Yang — one of the first group of Chinese pilots to check out on the U-2 in Texas — who made the longest-lasting impression on the Americans at Taoyuan. He was a colourful character who already had considerable flying experience over the mainland in an assortment of CAF aircraft. When he later moved up to take charge of the Black Cat squadron, he became known as 'Gimo' — short for Generalissimo, after Chiang Kai-Shek, the famous leader of the nationalists in exile.

When President Kennedy assumed office in January 1961, he confirmed the pledge given the previous year by his predecessor that the US would conduct no more manned overflights of Soviet territory. The pledge implicitly excluded overflights by satellite, and also manned overflights of other Communist Bloc countries. The People's Republic of China therefore remained a legitimate target for airborne reconnaissance as far as the intelligence community in Washington was concerned — and there was plenty that needed finding out. In particular, what effect was the Sino-Soviet split having on China's nuclear weapons programme?

It was clear that the Chinese communists were intent on pressing ahead with plans to acquire the bomb, despite propaganda statements to the contrary. Even before the rift with Moscow had deepened in 1959, Peking had laid plans for self-sufficiency in nuclear weapons production, and in the means to deliver them, namely ballistic missiles. In 1958, the Soviets had commissioned an experimental heavy water reactor for the Chinese at the Institute of Atomic Energy near Peking, having previously supplied a chemical separation plant capable of producing weapons-grade uranium-235 and plutonium. This had been set up for the two countries' joint use in the remote Sinkiang autonomous region, where radioactive ore products were to be found. Building on this base, the Chinese used Soviet blueprints to construct their own small research reactor at the Peking Institute, fuelled by enriched uranium-238 supplied by the Soviets before they left the country. Between 1959 and 1961, a further four small reactors were established elsewhere, including one of indigenous design at Nankiang.

Could the People's Republic replace the Soviet expertise which had now been withdrawn with its own home-grown talent? A large number of leading scientists led by Professor Chien Hsu Sen now left their posts in Peking and the other major cities. It soon became clear that they had been transferred to the remote Takla Makan desert in Sinkiang province, where a nuclear weapons testing site was established at Lop Nor.

The American-trained Chien was also put in charge of China's ballistic missile development. A launch site and test range for the missiles was set up a few hundred miles to the east in Kansi province. This site became a prime target for the Black Cat squadron, whose missions were planned jointly between Washington and Taipei, with both sides having equal access to the results. However, the missile test site and other nuclear weapons facilities were huge distances away — nearly 2,000 miles from Taiwan. These would be long, dangerous flights at the limit of endurance for both aircraft and pilot. Moreover, unlike a border surveillance flight, or one which made only a brief incursion, there would be no chance of the aircraft limping back to Taoyuan or another friendly field in the event of some mechanical failure over the deep interior. Lop Nor itself was out of range from Taiwan, forcing the Agency to consider alternative bases for the U-2 and other reconnaissance aircraft in countries surrounding the People's Republic, in order to reach this vital intelligence target. The only suitable candidates were Thailand and the nations of the Indian sub-continent. After the political furore of the Powers shootdown, however, there were formidable diplomatic problems involved in securing permission to operate U-2s from any of these.

The Taiwan operation moved into top gear during 1961, as the Chinese pilots embarked on some very long penetrations of the mainland, as well as shorter flights which brought back intelligence on the communists' military dispositions opposite Taiwan. There were as many as three overflights each month, some flown simultaneously. On the longest overflights, the pilots had to fly virtually straight as an arrow from Taoyuan towards the north-west for more than 1,800 miles, before turning round somewhere near Yumen and returning along a parallel route some 200 miles to the side. Because the fuel margins were so tight on these missions, the flight planners at project HQ and within the Black Cat squadron were unable to insert the dog-legs and other meanderings which were typical of a U-2 overflight, and designed to evade enemy radar tracking, or at least keep the air defence operators guessing over what was the real target objective.

Security around the small wooden hangar of Second World War vintage occupied by the two U-2s at Taoyuan was always tight, but through their radar tracking and (probably) their network of agents on Taiwan, Peking soon realized what was going on. In July 1962 the communists broadcast a message promising the equivalent of $280,000 in gold to any Chinese U-2 pilot who defected with his aircraft. Then, on 9 September 1962, the New China News Agency in Peking suddenly announced that a U-2 'of the Chiang Kai Shek gang was shot down this morning . . . when it intruded over eastern China'. No mention was made of the pilot's fate, but nothing was ever heard again from thirty-two-year-old Colonel Chen Huai Sheng, except for some unconfirmed intelligence reports that he had survived a crash-landing, only to die later. He was believed to have been hit by an SA-2 missile. In Peking a few days later, Premier Chou En Lai addressed a rally of 10,000 people called to celebrate the downing of the nationalists' aircraft.

The mainland's air defence system at the time was almost entirely of Soviet origin. Despite the Sino-Soviet split, which became final in July 1960, the communist Chinese had maintained good radar coverage, and now it appeared they had retained a working SAM capability as well. As far as fighter-interceptors were concerned, they had built up a large force of licence-built MiG-15 and 17 types, and the manufacture of more advanced MiG-19s had commenced at the Shenyang aircraft factory before the Soviet military advisors were withdrawn. These types had good altitude performance — better, in fact, than some of their successors. They could cruise with no problem at 40,000 feet — above the level to which a flamed-out U-2 would have to descend in order to get a relight. One of the new MiG-19s was thought to have accounted for the CAF RB-57D which was shot down in October 1959 after it was forced to make a premature descent. United States intelligence also credited the Chinese with a force of at least twenty-five MiG-21 interceptors, despite the withdrawal of Soviet military aid. The pilots of the Black Cat squadron were under no illusions: it was dangerous territory over there!

Chinese U-2 pilots were required by their government to complete ten overflights of the mainland before they could be transferred out of the unit. The exception was for those promoted to squadron commander; these would now know too much about the management of the operation to risk being captured (and perhaps tortured) by

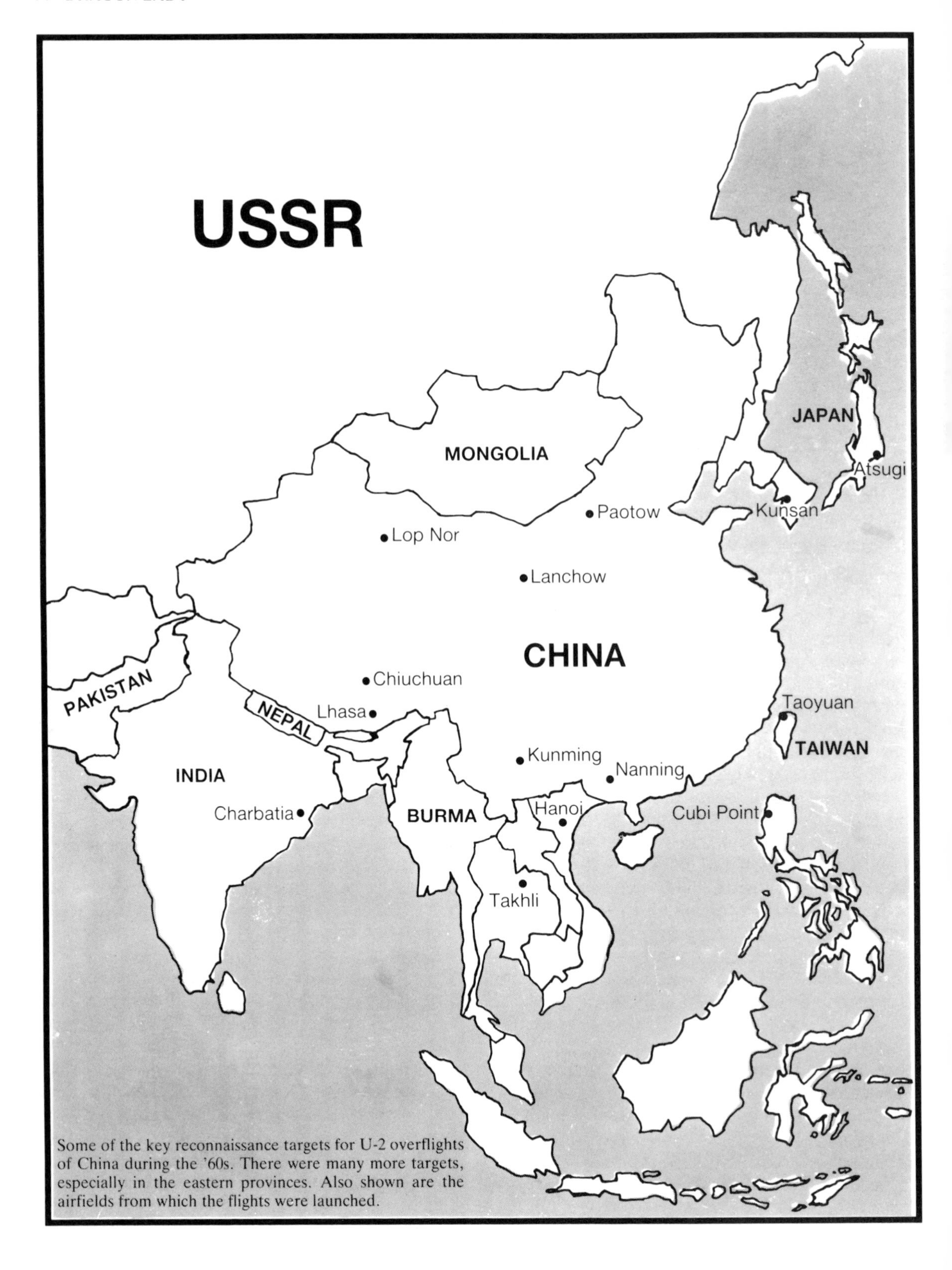

Some of the key reconnaissance targets for U-2 overflights of China during the '60s. There were many more targets, especially in the eastern provinces. Also shown are the airfields from which the flights were launched.

Unlike many of their colleagues, these three Chinese U-2 pilots all completed their combat tour of ten flights over the mainland. Left to right, Major Hua Hsi Chun, Colonel 'Gimo' Yang, and Major 'Tiger' Wang. All are wearing the uncomfortable partial-pressure suit and helmet, with faceplate removed.

the communists. So there was a steady stream of new recruits selected for U-2 training in the US. The fresh intake of pilots were also needed to replace those lost in training accidents — of which there were many. Shortly before Chen was shot down, another of the original group of trainees was killed in a spectacular crash on the runway at Taoyuan in early September. Major Chih Yao Hua was flying night touch and gos when he took a high bounce and failed to control the aircraft as it subsequently began to roll. It cartwheeled over and exploded in a fireball.

Landing the U-2 was the biggest problem for the Chinese pilots. Sometimes they failed to get the fuel balance right before making an approach. At other times they managed this, but simply couldn't control the final few feet before touch-down, when the plane had to be virtually stalled onto the runway. Lockheed suggested an adjustment to the fixed stall strips on the inboard wing leading edge for their benefit. But this adjustment made it difficult for the pilots to distinguish the stall from the Mach buffet at altitude. Eventually, small retractable stall strips were fitted halfway along the wing leading edge. When preparing to land, the pilot pulled a T-handle by his right knee which manually extended the strips by half an inch. This was sufficient to help keep the wings level during the crucial final moments of landing as the aircraft approached the stall. There were at least two more accidents in Taiwan attributed to this cause, and a number of other close calls. The squadron lost the services of another U-2 pilot when 'Tiger' Wang Tai Yu completed his ten mainland missions and moved across the base to take command of the RF-101 squadron.

Despite the formidable training problems and operational dangers, the CAF apparently had no difficulty in recruiting replacement pilots. Four more were despatched to the US for training in 1963. Following the loss of Chen's aircraft, Taipei and Washington both released the cover story about Lockheed having sold two U-2s to the nationalists, after being granted an export licence. In response to some sceptical questions from journalists, a State Department spokesman declared, 'We have said everything we are prepared to say on the subject of U-2s,' adding that, 'We have made it clear that it was a Chinese Nationalist affair.' This distinction didn't fool the communists in Peking, who declared that the US government was 'the chief culprit in this aggressive crime'.

All through 1963, a Chinese nuclear test seemed possible. New reactors were built at Xian and Chungking, bigger than the previous ones and perhaps therefore capable of yielding larger amounts of weapons grade plutonium, but they were also progressing down another route towards bomb-making capability. A gaseous diffusion plant utilizing a large reactor was under construction in secret at Lanchow in Kansu Province, and if the Chinese somehow managed to master the difficult techniques involved, this facility could separate the uranium-238 isotope from the naturally-occurring uranium ore (U-235), so that the scientists could use this instead of plutonium as the fissile material. Progress was also made with a nuclear delivery vehicle — the first Chinese ballistic missile was test-flown in October 1963.

Once SAC had taught the Chinese pilots to fly the U-2A model, they returned to Taoyuan and were checked out by the Agency's resident American pilot on the more powerful C-model. They were then ready to go. Major Yei Chang Yi had hardly been back in Taiwan for three months when he became the second victim of the mainland's air defences on 1 November 1963. He was shot down near Shanghai and presumed killed. A few days later, Peking announced in triumph that it could shoot down U-2s 'at any altitude, and whether they come by day or by night', which suggested that they were now quite confident in their SAM defences.

Five months later, another Chinese pilot died, when Major Liang Teh Pai lost control of a U-2 during a turn at high altitude just off the mainland coast. He was on a training flight which also served as a peripheral SIGINT mission. The aircraft crashed into the sea but the pilot's body was recovered. His death was a bit of a mystery. Liang had evidently made a successful escape from the aircraft — his parachute had deployed. He appeared to have drowned, so perhaps he had hit the water unconscious. But there were no bruises to suggest, for example, that he had collided with part of the aircraft on his way out. Someone came up with the idea that he had drowned because he was too small to negotiate the steep side of the inflatable life raft which formed part of the emergency kit in the seat pack. The life raft was therefore modified to make it easier for the Chinese pilots, who were physically smaller than their American counterparts, to clamber in.

There were more problems with the aircraft in early 1964. The Black Cat squadron experienced a rash of flameouts over Taiwan while climbing up through 'the Badlands'. Lockheed sent test pilot Bob Schumacher to Taoyuan, but as far as he could ascertain, the Chinese pilots were following the agreed procedure of easing back on the throttle while the aircraft climbed through the danger area between 40,000 and 60,000 feet. The problem was apparently caused by some unusually cold temperatures in the tropopause above the island. Because of the Reynolds Number effect, on a very cold day the pilot could retard the throttle to a lower EGT setting, and yet the engine would still produce the same amount of thrust. The solution was to add an Engine Pressure Ratio (EPR) gauge to supplement the EGT indicator. An EPR schedule for climb and descent was devised, and issued to all U-2 pilots; from now on, the engine could be operated at the higher altitudes with much greater precision.

While this solution was being worked out, an aircraft and crew was despatched to NAS Cubi Point in the Philippines, where no climb-out problems were encountered. Launched from here, Lieutenant Colonel Terry Lee made a successful penetration of the Chinese mainland in late June. Early on the morning of 7 July he took off from Cubi Point again for a long, meandering flight across China's southern provinces of Kwangsi and Kwangtung. This was the region through which China was sending most of its military aid to North Vietnam and the Viet Cong, who were now posing a serious threat to the US-supported regime in Saigon.

Lee penetrated the mainland north of Hainan Island and headed for Tsungtso, a town near the border with North Vietnam, and on the main railway line going south-west to Hanoi. He then turned north-east, following the road and rail links all the way to Henyang. Now followed another sharp turn to point the U-2 south-east towards Hong Kong. Before reaching the British colony, however, Lee turned back north-east so that the big camera could capture communist military dispositions along the heavily-defended lowland area opposite Taiwan. He was now on the last leg of the overflight, and was shortly due to 'coast-out' over the nationalist coastal enclave of Quemoy island before flying high over Taoyuan and setting course back to Cubi Point. It had been an uneventful flight, as these things went. The communists had been scrambling MiGs below him, and through the driftsight he had caught sight of the occasional contrail far below. Fortunately, the aircraft had functioned perfectly, giving the pilot no cause to worry about a premature descent to their level.

As Terry Lee approached Swatow, his System

12 indicator suddenly lit up for the first time in the flight. This was the warning device which alerted pilots to the tracking of their aircraft by an SA-2 missile's Fan Song radar. On the panel in front of him, Lee was seeing the 'Hi-PRF' light flash, an indication that the radar had already been switched to missile firing mode. In the command posts at Cubi Point and Taoyuan, his excited voice came over the SSB radio link: 'Hi light on 12!' he repeated three times. This was the last that was ever heard from him. He was shot down almost within sight of home. A ground radar site at the southern tip of Taiwan tracked a descending target over Swatow — it was obviously Lee's crippled U-2. An air and sea search was mounted as close to the mainland coast as the nationalists dared, in case the pilot had managed to get out of the stricken aircraft and had been blown out to sea. Nothing was found. The next morning, Peking radio announced another victory over the bandit U-2 planes.

Upon receiving a SAM warning, Lee would have initiated a descending turn, but only gradually at first or the aircraft's limiting Mach number would have been exceeded. The manoeuvre had evidently not been enough to avoid the SAM — the U-2 clearly needed more protection. In September 1964 the Black Cat squadron received its first active SAM countermeasure in the form of System 13. This was an adaptation by Sanders Associates of the deception jammer they had produced for the Navy's A-3 Skywarrior jet. It was less bulky than the 'brute-force' noise jammers used by the Air Force at this time, but it was nevertheless still too big to fit inside the U-2. Instead, it was housed in wing slipper tanks which had been modified with deeper, fibreglass bottoms. Carrying the ECM system in this way entailed the loss of 200 gallons of fuel-carrying capacity, but within a year the Skunk Works came up with a drop tank arrangement under each wing, attached by an integral pylon to a point just outboard of the slippers. These tanks exactly compensated for the fuel capacity lost when the ECM was carried in the slippers.

System 13 was completely automatic. Once the Fan Song radar signal was detected, it would begin re-transmitting the signal in a manner which was calculated to confuse the radar operators down below in the missile's tracking cabin. On their radar scopes, a false display of the U-2's position would be generated, so that the SA-2 missile was despatched to a rendezvous in the wrong piece of sky. Unlike a typical noise jammer, which would fill the scope with clutter, System 13 gave the missile crew no indication that they were being spoofed. On the U-2, the transmitter was housed in the starboard slipper pod and the inverter in the left pod. In order to prevent the other side from acquiring this high technology in the event of an aircraft being forced down, the U-2's destruct device was moved from the Q-bay to the modified pods.[2]

Overflights were resumed, and System 13 soon proved its worth over the mainland, but it was yet another addition of weight to the airframe. What with the HF radio, Birdwatcher, Systems 9 and 12, and other minor improvements, over 700 lb. had been added to the airframe since 1959, when the J75 engine had been substituted in an attempt to cure the aircraft's original weight problem. Once again, added refinements had eaten into the U-2's range and altitude performance. While the air-refuellable U-2F was now available, it couldn't help mission planners trying to stretch the legs of the U-2 all the way from Taiwan to Lop Nor and back. They could hardly send the KC-135 in to set up a refuelling track over the heart of mainland China! Because of the need for the U-2 to climb rapidly to near-maximum height before crossing the mainland coast, on account of the missile threat there, the more remote Chinese test sites were by now out of range from Taiwan, and had to be reached some other way.

Looking at the map, Peshawar in Pakistan would have been a suitable alternative launch base. Not only was it much closer to the target, but Chinese radar coverage of the approach routes across the Himalayas might not be so comprehensive. However, Pakistan had been badly 'burned' by the Powers incident in 1960 and, of greater significance, it had recently developed friendly relations with communist China. By contrast, its larger neighbour to the east had very poor relations with the People's Republic. India and China had fought a major war in two disputed sections of their mountainous common border in October 1962, and nearly two years later the two countries were still at odds, with occasional border skirmishes always liable to break out.

In the 1962 clash, India received a bloody nose and turned to the West for help, not only to

Ground crew rush to insert pogos as a U-2C rolls to a halt after a test flight. US civilian registrations were applied to aircraft returning from Taiwan and other overseas CIA detachments for overhaul. The jetpipe of this aircraft has been modified with the first version of the 'sugar scoop', intended to make it more difficult for an infra-red homing missile fired from an interceptor to find its target.

its former colonial masters in London, but also to Washington. The US had airlifted infantry arms to Delhi and agreed to provide the Indian Air Force with a complete air defence radar system for the Chinese border. Throughout 1963, the military relationship between Delhi and Washington strengthened, although India was officially determined to remain non-aligned. (It proceeded to demonstrate this by negotiating alternative arms deals with Moscow, including one for MiG-21 fighters.)

So the CIA suggested a temporary U-2 detachment in northern India. From here, the nuclear test site and other remote intelligence objectives in Sinkiang province and Tibet could be overflown and photographed to the mutual advantage of the US and its host. The White House approved the plan; protracted negotiations with Delhi ensued, and became embroiled with an Indian request for three squadrons of F-104 fighters. But a deal appeared likely in spring 1964, and the Indians allocated an old wartime base named Charbatia on the east coast near Cuttack for the U-2 deployment. For obvious reasons, both sides wanted the deployment to be kept top-secret. An aircraft and its American crew were despatched from Edwards to Cubi Point, where they were kept waiting for several weeks before the final permission to fly into India was granted.

Two or three U-2 missions over China were flown from the Indian base. The Indian Government received up-to-the-minute intelligence on Chinese military deployments along the border — and the US intelligence community got the pictures of Lop Nor that they so badly wanted. What an extraordinary view the pilots of these flights must have had, as they soared across the spectacular Himalayan mountains which form the roof of the world! One flight was routed over the Tibetan capital of Lhasa, and brought back superb pictures of the famed Potala Palace. The Lop Nor nuclear test site itself was in a 2,000 feet-deep depression, twenty miles wide and twenty miles across, in a plateau which was otherwise 4,000 feet high. A blast tower had been erected in the middle of the depression, and there were various support facilities visible nearby. But there hadn't yet been a test shot there.

Although the U-2 flights from India were handled in the Agency's classic low-profile operational style — a quick deployment of the U-2 and accompanying transport; fly the mission(s); pack up and rapidly depart — the new venture had an inauspicious beginning. When the the very first mission returned to Charbatia, the aircraft ran off the end of the runway! There was a substantial hump in the middle of the old airfield's brick runway, and the optimum end for landing was ruled out because it entailed overflying a populated area, which would reveal the operation to a wider audience. One of the Agency's most-experienced pilots was chosen for this first flight, and he made an uneventful flight until the aircraft was rolling out after touchdown, when the brakes failed to operate. The U-2 crested the hump and began to gather pace as it rolled down the other side. The pilot tried to drag one wing and then the other, to no avail, and the aircraft ran off the other end of the runway and stuck fast in the mud. The pilot was unhurt, and the photographic 'take' undamaged, but the U-2 had to be dismantled and returned to the Skunk Works for repair. The incident brought the first U-2 deployment to India to an abrupt conclusion.

What with these and some other very discreet operations, the Agency's U-2 pilots were kept busy. A couple more from the original groups now wanted to resume their military careers, so for the first time since 1956, the CIA needed to recruit some new American pilots. The SAC U-2 wing was the obvious source, and so in the summer of 1964 two of the 4080th's more experienced U-2 pilots — Robert 'Deacon' Hall and Dan Schmarr — were duly put through the 'sheep-dip' process, eventually emerging at Edwards North Base. A third Air Force pilot, Warren Boyd, was sent to Taoyuan as a combined maintenance officer and test pilot, but this post remained on the military payroll.

In September 1964 two aircraft and crews from the Black Cat squadron were deployed to Takhli airbase. The Vietnam War was hotting up, and the US wanted more frequent coverage of the communist supply routes through southern China. There was some fear in Washington that the Red Army would be thrown into the conflict on Hanoi's side, as had happened during the Korean War, when the US was taken by surprise. The SAC U-2 unit was flying over much of Indo-China from Bien Hoa, and had also been sending the Lightning Bug drones across southern China. Nevertheless, the hierarchy in Washington decreed that only CIA-manned U-2s should penetrate Chinese territory. Peking's support for North Vietnam and the Viet Cong now became a major focus of effort for the Black Cat squadron, which maintained a semi-permanent detachment at Takhli from hereon. Flights over the railroads and airfields of southern China were interspersed with training flights, during which most of Thailand was photo-mapped. Within a few months, the hitherto sleepy Thai base was swarming with Americans, since it became home to an entire Air Force F-105 fighter wing, brought in to strike against selected targets in North Vietnam. To maintain 'cover', the CIA security people told Gimo Yang and his Chinese U-2 pilots to masquerade as Hawaiian-Americans flying for Air America (an Agency-owned airline used for covert operations, which kept some aircraft at Takhli). The CAF pilots had a great time swapping fighter pilot's tales in the officers' club with the F-105 jocks, who probably didn't believe the Hawaiian story for a moment!

After India, the Thai base was also the next best launch point for a U-2 mission to Lop Nor, although this would be a very long haul indeed. At this time, there was some doubt whether Delhi would permit any more U-2 flights from its territory; a new administration had been formed there following the death of Prime Minister Nehru on 27 June. As it turned out, the call to fly all the way to Lop Nor from Takhli didn't come at this time, and a further U-2 deployment to Charbatia went ahead in December 1964. By then, the long-awaited first Chinese nuclear test had taken place, on 16 October 1964. The weapon yield was estimated at twenty kilotons, or about the size of the bomb dropped on Hiroshima in 1945.

Thanks to the excellence of US intelligence — due in no small measure to the earlier work performed by the Black Cat squadron — no-one in the West was much surprised by the first Chinese nuclear test. Just 17 days before the blast, US Secretary of State Dean Rusk predicted in public that it was about to occur. Strangely, the U-2s at Taoyuan were not tasked to conduct atmospheric sampling flights at this time, but the US Air Force did, and it was not long before the weapons diagnostics people reached a surprising conclusion about this first test. From an examination of the fallout particles, they concluded that uranium-235 had been used as the fissile material, instead of plutonium. How had the Chinese acquired enough of this highly-enriched material to create the explosion? Had they stockpiled the isotope from the days of Sino-Soviet co-operation, or had the gaseous diffusion plant at Lanchow gone into top gear with the separation process?

The latter explanation seemed more likely, but if so, it was a worrying development. If the diffusion plant was in production, it meant that the Chinese communists would soon have a substantial nuclear stockpile. For not only could they produce highly enriched uranium-235 as fissile material in its own right, but they could probably also speed production of weapons-grade plutonium by providing their reactors with slightly-enriched uranium as fuel. This was the more efficient method of getting good quantities of fissile material from reactors — and it was the course which the US had adopted twenty years earlier. More than this, China's apparent success in producing uranium-235 could quickly lead to a tritium-production capability, and therefore to the H-bomb.

The diffusion plant at Lanchow now became a prime intelligence target, along with a reactor at Paotow in Inner Mongolia, which was thought to be the major plutonium-producing plant. What was the rate of production at these facilities? Project HQ decided to equip a U-2 with an infra-red scanning camera in an attempt to find out. This equipment, designed by Texas Instruments, produced its best results at night. CIA pilots therefore flew the device out of Edwards on nighttime sorties over US nuclear plants whose production levels were known. By extrapolating this data, the results that the infra-red scanner would bring back from over the Chinese mainland could be better assessed.

If the flights were successfully completed, that is. The cities of Lanchow and Paotow were on the Yellow River, although separated by 500 miles, and both were about a 2,600-mile round trip from Taiwan. With the extra weight of those recent modifications, this was now close to the maximum distance that the U-2 could fly, when the maximum altitude profile was being flown. In recent times, none of the squadron's overflights had penetrated more than an hour's flying time from the coast, or a friendly border.

By early November, the infra-red camera was delivered to Taoyuan and mounted in the Q-bay of U-2C Article 358. The Agency always kept two of its approximately ten-strong fleet of U-2s there at any one time. The first target was Lanchow, and Major Johnny Wang Shi Chuen was chosen to fly there. He had been flying the U-2 for just over a year. He launched into the night and set course to the north-west. For over three hours he kept going, until an electrical failure knocked out all his cockpit lighting. With only a flashlight with which to read his instruments and map, Wang turned round and flew home. He recovered safely, but the flight had been painted by mainland radar, and the track pointed straight as an arrow towards the target. The communists now knew that the U-2 went travelling by night — had they also worked out the target objective?

Next it was Major Jack Chang's turn. He was a relative newcomer to the squadron, having returned from training at Davis-Monthan the previous July. The target this time was Paotow, but when Chang performed the obligatory test of the defensive systems just before crossing the mainland coast, System 13 failed to function and he too had to abort, although not before the mainland air defence system had been alerted again. A few days later, Johnny Wang was launched on another attempt to overfly Lanchow. In an attempt to spoof the radars, he flew south down the coast for a few hundred miles before turning back towards the north-west for the long run-in. The tactic seemed to work. He flew unopposed to within thirty miles of the target, until his SAM warner started functioning. The scope indicated a SAM site almost directly ahead and within lethal range of the Lanchow complex. At the same time, the new System 13 countermeasures pod began jamming. Wang started to turn away from the danger area, but almost immediately he was blinded by the brilliant rocket flame as an SA-2 missile roared past him. Luckily, it did not detonate, and Wang managed to re-engage the autopilot and set course for home, as his vision slowly returned. He was not troubled further by communist defences, and landed back at Taoyuan as dawn broke. The infra-red camera film indicated that at least three SAMs had been fired, but System 13 had apparently saved him.

So near and yet so far! Mission planners back at U-2 project HQ in Langley decided on a different approach. They would launch out of South Korea and cover both the Lanchow and Paotow targets on the same flight, recovering into Taiwan. Tension mounted in the Black Cat squadron, and everyone started calling this flight 'The Big One'. The same two Chinese pilots were assigned, with Chang as primary and Wang as back-up. Early one morning in mid-December 1964, they flew off to Kunsan airbase in a C-130 with the ground crew. As usual with these detachment operations, the mission airplane would follow after dark, this time in the hands of the resident American pilot. Things began to go wrong shortly after the advance party arrived in Korea: Chang went down with a stomach complaint and so the flight was reassigned to Wang. The mission bird landed at Kunsan around 8 p.m., witnessed by virtually no-one save a security police crew, and was quietly refuelled out of sight on a remote taxiway. The Agency pilot had flown all the way from Taoyuan at 40,000 feet, so as not to alert radar operators along the mainland coast to the fact

The sorry-looking remains of Major Jack Chang's U-2 on display in Peking. The communists pieced together the first four U-2s that they shot down, and put them all on display in mid-1965, but only this one now survives. Chang was attempting to photograph the Lanchow atomic weapons plant when he came to grief on the night of 10 January 1965.

that a high-flying U-2 was in the air.

At 11 p.m., the still Korean night was shattered as the J75 roared into action and Wang's plane leapt off the runway. But twenty minutes later, 'The Big One' had to be aborted when Wang's routine self-test of System 13 indicated another failure. The U-2 pilot set course direct for Taoyuan along the coast, and the disappointed ground crew packed up and reboarded the C-130 for the long flight home.

It was the eighth day of the new year before weather conditions were right for another attempt. This time, they reverted to Plan A, and Johnny Wang set course yet again from Taoyuan to Lanchow. Against all expectations, it was an uneventful flight. No SAMs, no equipment failures. Wang safely overflew this once-small town on the Yellow River which had expanded dramatically into a major industrial centre in the last few years as Peking threw massive resources into the nuclear weapons effort. The infra-red camera duly recorded the required information and the U-2 returned home safely. Now for Paotow!

Once again, it was Jack Chang's assignment, and he set off in the early evening of 10 January 1965. In order to fly the shortest distance across 'denied territory', his flight plan went north over water via the East China Sea and the Yellow Sea to a coast-in point near Tsingtao. He reached this point at about 8.30 p.m. and headed inland. The Agency operatives in the command post back at Taoyuan lost track of him as his Birdwatcher signal faded out. Reception of the selected HF frequency was degraded after dark, and he had flown out of range. To keep abreast of his progress, they monitored the nationalists' SIGINT station, which listened in to the mainland's air defence system. The communists were tracking the U-2, but about three-quarters of an hour after coast-in, when Chang was about 200 miles south of Peking, the tracking ceased. Five minutes later, the communists were heard to call off their air defence alert. After a long night of hopeless waiting, Peking radio announced the next morning that it had shot another U-2 down. Article 358 — and Major Jack Chang — had finally run out of luck.

Now the communists had bagged four U-2s, not to mention a few of the Lightning Bug reconnaissance drones being flown over southern China by the 4080th Wing from Vietnam. They collected all the pieces, assembled them together as best they could, and mounted a grand display at the People's Military Museum in Peking the following August. Pictures of the event were flashed all round the world. Once again, they confirmed that the Dragon Lady could be shot out of the sky and yet fall to earth in relatively large pieces. The vertical tails and outer wings of the reassembled aircraft were virtually intact, and although there was some mixing of parts between airplanes, all four aircraft were quite recognizably U-2s. Chinese Air Force markings were still visible on two of them — the nationalists' 12-pointed sun on the rear fuselage and a four-digit number on the tail. (It was the practice to apply these markings in soluble paint; they could thus be easily removed if the aircraft was returned to the US for overhaul, or deployed to another country.)[3]

The Black Cat squadron reviewed its tactics. From the listening posts monitoring mainland communications, the intelligence specialists knew that the communists were calling an air defence alert every time their search radars spotted a U-2 climbing out of Taoyuan to beyond 40,000 feet, where no other aircraft ventured. One of the newer pilots was assigned to try and evade the radars during a training sortie. Major Pete Wang took off and halted his climb at 20,000 feet by deploying the gear and speed brakes. He had a rough ride for some 150 miles down to the southern tip of Taiwan, where he began climbing to cruise altitude. Despite the ruse, mainland radars duly began tracking him as a suspected U-2 as soon as he passed through 40,000 feet. There was no fooling the communists — if anything, their radar-tracking was better than that of the nationalists: before the U-2s were equipped with an IFF transponder in late 1963, Taiwan radar had experienced some difficulty in identifying the aircraft!

The air defence situation was getting worse. In the first few years of Chinese overflights, the SAM threat had been downgraded by routeing missions around the known sites, and if an unknown site was inadvertently approached, System 12 would give the pilot a warning to steer clear of the danger area. Good as System 12 was, the communists had learned how to devalue its worth to the U-2 pilot by keeping their SA-2 missile radars off the air until the last minute. The Fan Song system had a long-range scan mode which swept 360 degrees until a target was acquired. During this phase, it operated at low pulse repetition frequency (PRF). Once the target was acquired, the radar switched its scan pattern to establish a 'window' within which both the target and the missile (once launched) could be tracked. As the target came within missile range, the PRF would increase so that the target could be more precisely tracked. System 12 could pick up the low-PRF signals at a hundred miles distance, and display a flashing white light in the cockpit, along with the approximate direction of the site as a strobe on a small scope. This 'low-light' gave the U-2 pilot plenty of time to adjust his course around the missile site. If he were to ignore this warning, a high-PRF light would eventually illuminate in red to indicate the switch in radar mode which presaged a missile firing, and a screeching noise would be relayed through the headset. But when Terry Lee had been shot down, and on at least one other occasion since 1962, the first warning of a missile attack had been when the high-PRF light had come on, with the length of the strobe on the display scope indicating that the site must be very close. The missile crews were evidently relying on long-range search radars to feed them target information, so that they could refrain from operating their own Fan Song system until the U-2 was within firing range. The communists had probably also got Major Yei with this technique in November 1963.

Fortunately, advance technology soon came to the rescue. A new package of electronic countermeasures became available in early 1965, and the first U-2 to be modified was deployed to Taiwan shortly thereafter. You couldn't mistake the improved aircraft — it had grown a large spine which extended the full length of the fuselage aft of the cockpit. This was nicknamed 'The Canoe', and in addition to the HF radio and (on U-2F models) refuelling receptacle, it contained a much-improved airborne intercept jammer (System 9B). (The canoe could also house a doppler radar, although this equipment was not required by the CIA, which wanted its overflights kept as stealthy as possible.) The System 13 SAM jammer was miniaturized and relocated from the modified slipper tanks to a pressurized compartment just ahead of the tailwheel, and the SAM warner was also upgraded as System 12B.[4]

In the spring of 1965, another missile warning system was installed in the U-2. This was a very sensitive receiver known as 'Oscar Sierra'. It was able to detect the faint emissions from the SA-2 SAM command guidance transmitter, which operated on a different frequency from the radar. This transmitter needed to be tested at full power prior to missile firing, and most of this output was operated into what was known as dummy load. This was supposed to prevent the signal leaking out until the missiles had been fired, but Oscar Sierra's sensitive saucer-shaped antenna, mounted on the lower Q-bay hatch, picked up the inevitable signal leakage from dummy load, and caused a red light to glow in the U-2 cockpit. If the SA-2 transmitter was switched from dummy load to antenna — thus indicating that a SAM was now on its way — a higher-strength signal would reach the aircraft's antenna and the light would begin to blink. The U-2 pilot now had to take last-ditch evasive action, preferably as smart a 180-degree turn as possible, although the tight margins of 'coffin corner' did not usually allow for fighter-style tactics. The evasive action could throw the missile guidance computer, making the intercept equation too complicated, and causing the SA-2 to miss its target. The red light in the cockpit was labelled simply 'OS'; almost inevitably, it became known to the pilots as the 'Oh Shit!' light.

With all this new electronic gear, the U-2 cockpit had to be upgraded. Even so, it was a tough job fitting everything into such a confined space. There was a new threat warning display on the upper right instrument panel, with a three-inch cathode ray tube displaying information from Systems 9B, 12B, and 'OS'. The scope produced lines indicating the type and direction of the threat — interceptors with a dotted line and SAMs with a solid line. The warning lights were clustered around the scope. Any U-2 pilot now faced with opposition would be assaulted by a barrage of visual and aural warnings. During a SAM attack, for instance, the 'Oscar Sierra' red light would be flashing, one or more green strobe lines from the System 12B warner would appear on the scope, together with more flashing red and white lights, and the System 13A jammer 'repeat' red light would illuminate. Through the headset, the pilot would hear the high-pitched bird-like 'Chirrrp' of the Fan Song radar signal, a 'beep-beep' from 'Oscar Sierra', and a 'Gronnnk' from the Bird-watcher telemetry system as it relayed a fresh burst of data every time one of the threat systems momentarily lost and re-acquired the signals. The poor U-2 pilot was supposed to take all this in while throwing the delicate machine into the tightest-possible turn in order to get the hell out of the area!

Although SAMs continued to be the main worry, an unwelcome new threat emerged in the early spring of 1965. Chinese MiG-21s began to appear. The Soviets had delivered up to fifty of their early MiG-21F models before the split in 1960, but these had not previously troubled the overflying U-2s. Now the aircraft factory at Shenyang had succeeded in turning out copies,

Rare mid-sixties shot of a CIA aircraft carrying drop tanks on underwing pylons. The long dorsal canoe is also evident, having recently been added to house the latest ECM equipment. The full 180-degree 'sugar scoop' in the jetpipe is providing enhanced protection against IR-guided missiles.

which they designated Jian-7. According to information made public by the Chinese in later years, the Jian-7 first flew on 8 January 1965, but if this is so, then it was pressed into immediate action against the U-2s. On Captain Charlie Wu's very first overflight, a roundtrip from Takhli in February 1965, his aircraft was intercepted south of Kunming as the fighter loomed up from below his right wing and then dropped out of sight beneath the nose after a near collision with the Dragon Lady. The Chinese were evidently having the same intercept problems as the Cubans three years earlier, with the fighter going ballistic and flaming out as it neared the top of the curve. However, on this very first intercept, the Jian-7 had apparently managed to loose off two infra-red missiles: Wu reported seeing two black puffs ahead of him just before the fighter popped up. The puffs were thought to be the missiles detonating at the end of their runs.

The improved System 9B did not provide any warning of this and some further Jian-7 intercept attempts experienced by the U-2 pilots over the next few months. It was designed to detect and counter airborne radar threats (and had a powerful transmitter which could be switched to any of four antennae covering each quadrant). But there were indications that the Chinese had stripped the radars out of these new fighters, thus presumably improving their altitude performance a little, and were directing them from the ground via a data link. They were also using infra-red, rather than radar-guided, air-to-air missiles (AAMs). Given the shortcomings of early Communist Bloc AAMs, however, the fighter would probably have to approach from the rear to have any chance of a successful lock-on. Even then, the 'sugar scoop' which masked the U-2's engine exhaust would hopefully reduce this possibility still further. To keep the overflight altitudes as high as possible, less fuel was carried. The slipper tanks were now rarely used except for ferry missions, and overflights were planned to arrive back at the descent point with only fifty gallons remaining. This led to some anxious approaches when the weather was poor and there wasn't enough fuel left for a go-around or divert. For certain missions, the refuellable U-2F could be used, and there was nearly always one on hand at Taoyuan, although only the Agency's American pilots were qualified to fly tanker-assisted missions.[5]

With the air intercept threat now taken seriously again, another round of experimentation in U-2 camouflage took place. It resulted in the adoption of the 'Black Velvet' paint scheme in late 1965. This very matt black coating was devised by the Pittsburgh Paint and Glass Co., and apparently made some contribution towards reducing the aircraft's radar return. Microscopic particles of ferrous material weakened the electromagnetic currents on the aircraft's skin, and hence helped to suppress sidelobes. Minute balls of glass which were also present in the pigment helped soak up light and suppress reflections in the visible spectrum. It didn't exactly result in a 'stealth' U-2 (that idea had been tried and rejected years earlier), but it did make the task of the opposition's radar operator or interceptor pilot more difficult. Before 'Black Velvet' was finally adopted, the CIA sponsored trials of various alternatives. One of these was 'Chameleon', a cream-coloured paint from NCR that turned to brown as the aircraft ascended into the cold upper atmosphere!

To accompany 'Black Velvet', the cockpit sunshade was repainted from white to black, the speedbrake and landing gear wells from silver to black, and the canopy was treated to reduce its reflective qualities. Heeding the lessons from aerial combat in earlier conflicts, the Skunk Works did all it could to reduce the possibility of a stray burst of reflected sunlight reaching a pair of unfriendly airborne eyes. This effort complemented procedures which were already employed by the U-2 squadrons. It had, for instance, been the practice for some years to pull the circuit-breaker from the aircraft's top and bottom anti-collision beacons before an overflight mission, and stick masking tape over the lenses so that they couldn't serve as a reflector.

The People's Republic exploded its second A-bomb on 14 May 1965 — another twenty-kiloton device made of uranium-235. Taken together with the intelligence brought back from Lanchow by the U-2s, there could now be no doubt that the Chinese were well-advanced in uranium separation techniques. Overflights continued throughout the year, with up to six flown each month. In July, Johnny Wang flew his tenth mainland mission and thus 'graduated' from the hazardous duty. He was promoted to Lieutenant Colonel and succeeded 'Gimo' Yang as squadron commander. There were no more losses to Peking's missiles this year, but yet

When MiG-21s flown by communist Chinese pilots began to challenge the U-2 again in 1965, the Lockheed Skunk Works re-examined the U-2 paint scheme to see if the aircraft could be further camouflaged. The 'polka dot' idea shown here was not adopted, but this aircraft is wearing the new 'black velvet' coating which became standard. Note that even the formerly white canopy sunshade has now been painted black.

another Chinese pilot died on 22 October when Major Pete Wang lost control during a high-altitude training flight.

He had just completed a 180-degree turn off the north coast of Taiwan. Eyewitnesses on the shoreline saw the plane hit the water, but no parachute. Analysis of the strip chart from Birdwatcher showed that the plane had first exceeded maximum Mach and positive G limits. Then came alternating negative and positive G readings, and fluctuating fuel pressure, which were symptoms of a wildly-tumbling aircraft that had probably lost its horizontal tail and wings. Loss of cockpit pressurization and other indications of a flame-out followed as the altitude channels showed a rapid descent through 60,000, 50,000 and 40,000 feet. After forty seconds of this, the transmissions had ceased, probably because the wire antenna had been ripped off as the plane spun towards the sea. There had been no indication that the pilot had ejected, nor that he had jettisoned the canopy prior to a bale-out. An extensive search was nevertheless mounted, but all that was discovered were two of the aircraft's oxygen bottles. Wang had presumably been pinned in the cockpit by heavy G-forces. As the volume of data from accidents like this increased, it became ever-more apparent that the chances of survival from a high-altitude break-up were slim if one didn't get out at the first sign of trouble.

A classmate of Wang's at Tucson had clearly been thinking along these lines almost a year earlier, when he had hurriedly baled out of a U-2 during a training sortie from the SAC base. It was 19 December 1964, and twenty-eight-year-old Captain Steve Sheng Shih Hi had only recently resumed flight training. This was because he had hurt his back during an earlier accident, when he ejected from another U-2 at altitude over Boise, Idaho. In SAC's training regime, one normally didn't get a second chance at qualifying after an incident like that, when pilot error was clearly a major factor (he had allowed the fuel to become unbalanced). But the Americans were dealing with a different culture here, and Sheng's father was a big wheel in the Taiwan military, so he got his second chance. There were thunderstorm cells about on that December day, but none were showing up near the field when Sheng taxied out for a local flying sortie. He didn't get very far. Taking off from Davis-Monthan's Runway 12, he was supposed to climb straight ahead towards the VORTAC beacon before turning to make a standard approach back to the field. The U-2 would reach the VORTAC in four minutes at an altitude of 20,000 feet. Sheng had just reached this point when he suddenly transmitted 'Me get turbulence! Me get turbulence! Me bale out!' Before the instructor-pilot or Chinese interpreter in the D-M command post had a chance to pick up the microphone, he had gone over the side. Sheng landed rather uncomfortably amid the desert cactii, and was taken to hospital to have all the thorns removed! This time, he was scrubbed from the programme. Back on Taiwan, he resumed flying F-104s, but was killed in a crash some time later.

Yet more Chinese pilots were selected for U-2 training. At the time of Sheng's second accident, there were a further three at Tucson preparing for their first flights. These were followed a year later by three more, including Captain Andy Fan who managed the dubious distinction of ejecting (successfully) from his very first flight in the U-2 on 22 March 1966. By the time that this class of 1966 had completed the programme and returned to Taiwan, another pilot had perished there.

During a high-altitude training flight, Captain Charlie Wu Tse Shi shut down the engine when he received an overtemp warning. Much of Taiwan was covered in cloud, complicating his task of planning a deadstick landing somewhere. Wu turned south towards Kung Kwan airbase near Taichung, which had very long runways and was the divert field for Taoyuan. As he neared the field, he noticed a smaller airstrip through a tempting gap in the clouds, and decided to put down there instead. The strip was too short, and Wu landed halfway down it, bounced, and crashed over a ditch at the far end and into some houses. The U-2 exploded, and in addition to the pilot, five people were killed on the ground. This incident brought the number of Chinese U-2 pilots who had either been killed or gone missing over the mainland to eight. Out of the eighteen who had been trained to fly the Dragon Lady, only four had completed the ten overflights required before one could graduate from this dangerous assignment.

1 In common with widespread practice, Chinese names are presented here with surname (family name) first, followed by forenames. Most of the Chinese pilots adopted (or were given) a Western christian name; when these are cited, the christian name is given first, followed by the surname and Chinese forenames. Throughout this book, the old Wade-Giles system of rendering Chinese names into the roman script has been used. This system was the accepted method of romanization during the period under discussion, and remains in use on Taiwan to this day.

2 As far as the author can tell, System 13 was the first active jammer to be deployed on the U-2. Two years earlier, however, in the middle of the Cuba missile crisis, Lockheed was asked by the USAF to try and accommodate an ECM system from the B-52 on the U-2, and this resulted in the first use of modified slipper tanks for this purpose. A modified aircraft was test-flown, but not until after the crisis had been resolved, and was apparently not used operationally.

3 On the subject of U-2 markings, when CIA planes were test-flown by Lockheed in the US after overhaul, they were painted with an American civil registration taken at random from the series N801X to N810X. In 1963, the Lockheed U-2 depot was relocated to the Van Nuys airport some ten miles west of Burbank. The strange black aircraft seemed a little out of place amid the hundreds of lightplanes scattered about this busy field. One day, an FAA inspector doing the rounds passed by the Lockheed ramp and noticed one of the U-2s with its 'civil' identity. He approached the Skunk Works crew working on the aircraft and demanded to know how they had got permission to use the FAA-allocated registration, which didn't appear on his list of records. Receiving a non-committal and unsatisfactory answer, he drove off, vowing to pursue the matter with higher authority, and return. He never did!

4 One feature of particular note in the latest ECM package was a new type of destruct device consisting of non-explosive incendiary panels which were mounted against the circuit boards of System 13A. These boards contained the really secret technology, and would be burnt up when the destructor was operated, but the combustion would be contained within the pressurized box so that the aircraft could still be flown.

5 The classic intelligence method of protecting the secrecy of operations, known as 'compartmenting,' was practised extensively within the Taiwan U-2 operation. For instance, when the Agency's American pilots flew south from Taoyuan on reconnaissance missions over Southern China and Vietnam in 1963, the Chinese members of the squadron were told that these were test and training, not operational, flights. The CAF people were also not supposed to know about the carrier missions, or those flown out of India. To reduce the chances of compartmented information being inadvertently released, the Agency frowned upon social fraternizing between the Chinese and American contingents at Taoyuan. This stricture was not taken entirely seriously: pilots and wives from the two camps would arrange to meet off-base in Taipei instead.

Another example of the compartment theory in action occurred at Edwards later in the sixties, when CAF flyers were training on the U-2 at North Base at the same time as some British pilots. Headquarters decreed that the two groups of foreign nationals should not meet — and the Chinese were banished to off-base housing at California City, a remote community in the desert some twenty miles from Edwards. (The British got the best of the deal — they were housed in a motel in downtown Lancaster.) The Edwards squadron was supposed to schedule the two groups for training on different days — an irritating complication!

6 Some of the Agency's stipulations seemed particularly irrelevant, such as that which affected those people transferred from the 4080th Wing to the Agency operation. No-one at Laughlin (or later, Davis-Monthan) was supposed to know where they had gone, or why. This was ridiculous! Many of these two groups of U-2 pilots had known each other for years, and they had worked alongside each other during Cuba and other events. SAC people knew exactly what was going on at Edwards, Taoyuan or wherever. Nevertheless, the Agency decreed that all mail from the ex-4080th people to their friends back at Laughlin or Davis-Monthan had to be sent first to Washington for re-forwarding, so that its origin remained unknown. Eventually, the powers-that-be were persuaded to drop this requirement.

Chapter Seven
A New U-2

Without the authority assigned to the Skunk Works by our customers and the Lockheed Corporation, we would not have been able to accomplish many of the things we have done, things about which I felt we could take a risk — and did.
Kelly Johnson, 1985

One night in August 1963, one of the CIA's black U-2 models was flown into North Island Naval Air Station at San Diego and quickly craned onto the flight deck of the USS *Kitty Hawk*. Under cover of darkness, the big aircraft carrier headed out of the harbour, while the U-2 was positioned on the stern. It was the start of an ambitious project to 'navalize' the Dragon Lady. The Agency figured thus: with all the political and logistical complications of setting up U-2 bases in other countries, in order to fly missions over vital but remote targets, why not utilize instead those floating chunks of US real estate, the big aircraft carriers? It would be easier to convince a US Admiral that a U-2 flight was essential than some unreliable foreign politician! The Skunk Works believed that U-2 operations from aircraft carriers were feasible. Now they were going to find out for sure.

By daybreak, the USS *Kitty Hawk* was well out to sea. Lockheed test pilot Bob Schumacher took off and made a number of practice approaches to check basic performance parameters, such as wind-over-deck requirements and the effect of turbulent wake from the carrier's island structure on the U-2's delicate approach profile. It soon became apparent that unless there was a gale blowing, the carrier did not need to turn into wind before allowing the U-2 to approach, as was the usual procedure for recovering naval airplanes. If, for instance, the ship was to turn into a thirty-knot wind and then pump its own speed up to twenty-five knots, the poor old U-2 had virtually no forward speed as it approached the deck. Furthermore, the wake turbulence from the island became dangerous. The ideal speed for the aircraft carrier to maintain was fifteen knots. With some valuable lessons learned from this first trial (in which no touchdowns were attempted), the Skunk Works engineers returned to Burbank and set about designing the modifications necessary for full U-2 carrier capability.

The aircraft which emerged was redesignated U-2G. The tail was beefed up to cope with the additional structural loads imposed by an arrested landing, by extending the three longerons in the main fuselage into the tail section. A neat little tail hook was mounted ahead of the tail wheel, attached to the strong framing which formed the

A U-2G model approaches the deck in one of the very first carrier landings by a Dragon Lady. The hook is in the fully extended position, as is the generous area of flap, as the moment of truth approaches.

U-2 safely down on the USS *Ranger* during the series of qualification flights conducted in early 1964. The flights were supposed to be top-secret, and all extraneous personnel confined below deck while they took place, but the carrier's island is jammed with onlookers in this picture! The aircraft carries titles which read 'Office of Naval Research', but this was in reality another CIA scheme.

engine mounts and wing attachment points. The pilot released it from the stowed position by pulling a T-handle in the cockpit. Small fairings were added in front, behind and on both sides of the hook. These reduced aerodynamic drag, and coincidentally provided an anchor for a small radome which could be fitted over the modification when not in use, so that the hook would remain hidden from view to preserve the security of this top-secret programme. To deflect the carrier's arresting cable from the tail wheel in the event that the hook failed to connect, a metal tubing structure was added to the gear assembly and doors. The main landing gear was also modified; because of the restricted space available on the carrier, a swivel bar was added so that the plane could be moved sideways. With its eighty-foot wingspan, there was no way the U-2 could be moved off the elevator and into the hangar deck without this innovation. A fuel dump system was introduced, so that the pilot could quickly reduce his fuel load to achieve the required weight for a carrier landing. The system consisted of a dump valve and float switch in each of the four wing tanks, and an overboard chute on the trailing edge of each wing between the flap and the aileron. This innovation was subsequently extended to the rest of the U-2 fleet, and was much appreciated — if a mission was aborted shortly after take-off, it eliminated the long, boring hours of droning around while burning off enough fuel to bring the aircraft down to landing weight.[1]

In early 1964, Bob Schumacher and a few of the Agency's pilots were despatched to the Naval Air Facility at Monterey, California where instructor pilots checked them out on the T-2A Buckeye jet trainer. The group then transferred to NAS Pensacola for carrier landing qualification flights on the USS *Lexington*, also using the Buckeye. Then it was back to California, where the USS *Ranger* was cruising off the coast on a shakedown trial. With most of the sailors confined below deck, Bob Schumacher flew out from Edwards to make a series of touch and gos, to be followed by the first U-2 carrier landing. It was the start of a hairy process of trial and error, before a satisfactory technique was developed.

From the earlier approach trials on the *Kitty*

Hawk, it had been realized that the U-2's tremendous wing lift would have to be dumped more efficiently, if the aircraft was to be put down precisely where the arresting cables were situated on deck. A set of wing spoilers was therefore fitted, but the U-2 pilot was still required to approach the fantail at near to stalling speed, around seventy-eight knots, which meant flying at a high angle of attack since the power was still on. Eventually, the approach technique was refined with the help of John Uber, a US Navy Landing Signals Officer (LSO) who was specifically assigned to the programme. As the U-2 approached the fan, Uber would command 'First cut!' — the signal for the pilot to reduce the power to idle. Upon his command, 'Second cut!' the spoilers would be deployed, whereupon the aircraft was supposed to hit the deck and engage the cables. Of course, regular naval aviators would throw a fit if asked to cut the power *before* engaging a cable!

Schumacher's very first carrier landing went wrong. He successfully engaged the hook, but the rear of the U-2 tipped up and the nose was dug into the deck. The Skunk Works crew who had embarked on the ship fixed the broken pitot mast, and a somewhat shaken Schumacher flew the aircraft back to the mainland for more extensive repairs. (Taking-off from the carrier was the easy part!) The Navy determined that the accident was due to the tension of the arrester cables being too tight. The Skunk Works weren't taking any chances — they modified a landing pogo and fixed it under the nose as protection in case the same thing happened again. Springs were also added to the base of the wingtip skids.

A few days later, Schumacher flew out again and completed some half-dozen successful landings. Now it was the turn of the Agency pilots. All went well until Jim Barnes snagged one of the arrester cables with his left wingtip while performing a touch-and-go. The cable dug into the leading edge of the skid and the aircraft slewed off the port side of the flight deck. By some miracle it kept flying, and Barnes managed to climb away and head back to Edwards, but he experienced great difficulty in controlling the aircraft in the roll axis. After he landed, it was discovered that the left aileron had been almost completely jammed in the collision with the cable. In fact, the pilot's only lateral control during the flight back to Edwards may have been through wing-warping as he exerted maximum force on the control yoke, in order to get anything to happen. To prevent a similar accident in future, a metal plate was added to the front of the skid as a reinforcement.

Eventually the U-2 pilots were declared to be 'carrier-qualified', and the intelligence planners in Washington came up with the perfect target for a carrier-launched U-2 mission: Mururoa Atoll in the middle of the Pacific Ocean. This was part of French Polynesia, and had been selected by France in 1963 as its new nuclear test site, following the loss of the original site in the Sahara desert when Algeria became independent. The French were going it alone in nuclear weapons development, and also pursuing an increasingly independent foreign policy under President Charles de Gaulle. There was little love lost between Washington and Paris during this time; the following year the Americans were told to remove their NATO bases from French soil. Mururoa was thousands of miles from anywhere, and was little more than a twenty-mile by ten-mile strip of coral rising out of the ocean. In May 1964, a U-2G was embarked on the USS *Ranger* which set sail for the mid-Pacific. The U-2 flew off and secretly photographed the test site.

For the next four years, a few pilots were always carrier-qualified and three of the Agency's ten-strong U-2 fleet were carrier-capable. Since carrier missions were few and far between, the hook was removed for most of the time, so that these aircraft could be used elsewhere, such as Taiwan. But the G-models were not often favoured for overflight missions on account of the extra weight that could not be removed, such as the spoilers. One of them, Article 349, was a really heavy aircraft, for it was given the inflight-refuelling receiver as well as a hook, and became the one and only U-2H model.

Because this aircraft had a low ceiling, it seldom left Edwards on serious business. However, it nearly grabbed its share of glory over the Middle East during the 1967 Arab-Israeli War. Agency U-2 aircraft and crews were deployed across the Atlantic to the US base at Upper Heyford in the UK, to await the go-ahead for an intelligence-gathering flight over the war zone. But the planned mission encountered all sorts of political difficulties: the Spanish and French refused permission for the U-2 to overfly, and the British were reluctant to let the plane land at RAF Akrotiri on Cyprus after the flight. Thought was given to using the unique U-2H: it could take off from Upper Heyford, be refuelled in flight as it made the long detour round France and Spain, fly over the target area, and recover on a US Sixth Fleet carrier in the Mediterranean. But Washington tired of all the political complications, and when the Admirals started objecting about the likely disruption to naval air wing operations, the whole plan was called off.

When a deployment to the aircraft carrier was imminent, the pilots were required to complete nine practice carrier landings. A training site was set up on the lakebed at Edwards, with a deck pattern marked out and a portable mirror and other equipment set up to replicate the carrier landing aids. On 26 April 1965, two pilots were scheduled for this Field Carrier Landing Practice (FCLP). Although the first of these two completed his nine FCLPs, he reported that the aircraft was falling off on the left wing, meaning that it would stall a couple of knots ahead of the other wing. Each U-2 had its individual traits, but the pilot felt this was serious enough to call in the Lockheed people from Burbank. A call was made, and test pilot Bob Schumacher was despatched from the Skunk Works. In the meantime, Eugene Edens took off in the U-2 to complete his FCLPs as originally scheduled.

'Buster' Edens was one of the originals from the second class at Groom Lake, a contemporary of Frank Powers. Like all the long-serving Agency pilots, he had needed a share of luck to get this far. Three years earlier, he had written off a U-2 in a landing accident at Edwards. On final approach, the plane had stalled some fifty feet short of the lakebed runway, and slammed onto the ground. The main gear was driven up into the sump tank, and the fuel caught fire. Jim Cherbonneaux was on mobile duty that day, and rushed over to the wreckage. Finding Edens in a state of shock, he pulled him out of the cockpit just as the fire spread to the left wing tank. Both men beat a rapid retreat with singed legs and arms, and turned to watch from a distance as the left wing blew up.

Today, Edens' flight in the misbehaving U-2G also ended with the aircraft a wreck on the dry lake bed, but this time he did not walk away. He made an approach, and the wing dropped as advertised. Edens applied power and climbed away. He was advised to take it up to a safe altitude and check the stall characteristics, but for some reason the pilot elected to perform the check in the pattern. He lost control at about 2,000 feet and the U-2 spiralled into the ground. Edens was too low to complete a successful ejection, and was killed.

In the Far East, China exploded into the Cultural Revolution in mid-1966, and the Vietnam War intensified. Hanoi's MiGs rose to challenge US fighters on an almost daily basis, and by the end of 1966 some 170 separate SA-2 sites had been identified, with the communists switching missiles between sites to keep the attackers guessing. The cat-and-mouse electronic war in the skies over North Vietnam was now in full swing, with the first Wild Weasel fighters making attacks on the SAM sites with Shrike anti-radiation missiles, and ECM pods appearing on F-4 and F-105 strike aircraft.

Although they did not know it, the US fighter pilots who flew up North had the U-2 to thank for much of their electronic protection. The APR-25 SAM radar warning device, which was designed by Applied Technology[2] and rushed into service on the fighters in south-east Asia, was a direct descendent of the same company's Systems 9B and 12B which had been developed for the U-2 in the early sixties. If a U-2 made it back home after successfully evading a missile, its ELINT recorders could yield valuable data on the enemy's firing and fusing techniques. In the early days of the Vietnam War, much of the intelligence on the SA-2 which was used by the US to develop countermeasures, had been acquired at great risk and with some sacrifice by U-2 pilots over Cuba and mainland China.

At Bien Hoa, the codename for OL-20's U-2 operations changed from 'Lucky Dragon' to 'Trojan Horse'. Not content with that, the faceless military bureaucrats who decide these things now decreed a change in the wing and squadron numbers. In June 1966, the 4080th SRW, 4025th RS (Lightning Bug) and 4028th SRS (Dragon Lady) passed into history, replaced by the 100th SRW, 350th SRS and 349th SRS respectively. The official 100th Wing insignia of two rampant lions was adopted, but the subordinate units managed to retain their treasured Dragon and Black Knight symbols in new squadron patches.

U-2s from OL-20 were still flying photo missions over the heavily-defended North, and about halfway through the year the black boxes of missile ECM System 13 finally became avail-

One of the last U-2A models in SAC service stands on the ramp at Davis-Monthan AFB. This aircraft, serial 66953, was one of the five extra machines produced from spare parts after the initial production run was completed. It was subsequently converted to U-2C standard, and later became the first U-2CT trainer version.

able to the Air Force flyers. This was at least a year and a half after the system had been issued by the CIA to the Black Cat squadron on Taipei! On 8 October 1966 Major Leo Stewart of OL-20 was returning to Bien Hoa from a sortie up North when he encountered severe buffeting and overspeed during the descent. He disconnected the autopilot, but even with the speed brakes and landing gear out, the aircraft kept accelerating. He ejected from the aircraft and landed in the jungle not far from the base. A rescue helicopter soon picked him up and returned him to base.

The wing had lost their first U-2 in south-east Asia; when the wreckage was located, some officers from OL-20 accompanied Army special forces to the area and supervised the blowing up of the remains. Unfortunately, the tail section had separated inflight and was not to be found, and it contained the precious System 13. It was vital that the other side didn't get their hands on this exotic hardware; that was why the battery-powered destruct device had been included to burn up the system's logic circuits. The trouble was, Stewart's plane had been sent out *without* the destruct system onboard. Because the proper technical data had not yet been issued, no-one in the electronics shop at OL-20 had wanted to take responsibility for wiring it up, in case something went wrong and they were blamed for writing-off such a valuable piece of equipment.

That tail section had to be found! A U-2 was sent up on a low-level photo mission over the area where Stewart had ejected. The film revealed nothing. Some RF-101s were despatched to cover the same area with their low-level oblique cameras, but again nothing was apparent on film. Eventually, eight US special forces soldiers and 120 Vietnamese soldiers were inserted into the area by helicopter. After combing the dense jungle for three days they found the missing ECM system, and the whole of OL-20 breathed a sigh of relief.

About 600 miles north-west of Bien Hoa, CIA U-2 flights continued out of Takhli, Thailand. As they flew across China's Kweichow, Kwangsi and Kwangtung provinces, the nationalist pilots from Taiwan met increasing opposition from China's missile forces, and 'Oscar Sierra' or System 13 were often their saviours. On one flight in July 1966, Major 'Spike' Chuang Jen Liang was ambushed by a SAM battery when turning over Kunming airbase, and no fewer than eight missiles were fired at him in two salvos. Two of them were captured on film as the pilot steered the aircraft into an evasive turn and escaped destruction.

On another penetration to the Lanchow diffusion plant, Major Billy Chang Hseih watched two SAMs go by as he passed over the target. Turning round a little further to the north, his route brought him back across the missile site. The camera captured a frantic scene on film as the missile crews attempted to reload the launchers and fire another salvo.[3] A U-2 pilot flying at 72,000 feet who detected a missile launch would have about forty seconds to get out of the way, before the SAM reached his level. He needed to get well clear in this time, in case the missile exploded nearby. In the rarified atmosphere of the upper altitudes, the scatter pattern from an SA-2 explosion was about 800 feet. If the U-2 was hit by even the smallest fragment of debris, it might be downed. That was how Major Anderson had died over Cuba.

Chinese-built MiG fighters constantly snapped at the heels of the U-2s as they cruised above, thereby representing another threat to the successful completion of the mission. More air-to-air missiles were launched by the MiGs during speculative pop-up attacks. As a counter, the U-2 was equipped with System 20, an infra-red warner able to detect an approaching fighter or air-to-air missile in the rear quadrant. The device was faired into the trailing edge of the right wing, between the flap and the aileron.

The Chinese pilots were now flying over a country in turmoil, as Chairman Mao's Red Guards ran riot against 'bourgeois reactionaries'. These young radicals thought that much of the country's ruling establishment — including the top military leaders — fell into that category, although Defence Minister Lin Piao seemed to be on their side. At first, the nuclear weapons scientists and their research establishments were protected from the upheavals by the moderate faction led by Prime Minister Chou En Lai. A third Chinese bomb was exploded on 9 May 1966, this time from an aircraft flying at 10,000 feet over the desert, and it was found to contain some thermonuclear material. More ominously, ballistic missile tests had been stepped up, and US intelligence predicted that China would have an operational deterrent capable of reaching 700 miles within a couple of years. A U-2 from Taoyuan photographed a Soviet Golf-class ballistic missile-carrying submarine that the Chinese had managed to complete at the Dairen shipyard.

But the toll on both men and machines from the Black Cat squadron continued to increase. On 21 June 1966 Major Mickey Yu fell victim to an engine failure on a high-altitude training flight. He was over the East China Sea on his way back from Okinawa to Taoyuan and at first he reckoned he was close enough to stretch the glide to Taiwan. Then he changed his mind, and turned back towards Okinawa. The hesitation proved fatal. He no longer had sufficent altitude to make it as far as Kadena AB. Yu's last desperate option to save the U-2 was to deadstick onto a small island which was visible in the ocean below, but as he neared the ground, he realised that the terrain was unsuitable. He ejected, but the aircraft was too low for him to survive. The final moments of the drama were witnessed by the crews of three fishing boats in the vicinity of the island.

A fifth bomb was set off at Lop Nor in December 1966, and on 17 June 1967, Peking triumphantly announced that it had exploded its first H-bomb. Once again, the event came as no surprise to US intelligence analysts. Just twelve days earlier, Spike Chuang had flown a U-2 over the test site on an epic 3,700-statute mile roundtrip from Takhli. Climbing out of the Thai base in the dead of night, Chuang flew north-west across Burma, Assam, and Tibet to reach his objective at the ideal time for photography — early morning. He returned along much the same route, and brought the aircraft safely back to Takhli after a nine-hour flight. After the blast, another Chinese pilot took the same route to conduct a sampling operation. The two missions to Lop Nor could only be flown using the maximum range profile; the aircraft didn't rise above 60,000 feet in the first few hours of the flight, but the air defence threat over the Irrawaddy valley and the eastern Himalayas was correctly judged to be negligible.

A SAC U-2C model in the late sixties, after application of the 'black velvet' paint scheme. The dorsal antenna shared by Systems 3 and 6 is prominent, while a recent innovation is the data-link, visible in the form of a small fairing behind the tailwheel.

On 9 September 1967 the Chinese communists claimed their fifth U-2 when Captain Tom Hwang Lung Pei fell victim to an SA-2 over Chuhsien during an overflight from Taoyuan. According to the information relayed by Birdwatcher, Hwang had received a warning from 'Oscar Sierra', but it clearly hadn't been sufficent to save him.

The number of U-2s that had been destroyed in accidents or shoot-downs now topped forty. The few airframes that remained were beginning to show their age. Considering that they had originally been built for a one-time overflight project which no-one envisaged lasting more than two or three years, they had done remarkably well. They were overhauled and maintained by some of the finest engineering talent that Lockheed and the Air Force could muster. Even so, the effect of long flying hours and increasing equipment loads was taking its toll. Cracks were discovered in a SAC aircraft at OL-20 which had just returned from a particularly turbulent flight. Accident investigators reviewed two other recent crashes where structural failure was involved.

The first of these had occurred on 25 February 1966 at Edwards. A new pilot was being trained by WRSP-4, having been drafted in from the US Navy as a quid pro quo for making the aircraft carriers available to the CIA for U-2G operations. He took off in one of the squadron T-birds to observe a U-2 in-flight refuelling demonstration to be carried out by Robert 'Deacon' Hall. It turned out to be quite some demonstration!

Deke Hall had practised refuelling on numerous previous occasions. After the hook-up was completed, and the U-2 had safely withdrawn abeam the tanker, he was in the habit of making a spectacular, fighter-style pull-up to impress the onlookers watching from the KC-135. On this occasion, someone onboard the tanker requested that Hall do his usual party piece. The manoeuvre was within limits if conducted at the refuelling speed (210 knots) and with control surfaces placed in the gust position. Unfortunately, with his eyes out of the cockpit to ensure he kept a safe distance from the KC-135, the U-2 pilot failed to notice that the tanker he was formating on had increased speed to about 250 knots. As he pulled up, he heard a loud crack and the aircraft came apart! The wings separated and the tail fell off, and there was an explosion as fuel spilled out and was ignited. Watching the various pieces of wreckage come floating by, the boomer on the KC-135 was amazed to see the U-2 pilot amongst it, safely strapped to his ejection seat. The Navy pilot in the T-33 followed him down and reported that Hall's parachute had opened at 22,000 feet. He made a safe landing, and the aircraft wreckage was gathered together for analysis.

The initial verdict was that it was the pilot's fault for pulling up at too high a speed. But on 1 July 1967 Captain Sam Swart of the 100th SRW was forced to eject from a U-2C only forty miles out of Barksdale AFB when the wing came off. Once again, cracks were discovered and the remaining U-2 fleet was subjected to inspections every twenty-five hours of flight. Lockeed came up with a fix. The fatigue was starting at countersunk holes in the structure. These could all be drilled out oversize and then bevelled, so that a rosette fitting could be inserted. But was it worth it when so few aircraft remained?

It might have been curtains for the U-2 there and then, had an important decision not been taken a year earlier. In August 1966, the Skunk Works received the go-ahead for a new and much-improved version of the Dragon Lady. Kelly Johnson convinced the government that despite the advent of the supersonic Blackbird, there was still a valuable role for the U-2 to play in the gathering of airborne intelligence. The new Mach 3 spyplane was now in service with the CIA, and had been deployed to Kadena AB, Okinawa for flights over the Chinese mainland. The Air Force was also working up on the definitive SR-71 version at Beale AFB. It was capable of reaching well above 85,000 feet, and of outrunning anything that the other side could throw at it. If photo coverage of a high-threat area was required, the Blackbird would henceforth be the platform of choice. But it was a very expensive aircraft to operate, and there were many other missions which did not require that degree of performance. In particular, the demand for long-duration SIGINT-gathering flights was growing, and the U-2 was an ideal vehicle for these. Moreover, a new generation of long-range oblique reconnaissance (LOROP) cameras capable of photographing many miles to the side of the host aircraft were being developed. Equipped with one of these, a U-2 could still perform many photo missions by operating outside the perimeter of the opposition's defences.

So government approval for a limited production run of a new U-2 version was given, with

Line-up of U-2R models during the flight test programme at Edwards North Base. This new version differed in a number of important respects from the original machines, not least of these being its size — about one-third bigger. Aircraft N803X in the foreground has yet to be painted.

the CIA once again funding the airframes, and the Air Force providing the engines. In classic Skunk Works tradition, the U-2R was ready for flight test just one year after the go-ahead, and in operational service less than a year after that. Although the basic design principles of the original U-2 were retained, the U-2R was effectively a new aircraft built from a completely new set of drawings. It was about one-third bigger than the earlier U-2s, but its fuel capacity and payload was doubled. The cockpit was enlarged forty-five per cent, which enabled the pilot to wear the latest and much more comfortable full-pressure suit, and also permitted the installation of a larger, zero-zero ejection seat. As far as its flying qualities were concerned, the U-2R was a much less capricious beast than its predecessor. One still needed to be an extremely good pilot to fly the U-2R, but it flew better at altitude and landed more easily than the early-model Dragon Ladies.

Pratt & Whitney's high-altitude version of the J75 powerplant was retained in the new model. This engine had grown in thrust from 15,800 lb. in the original -13, fitted to the U-2C in 1959, to 17,000 lb. in the -13B which was now fitted to nearly all the surviving early-model aircraft. The U-2C's performance had always been airframe-limited, rather than powerplant-limited, and so the challenge for the Skunk Works team (headed by Ed Baldwin as U-2R project engineer) was to design an airframe to coax maximum performance out of this reliable motor. For the wing, the team retained the original aircraft's stock NASA 64A airfoil (a new supercritical section was considered and rejected), but they scaled it up to achieve a 104 foot span and an area of 1,000 square feet. Lift over drag ratio remained twenty-six, but with gross take-off weights of 40,000 lb. soon to be approved for the U-2R, the wing loading remained close to that of the C-model. The fuselage was stretched to sixty-three feet in length and expanded to a width of 101 inches and a depth of sixty-two inches. The longer fuselage with its greater moment arm would have exerted even greater loads on the horizontal tail, had it not been for one of Kelly Johnson's innovations, the all-movable tail where both the horizontal and vertical sections were pivoted as one about a central point. This had made its first appearance in 1957 on the Lockheed Jetstar executive transport. On the U-2R, this hydraulic action effectively distributed the pitch forces to the entire horizontal tail, not merely the elevators, and the whole slab trimmed. The arrangement also widened the aircraft's cg limits, and eliminated the need for ballast.

Within the enlarged fuselage, room was found for the engine oil cooler intake which had been placed in an external scoop on the starboard side of the fuselage on the earlier model. Kelly Johnson was keen to eliminate all such draggy protuberances from the new design. Instead of an externally-strung wire, the HF antenna became a slot type faired into the leading edge of the vertical tail. The long above-fuselage canoe was eliminated and its contents redistributed elsewhere (the HF radio amplifier and receiver remained on top of the fuselage, however, in a fairing ahead of the tail which also housed the IFF transponder). Although the U-2R emerged with an exceptionally clean external appearance, it was not long before Johnson's 'clean-it-up' edict was countermanded, and lumps, bumps and wires began appearing. The first of these was the ADF sense antenna. It had to be relocated to the top of the fuselage after the initial design which was faired into the undercarriage doors proved unsatisfactory.

Since the engine exhaust now travelled through a longer tailpipe before emerging into the atmosphere, there was no need to retain the anti-infra-red missile countermeasure known as the 'sugar scoop'. Certain navigational aids which had been denied to the early U-2 pilots and only grudgingly incorporated later, such as ILS, TACAN and doppler, were now included as standard. The sextant disappeared, but the driftsight — so handy for spotting those MiGs below — was retained. There were new avionics for flight reference and flight direction, and the pilot now even had the luxury of a fuel quantity indicator for the sump tank, to supplement the crude fuel-remaining counter of the earlier U-2 models.

Flight control was considerably improved in the U-2R. The moving surfaces were larger for a start, and two hydraulically-actuated spoilers were added on the top surface of each wing. The inboard pair dumped unwanted lift in the descent phase, while the outboard pair improved roll control at low altitude. The retractable leading-edge stall strips which had been added to the earlier models in 1964 at the mid-wing point of the leading-edge were retained, and all this made landings and other critical manoeuvres much easier. During the most critical portion of the cruise-climb, 'the throat' or 'coffin corner' widened to give an approximate fifteen knot tolerance between the stall and maximum allowable airspeed. At altitude, the R-model could also be banked to forty-five degrees with safety, a fifteen-degree improvement over the earlier models.

The most significant feature of the enlarged wing was the huge increase in fuel capacity. Compared with the earlier models, the fuel tanks were re-arranged: instead of the main tank being in the forward section of the wing with the auxiliary behind it, the wings tanks were now arranged inboard and outboard. Fuel stored in the larger (1,169 gallon capacity) inboard tanks would be used first during a mission, so that the weight of the remaining fuel in the smaller (239 gallon capacity) outboard tanks would help dampen wing bending and torsional loads. With another ninety-nine gallons of fuel in the sump tank, this brought the total U-2R fuel capacity to 2,915 gallons, or nearly 18,500 lb. There was no more need for extra slipper or drop tanks — this aircraft could fly for fifteen hours on internal fuel alone, if the pilot could stay awake that long! The landing gear was beefed up to cope with the increased gross weights.

On operational missions, the weight of the sensors and other factors would reduce the time aloft, but ferry flights of thirteen hours or more would soon become routine. A typical U-2R operational mission could consist of a 3,000 lb. payload as part of a 22,500 lb. zero fuel weight. By adding 12,250 lb. of fuel, a 3,000 nautical mile mission lasting seven and a half hours could be flown, most of it above 70,000 feet. The maximum achievable altitude remained subject to a host of variables such as gross takeoff weight and outside air temperature, but with the J75 red-lined at 665 degrees centigrade, it was usually in the mid-seventy thousands of feet.

Lockheed test pilot Bill Park took the U-2R into the air for the first time from Edwards North Base on 28 August 1967. There was no prototype; Article 051, the first aircraft off the line, eventually went into operational service. Only twelve R-models were built, surely one of the shortest production runs ever mounted for an operational US military aircraft. Of these twelve, six were allocated to the CIA and six to the Air Force. The Agency sent the first of its new aircraft into service at Taoyuan in mid-1968, and retired all its early-model U-2s. The first U-2R allocated to the Air Force reached the 100th SRW at Davis-Monthan in September 1968. SAC had eventually turned in its old J57-powered aircraft in 1966 for re-engining, and had since operated U-2C and U-2F models painted in the 'Black Velvet' finish. Now, the new R-models were sent to OL-20 in Vietnam and to OL-19, which moved back to McCoy from Barksdale in 1968, and was still responsible for the Cuba mission. Only some fifteen U-2s remained from the original production. A few of these were stored; others were reworked and continued in use with the 100th SRW, mainly for training at Davis-Monthan. One of them — Article 347/Serial 56-6680 — had now flown more than 5,000 hours, and had become known as 'old sick-eighty' to those responsible for keeping it airworthy.

By now, OL-20 had clocked up 1,000 U-2 missions in south-east Asia. The deployment codename had been changed yet again, and was now 'Giant Dragon'. Drones were assuming an ever-increasing workload, as low-level and night photo reconnaissance versions were deployed. They even started putting an SA-2 jammer on the little bugs. From April 1968 the SR-71 Blackbirds based on Okinawa made twice-weekly photo runs across North Vietnam, and the U-2s of OL-20 were assigned virtually full-time to the SIGINT, and especially COMINT, mission. This required long on-orbit times, as the new suite of antennae and recorders onboard the U-2R did their snooping. Eight-, ten- and twelve-hour missions over the Gulf of Tonkin became routine. For the pilots, it could be very boring up there, with only the occasional course correction on a racetrack pattern to worry about. Eventually, they even had much of the responsibility for operating the receivers taken away from them; a data-link had been fitted in a compartment under the tail of the U-2, and some of the intercept systems could be switched on and off and retuned by SIGINT operators sitting miles away in ground stations. This operation was codenamed Senior Book.

Of all the intelligence-gathering activities performed by a great power such as the US, COMINT can provide the greatest pay-off. Even when the enemy's communications are encrypted — as they most likely will be for any items of significance — the other side can make significant deductions from the type of transmission, its addressees and signatures, the number of messages sent, their length, the type of transmitters used, their powers, ranges, type of modulation, type of code, and so on. If a COMINT analyst can track the relative frequency of multiple address messages to a group of users, he may establish their common interest. Later, information acquired from another source about one of that group may indicate the interests and missions of the others. Order of battle information can be built up, and the other side's intentions deduced. The long hours put in by the U-2 flyers over the Gulf of Tonkin and, increasingly, further north along the coast of China, helped to gather the raw materials that the analysts back at the Air Force Security Service (AFSS) and the NSA needed to piece together this jigsaw puzzle. Of course, those analysts had other sources of COMINT available, from satellites and ground stations in particular, but the U-2's sensitive receivers operating at great height helped hoover up a lot of transmissions that might have otherwise escaped into the ether unheard, particularly from the shorter-range VHF and UHF bands.

Marshall Lin Piao, the Chinese Defence Minister who led a failed coup against Mao Tse Tung in September 1971. He fled the country in a commandeered airliner, but loyal fighter pilots pursued him and shot it down. The story was pieced together by US intelligence, largely thanks to SIGINT supplied by a high-flying U-2.

The analytical work was painstaking, but sometimes there were dramatic, almost instant, successes. In 1971 a new COMINT sensor produced by Melpar and codenamed 'Senior Spear' was added to the U-2R. It was carried in a podded, mid-wing installation similar to, but bulkier than, the old slipper pods on the U-2C. It had hardly been operated in the field for a month when, on the night of 12-13 September, a U-2 carrying the sensor picked up signs of a major commotion in the Chinese air defence system. They were grounding all the airliners and launching interceptors in every direction, but the U-2 was flying safely offshore on a well-established pattern in international airspace, and no other US planes were penetrating over Chinese territory at the time. The tapes were analysed, and with input from other SIGINT sources, the fragments of an amazing story began to unfold. Chinese Defence Minister Lin Piao had apparently launched a coup against Mao Tse Tung, but it had failed. He had then attempted to flee the country in a Trident airliner used by top officials as a VIP transport. Communist Chinese interceptors had pursued the plane, and it had been shot down over Mongolia.

To the China-watchers in the State Department, this was astounding news. Lin Piao was nominated as Mao's chosen successor, and there had been no firm indication from diplomatic sources that the relationship had turned sour. Of course, it was difficult to make any sense at all about what was going on behind the Bamboo Curtain, especially since the upheavals during the Cultural Revolution, when many foreign residents and diplomats had left the country, and were only now beginning to trickle back. The order grounding all Chinese aircraft soon became public knowledge; all army leave was cancelled; and the annual military parade scheduled for 1 October in Peking was cancelled, although no-one knew why. But the coup was hushed up, to such an extent that Lin Piao remained officially listed as Minister of Defence for almost an entire further year, despite having perished along with his wife and son and the six other people onboard the downed airliner. The true story emerged only gradually over the next year. It transpired that Lin had been manoeuvring behind the scenes for years to re-establish friendly relations with Moscow, with the support of top Chinese Army and Air Force commanders. The Defence Minister opposed the overtures towards the US that Premier Chou had been proposing since early 1970, and which had already led to Henry Kissinger's secret first visit to Peking in July 1971. But Mao had refused to lend his support, and Lin realised that his only chance of assuming the reins of power was to stage a coup.

Since most senior Chinese officials or cadres had little idea of the extent of this doctrinal conflict in the leadership, it was small wonder that the West also remained uninformed. Thanks largely to the U-2 and the NSA's ground stations, however, the US had hard intelligence on the upheavals in Peking at a crucial time during the diplomatic opening to China. Six weeks after Lin's death, Dr Kissinger made a second visit to Peking. Premier Chou now presided over a limited opening-up to the West, although the position of Mao and the radicals associated with him on improved relations remained equivocal. United States diplomacy continued to be informed and influenced during these watershed times by SIGINT intercepts, much of it provided by the U-2R.

Immediately following Dr Kissinger's second visit, the United Nations voted to admit the People's Republic of China. President Nixon made his famous visit to Peking in February 1972. The nationalists were expelled from the UN as a consequence of Peking's admittance, and during Nixon's talks with the Chinese leadership, he promised to suspend manned

Once in operational service, the U-2R proved to be a reliable workhorse in the SIGINT role, which now became the primary mission. As it heads out for yet another long patrol, this aircraft carries the Senior Spear COMINT sensor in wing-mounted pods.

Chinese pilots also transitioned to the U-2R model, hence the full-pressure suit. This group photo was taken in the late sixties, and shows (left to right) David Lee, Johnny Shen, Tom Wang, 'Gimo' Yang, 'Spike' Chuang, 'Tiger' Wang, Andy Fan, Terry Liu and Denny Huang. Wang and Yang had left the Black Cat squadron by then, but borrowed the appropriate gear for this reunion line-up.

reconnaissance flights over China.

With all the disruption caused by the Cultural Revolution, and then by Lin's attempted coup and the forced resignation of 250 senior commanders which followed, progress towards achieving a credible Chinese nuclear deterrent had slowed down. The pace of testing of both bombs and missiles slackened, and the predictions were now that the Chinese would not have an ICBM capable of reaching the US until at least the mid-seventies. With the US President in Peking and American forces on their way out of Vietnam, it seemed a different world from ten years earlier. Peking was no longer considered a major threat to US security. China had security worries of its own — forty-odd Soviet divisions massed along the common northern border — and there now seemed little likelihood of Taiwan falling victim to communist aggression.

As the US withdrew its military units from Taiwan, in accordance with understandings reached with Peking, the joint U-2 operation on Taiwan was wound down. Since the introduction of the U-2R and the end of mainland overflights in 1968, there had been only one fatal incident. In the autumn of 1970 Major Eddy Huang had been killed at Taoyuan in yet another landing accident when he bounced, tried to apply power, but veered off the runway and cartwheeled over in the mud. There now followed an anticlimactic end to the decade-long marriage between the CIA and the Chinese Air Force. The aircraft remained shut inside the 35th Squadron's recently-built hangar at Taoyuan, on non-flying status, while the US figured out how to achieve an amicable divorce. They were eventually returned to the US in 1974.

Elsewhere, the new version of the Dragon Lady was also proving very reliable, and a lot easier to fly than the U-2C. The Air Force unit was achieving a remarkable accident-free record over south-east Asia. As for the CIA, it continued to operate a headquarters squadron at Edwards North Base. In mid-1969 it was redesignated from WRSP-4 to the equally obfuscating 1130th ATTG (Air Technical Training Group).

Later that same year, the CIA demonstrated that the U-2R model would be able to emulate its much smaller predecessor by landing on aircraft carriers. Lockheed test pilot Bill Park and some of the Agency pilots took the Navy's carrier qualification course at Pensacola. The Skunk Works fitted the old U-2G model hook to a U-2R. They had already taken a prudent decision at the design stage to include a wing-fold on the longer-span R-model. The outer six feet hinged upwards, reducing the wingspan by enough to accommodate the aircraft on a carrier's elevator. Park returned to California and flew FCLPs with the modified aircraft at Edwards, with a Navy LSO assisting. The team then migrated to the East Coast, since the

More carrier landings, this time with the R-model on USS *America* in 1969. The first trap was made by Lockheed test pilot Bill Park, but some CIA U-2 pilots also qualified at this time, although there is no record of subsequent operational flights.

Atlantic Fleet carrier USS *America* had been selected for the demonstration.

The assigned day of 21 November 1969 dawned clear, although the sea was fairly rough. Park flew out from the NASA airfield at Wallops Island and began his first approach. At Edwards, he had decided upon a forty-five-degree flap setting and an approach speed of seventy-two knots with a wind-over-deck of twenty knots. Like the earlier G-model pilots, Park had to rely on the airspeed needle — he had no angle-of-attack indicator in the cockpit. As Park neared the ship, he pulled the lever to lower the hook, but nothing happened. Someone had forgotten to remove the locking pin before he took off! He returned to shore for a quick fix, and was soon out at the boat again. A number of landings and wave-offs were conducted without further incident over the next couple of days by the test pilot and the Agency flyers. In Park's opinion, the hook was hardly needed at all.

The next major adventure for the CIA pilots came in August 1970, after yet another flare-up in the Arab-Israeli conflict. Egypt had pledged to retake Israeli-occupied Sinai by force, and the two sides were slugging it out in a war of attrition across the Suez Canal. To stop the Israeli Air Force from attacking targets deep inside Egyptian territory, the Soviet Union sent a complete Air Defence Division to Cairo's aid in late February 1970. It consisted of early warning radars, MiG-21M fighter regiments and SA-2 and SA-3 SAM batteries. After first being set up in rear areas, the mobile missile batteries began rolling east in June, threatening Israel's hitherto unchallenged air superiority over the canal itself. Two of Israel's F-4s were shot down, and a large dogfight followed over the Gulf of Suez on 30 July. The US and the UN tried to open a dialogue between the two sides, and a thirty-day ceasefire was agreed, to take effect from the night of 7 August.

Two U-2R models were despatched in a hurry from Edwards to the trouble zone. They staged through Upper Heyford in the UK and eventually landed at RAF Akrotiri in Cyprus. Washington's idea was that they be used to monitor the ceasefire, by flying down the Canal and photographing the strip of territory — thirty-two miles from each bank — within which the two sides had agreed not to make any further deployments.

United States Air Force transports flew into the Cyprus base with the latest photo-processing vans complete with a satellite link so that the film from the U-2s could be developed and then immediately transmitted back to the US. The monitoring flights from Akrotiri began on 9 August, and they were a twitchy affair indeed for the pilots that flew them. For a start, Egypt had not consented to them, and even the Israelis were suspicious of US motives. Extreme care had to be taken to fly precisely along the allotted track, which was ten miles inside Israeli-held territory. Then there was the SAM threat. The Egyptians apparently moved some SA-2 batteries into the zone after the ceasefire took effect; they were definitely close enough to threaten the high-flying spyplanes. Soviet-manned SA-3 batteries were only just outside the zone. Both sides tracked the overflying US spyplanes with their missile radars: the low-light in the U-2 cockpit was almost constantly illuminated! Sometimes, the hi-light illuminated, and the aircraft's active jammer went into action.

The U-2 flights continued every three or four days for the next three months. The ceasefire held, but the provision that neither side would move fresh military equipment into the canal zone broke down. That much was evident from the U-2 photography. Detailed coverage of the sixty-four-mile wide zone from only the one flight track was made possible by the LOROP cameras now carried by the aircraft. In the mid-1960s, Hycon's Type H had been the first of these to be introduced on the U-2. It was a sixty-six-inch focal length monster with folding optics which weighed 700 lb. It only just fitted in the Q-bay, and took pictures from horizon to horizon through a new hatch with three flat glass windows which protruded a small way below the fuselage contour. The Type H was more easily accommodated on the U-2R, since the Q-bay was bigger and the tracker camera was relocated to the nose.

Egypt's objections to the U-2 flights became more vocal as time wore on, and they were halted on 10 November. The aircraft were withdrawn in early December, and staged through Upper Heyford again on the way home. Here, the RAF's Flight Lieutenant Harry Drew climbed into one U-2R and flew it nonstop across the North Atlantic and North America to Edwards in fourteen hours twenty minutes. The flight was an endurance record for the type which apparently still stands to this day.

Drew was one of the last two British pilots to be trained on the Dragon Lady, although the British had kept a foot in the door of U-2 operations throughout the 1960s. Following the first group of five officers (one of whom had been killed during training), six more RAF flyers were sent to the US in pairs for three-year tours in 1961, 1964 and 1967. After they were checked out in the U-2, there was little for them to do except hang around the CIA squadron at Edwards. The call from London for them to perform an operational mission never came. The Agency allowed them to fly U-2s being ferried to Taipei as far as Hawaii. Otherwise, they kept current on the U-2 with round-robin training flights from Edwards, and filled up their log-books by flying the squadron's T-33 proficiency trainers.

The British connection was renewed in 1973, when following the Arab-Israeli War, the Agency squadron once again deployed a U-2R to Akrotiri for flights along the Suez Canal. This time, the operation had the blessing of both Israel and Egypt, since it had been arranged under the auspices of the United Nations. On a monthly basis, both sides were provided with the aerial photographs and reports outlining the disposition of each other's forces. This confidence-building measure helped ensure that the latest ceasefire held, and was eventually converted into a peace agreement.

Following the demise of the Black Cat squadron in mid-1974, the curtain finally came down on eighteen years of CIA U-2 operations. The 1130th ATTG at Edwards was closed down, and its handful of U-2R models were turned over to the 100th SRW, which also took charge of the detachment at Akrotiri. On the other side of the world, operations over south-east Asia continued. OL-20 had moved from Bien Hoa to Utapao airbase in Thailand in July 1970, becoming OL-RU in the process, but in November 1972 the unit achieved full squadron status as the 99th SRS. The Ryan drones in the 350th SRS were playing a more important role than ever; at times between the halt to bombing in October 1968 and the start of the Linebacker I raids in May 1972, they were the only US aircraft allowed over North Vietnam — even the Black-birds were sometimes restricted by the politicians in Washington to an offshore patrol in the Gulf of Tonkin. Enemy air defences forced OL-20 to fly the little bugs lower and faster, as the loss rate increased. Even so, they proved invaluable during the Linebacker operations, especially for post-strike photography, some of which was released onto the front pages of the newspapers in the US.

In January 1973, the 100th SRW flew 500 U-2 hours in one month for the first time in its history. Two months later, the wing was awarded the General Paul T. Cullen Memorial Award for its 'outstanding record in support of SE Asia requirements and unique contributions to Line-backer II operations'. This was the intensive interdiction campaign waged with B-52s over North Vietnam which is credited with finally bringing the North's negotiators back to the conference table for serious discussions concerning the US withdrawal. An unprecedented sixth outstanding unit award soon followed. Apart from its operation of the drones, the 100th's outstanding record included a U-2R maintenance readiness of ninety-eight per cent, and accident-free operations from the introduction of the U-2R right through until August of 1975. On 15 August, Captain John Little ejected from an aircraft he was ferrying back to the US from U-Tapao. In company with another U-2R and a KC-135, Little took off from the Thai base on a very dark night. The group had not been airborne long when his aircraft developed autopilot problems, and Little lost control as it oversped. The tail came off and Little ejected into the Gulf of Thailand, where he bobbed around in his liferaft until being picked up by a fishing boat the next morning.

Although the US involvement in Indo-China was drawing to an unhappy close, the 99th SRS was kept busy monitoring the North Vietnamese advance towards Saigon, and other tasks. In April 1975 they deployed a U-2R to Diego Garcia in the Indian Ocean, from where it took off to overfly Somalia and photograph the Soviet military build-up there. In May, another 99th SRS U-2R was pressed into service as a communications relay platform during the *Mayaguez* incident, when the Cambodian Khmer Rouge communists captured a US cargo ship and imprisoned the crew. With a Defense Systems Communications Satellite out of commission, the U-2R flew at altitude for twenty-seven hours over a three day period to relay messages to and from the rescue force, which succeeded in recovering the boat and its crew intact. Having been one of the first US Air Force outfits to arrive in the region, the U-2 unit was one of the last to depart, in April 1976. (Drone operations were wound down the previous year.)

Although the Dragon Lady was now out of Taiwan and Thailand, it had recently established a new, permanent home further north in Asia. In March 1975, the 100th SRW was instructed to open up a new OL for the aircraft at Osan airbase, South Korea. They were to replace the high-altitude drones which the wing had been operating from there as COMINT-collection platforms since February 1970. The drones were drafted in as an unmanned replacement for US Navy EC-121M intelligence-gathering aircraft, one of which had been shot down by the North Koreans over international waters in April 1969 with the loss of its thirty-one crew. This disaster followed the North's capture in January 1968 of a US Navy ferreting ship, the USS *Pueblo*, and proved beyond doubt that the belligerent regime in Pyongyang would strike out at random if it saw a chance to embarrass the US, and especially the intelligence community.

The solution was an unmanned collection platform, with COMINT receivers linked to a ground station via a wideband data-link. Ryan came up with a special big-wing variant of the standard drone which could cruise higher and, with an external fuel tank added, was capable of eight hours endurance. The operation was code-named 'Combat Dawn' and it was a mixed success, according to the Air Force verdict, because of low reliability and high mission costs, once you took into account all the support paraphernalia such as C-130 launch platforms and the helicopters and boats needed to recover the drones. Ryan and the RPV lobby disagreed, citing the almost 500 flights made during Combat Dawn and the return rate of 96.8 per cent which had now been achieved. But with the U-2R proving to be such an outstanding success elsewhere, SAC headquarters hastened to replace the drones as soon as surplus aircraft became available. The last Combat Dawn mission was flown from Osan on 3 June 1975, and one or two U-2s have been in residence there ever since.

Photo-reconnaissance drones also fell out of favour with SAC as the Vietnam War drew to a close, and it was decided to turn the remaining vehicles, helicopters and C-130s over to Tactical Air Command (TAC). This stripped the 100th SRW of over half its personnel, and made the continuation of a wing structure for just one U-2 squadron unfeasible in the budget-cutting atmosphere of the post-war years. Shortly before the U-2 wing celebrated its twentieth anniversary on 1 April 1976, the Pentagon announced that it would be disbanded. The U-2 squadron would be moved from Davis-Monthan to Beale AFB as the 99th SRS, to become part of the 9th SRW, operators of the now-famous Blackbird. The two Lockheed machines, dissimilar in almost every respect save that they both performed strategic reconnaissance, were going to become stablemates for the first time.

1 One slight disadvantage of the fuel dump chutes was that

they reduced the aircraft's internal fuel capacity by twenty-five gallons to 1,320 gallons.

2 Applied Technology was originally established in 1959 with CIA funds. In 1967, the company became part of Itek, short for Intermediate Technology. Itek was founded in 1957 by Richard Leghorn and Duncan Macdonald, when their old Boston Optical Research Laboratory was closed down. The company eventually supplanted Hycon as the main provider of reconnaissance cameras for the U-2.

3 The film from this sortie also revealed a previously-unknown airbase with a 16,000 foot runway. The puzzle was, it was located 6,000 feet up in the mountains of Kansu, and there were no road or rail links to the site. How it had been constructed remained a mystery!

Chapter Eight
Swords Into Plowshares

We have one resource which is not being used to its full potential. If it were properly employed it could save countless lives and billions of dollars in property damage each year. That resource is the nation's aerial reconnaissance and interpretation technology.
Dino Brugioni,
former CIA U-2 photo-interpreter, 1985.

While the CIA U-2 detachments in Turkey and Japan were patrolling (and occasionally penetrating) the borders of the USSR, and the regular SAC squadron was pounding the Crow-flight beat throughout North and South America, a third, quite separate Dragon Lady unit came into being in a remote corner of Edwards AFB. For the next twenty years, this outfit would make a major contribution to US security, though it never flew for so much as one minute across denied territory during all that time. The unit became known as the Special Projects Branch of the USAF Flight Test Center. During thousands of flights in support of highly classified projects, it helped perfect the equipment and techniques for satellite surveillance which were

Two aircraft from the Special Projects Branch fly in formation over Edwards AFB on 18 January 1961. In the foreground, serial 66722 is a U-2D with an aluminium turret housing the MIDAS sensor clearly visible behind the cockpit. The forward top half of the aircraft is painted black to prevent sunlight from reflecting off the U-2's shiny metal surface and confusing the sensor. The other aircraft is a U-2A, serial 66701.

to revolutionize US intelligence-gathering in the sixties and seventies.

It was, perhaps, inevitable that the Air Force would find a reason to create a U-2 test-flying operation of its own. Throughout the first three years of the programme, the Lockheed Skunk Works had taken care of all the testing and development work, thanks to the unique nature of its CIA-controlled contract with the government. Unlike every other type of aircraft in the Air Force, this one had not been subjected to the usual — and sometimes protracted — process of evaluation and clearance conducted by military test pilots. But while the U-2 had got into the Air Force inventory via the back door, the generals weren't prepared to let Kelly Johnson make all the running in the future. Shortly after the Lockheed U-2 flight test and overhaul operation was transferred from The Ranch to Edwards in 1957, the Air Force created a U-2 test unit of its own.

One of the earliest tasks for the Special Projects Branch was to investigate the aircraft's structural integrity, for there was a widespread suspicion among the blue-suiters that the U-2 airframe simply wasn't going to stand up to more than a couple of years use. Unknown to Johnson, the test unit embarked upon a series of these flights. While the Lockheed team was ensconced in secure quarters at Edwards North Base, the Air Force people took up residence in a hangar on the equally remote South Base, almost ten miles away. Major Robert Carpenter headed a small, select team of four test pilots, who all checked themselves out in the U-2 at Edwards, although they attended ground school alongside SAC's rookie U-2 pilots at Laughlin AFB.

The first and only fatality to be recorded during Special Projects U-2 operations occurred in the early days, on 11 September 1958. Captain Pat Hunerwadel suffered a heavy landing at Edwards, possibly as a result of gusting wind conditions. He applied power and tried to climb away, but apparently failed to retract the flaps. The aircraft stalled and spiralled into the ground. The wreckage was subsequently buried on the site.

The Skunk Works team at North Base eventually performed a complete U-2 structural test programme, before moving on to flight test the re-engined U-2C model. In the next few years, the Air Force U-2 unit at Edwards was tasked with the first of the many sensor test programmes which became its *raison d'être*. Its early history, however, is shrouded in mystery, and no official record of its activities in the late fifties appears to exist.

Whether this has anything to do with the equally mysterious U-2B project remains a matter for conjecture. Only a handful of people appear to have been involved with the B-model; senior Skunk Works people today deny that this version was ever built. The author has been unable to prove otherwise, but evidence exists to support the contention that the B-model was originally conceived as — a bomber!

Studies into the feasibility of adding an offensive weapon to the U-2 are known to have been carried out by Kelly Johnson in the late fifties. Moreover, at least five projects which involved the dropping of payloads from the U-2 reached the flight-test stage from the late fifties onwards, although none of these could strictly be interpreted as making the U-2 into an offensive weapons system.[1]

The rationale for a U-2 bomber was clear, however. While SAC had built a huge bomber fleet, partly against the expectation that only a proportion of them would get through to the target, here was an aircraft which soared above the Soviet air defences with impunity. Of course, it could carry only a fraction of the bomb load that a B-47 or B-52 was capable of hauling to Moscow, Leningrad or wherever, but maybe the U-2 could be a weapon of limited response. Moreover, US scientists were developing smaller warheads for the next generation of nuclear weapons . . .

It therefore seems entirely possible that an offensive version of the U-2 figured on the US Air Force wish-list in the late fifties. Whether President Eisenhower was appraised of the proposed development is unknown, but his reaction to the idea would surely have been emphatically negative. More than most of his high-level advisors, Eisenhower harboured a fear that the Soviets would mistake a U-2 on a penetrating reconnaissance flight for a nuclear bomber. Soviet spokesmen made exactly that point when they protested about other intelligence-gathering missions staged by the US Air Force and Navy in the fifties, some of which undoubtedly crossed the border. At the CIA, Richard Bissell had also contemplated this disturbing scenario. And the President's desire to draw a clear distinction between 'offensive'

Another view of serial 66722, this time circa 1962 in a new configuration. Protruding from the Q-bay are two spheres designed to represent payload re-entry capsules from an early US photographic reconnaissance satellite. While this U-2 test programme has been declassified, mystery still surrounds some other payloads which were designed to be dropped from the high-flying spyplane.

flights of the sort that might be carried out by the military, and the 'non-offensive' nature of the U-2 flights, had been part of the rationale behind assigning control of the project to the CIA, rather than the Pentagon.

After Powers was shot down in 1960 and the Soviet propaganda offensive began, the US laid much emphasis on the fact that the U-2 was an 'unarmed, non-military plane' (to quote the President's own words). The ground would have been cut from under Eisenhower's feet if it had become known that a bomber version of the U-2 was in the works. As it was, the May Day incident created the biggest political shock-wave ever caused by an aircraft. Reason enough, therefore, for all mention of the U-2 bomber idea to be suppressed — preferably for ever.

It remains unclear whether a U-2 bomber prototype ever took to the air, and if it did, whether Lockheed, the CIA or the Air Force sponsored the flight tests. However, the following is beyond dispute. By March 1959 Lockheed was test-flying U-2s which dropped 'payloads' from the Q-bay, which was modified with bomb-bay type doors for the purpose. In one set of trials flown by CIA pilots at high altitude, a downward ejecting seat was fired from the Q-bay. Incredible as it may seem, the idea may have been to provide a new way of inserting CIA agents into hostile territory! In another series of flights, this time conducted by Lockheed test pilots at low altitude, the payload was leaflets — hundreds of them, which were stacked inside the Q-bay in bundles and released at the desired moment by a cutting device controlled from the cockpit. After one test flight of this system, the entire lakebed at the north end of Edwards was covered with litter! The U-2 never made it into service as either a bomber, agent-inserter or propaganda-dispensing aircraft, but the drop system was later used by the Edwards unit to dispense radar-spoofing bundles of metallic chaff from high altitude in connection with a number of US radar research programmes.

By 1961, the Air Force Special Projects Unit at Edwards was also flying an aircraft which had been modified to release two spherical items from the Q-bay. These represented the payload capsules which were ejected from each of the 'Discoverer' series of satellites as their orbital path crossed Alaska heading south. Tracked by a variety of ground, ship and airborne sensors, the capsule was designed to re-enter the atmosphere protected by an ablative shield. At 52,000 feet, the heat shield was supposed to separate and a parachute deploy. The capsule could then be snared in mid-air at a lower level, by a patrolling Air Force C-119 transport from which was trailed a special wire contraption. Otherwise, it would have to be recovered from the ocean surface by naval vessels.

What was so vital about these capsules, that such trouble was taken to recover them? The public was told that they contained 'instrumentation packages', but the reality was that Discoverer was the prototype US photo-reconnaissance satellite, and the capsules contained the exposed film which was being returned to earth for intelligence analysis. General Electric had received an Air Force contract to develop the system in October 1956. The military had later lost interest, believing that the complicated recovery system would not work, and preferring instead to sink resources into what became the SAMOS (Satellite And Missile Observation System) series of satellites, in which the exposed film was processed in space and the images then linescanned and relayed to earth via a radio link.

But the CIA kept faith with the capsule recovery method, perceiving correctly that it alone was capable of returning images with the required clarity for military photo interpretation. With the ubiquitous Richard Bissell in charge, the Agency took control of Discoverer in February 1958, and the first flight followed a year later. Throughout 1959 and the first half of 1960, the system promised much but delivered nothing, thanks to a long series of technical failures. As we have seen, President Eisenhower sharply reduced the number of U-2 overflights of the Soviet Union at this time, in the hope that Discoverer would soon prove to be a risk-free alternative. But the complex series of events which comprised a Discoverer mission proved difficult to master. There were launch rocket failures, second stage failures, telemetry failures. Even if all went well and the satellite took the required pictures, the capsule re-entry procedure required split-second timing of five separate operations in sequence, if the small prize was to float gently down into the recovery area on the end of its parachute. Even then, the waiting Air Force transport might fail to snare it.

This is exactly what happened on 11 August 1960, when Discoverer 13 became the first in the series to correctly return a capsule to the recovery area. Luckily, it was retrieved from the Pacific Ocean some 330 miles north-west of Honolulu by Navy frogmen. The Air Force transport crews needed practice in the fine art of capsule-snatching, and they got it at first by courtesy of a B-47 from Edwards, which flew as high as it could to drop dummy capsules into the practice recovery area. Unfortunately, the big bomber could not get high enough to replicate the true height of capsule parachute deployment — 52,000 feet. The U-2 could do this, however, and had the additional advantage of flying high enough to preclude visual detection from below — the transport crews were suspected of cheating, through being able to observe the contrail of the converted bomber as it cruised above them before releasing the practice capsules!

The Edwards-based U-2s were therefore modified to drop dummy capsules for recovery practice, and deployed to Hickam AFB on Hawaii to work with the recovery crews. Often, a second U-2 would cruise alongside to photograph the drop so that the capsule's aerodynamic behaviour and parachute deployment sequence could be studied. With this and other technical problems solved, the Discoverer success rate improved dramatically in 1961. The twenty-ninth satellite in the series is credited with returning the first good coverage of the Soviet Union's suspected northern ICBM launch site at Plesetsk, on 30 August 1961. At last, fully sixteen months after Frank Powers had been launched from Peshawar in a famously unsuccessful attempt to photograph the site, the US had confirmation of an operational deployment for the mighty SS-6. Thanks to further successful Discoverer missions, and some less detailed coverage provided by the early SAMOS (image-transmission) satellites, the US soon knew that all Khrushchev's bold talk of massive missile deployments was mere sabre rattling. There were less than ten SS-6s deployed at only four sites, and the cumbersome liquid-fuelled missile posed little threat to US security. The much-touted 'missile gap' did not exist, and this knowledge helped strengthen President Kennedy's hand during the Cuban missile crisis a few months later.

As the American conquest of space gathered pace, demand for the Dragon Lady's services as a testbed grew. Through its Missiles and Space Division (LMSD), Lockheed was the prime contractor on the first three reconnaissance and surveillance satellite programmes. LMSD provided the Agena second-stage device which was carried aloft by the Thor and Atlas launch rockets. This bullet-nosed, nineteen foot-long cylinder became the US workhorse in space, carrying a variety of payloads into low and medium-altitude orbits. It was natural therefore for LMSD engineers at Sunnyvale to enlist the help of their Skunk Works colleagues, especially since the stable, high-altitude cruise of the U-2 was the best available replica of conditions that the satellites would encounter.

For a major flight research project in support of the MIDAS satellite programme, the LMSD engineers called for an observer to be carried to high altitude, so that he could operate sensitive

One of the observers attached to the Edwards test squadron poses atop the two-place U-2D model. He occupied the modified Q-bay behind the cockpit, but had only a poor view of the outside world once the hatch above his head was secured. On top of the hatch is a fairing to smooth the slipstream behind the MIDAS sensor barrel, which is unpainted.

infra-red radiation detectors to be mounted in the aircraft. MIDAS (Missile Detection and Alarm System) was supposed to provide early warning of a Soviet nuclear attack by detecting and tracking the rocket plumes from missiles' boost phase as they ascended into the stratosphere. It was an ambitious project, given the state-of-the-art in infra-red technology, and the huge distances over which the sensors would be expected to detect missile activity. A good deal of testing was envisaged, and the sensors would first be flown in the vicinity of US rocket launches as part of the development programme.

The Skunk Works therefore created the unique two-place Dragon Lady configuration which was designated U-2D. They stripped out the Q-bay of a U-2A and installed the MIDAS instrumentation around and below a second seat. The upper Q-bay hatch was replaced by a rearward-hinged cover with a tiny triangular window on the right side. In the small fuselage cavity between the cockpit and Q-bay, a complicated mirror system which focused infra-red activity onto the detecting instruments was housed. This system protruded from the top of the aircraft in an aluminium turret. Above the turret a revolving sensor head was fitted, so that the ascending missile's trajectory could be followed. Apart from his ejection seat, the second crew member was provided with a few basic flight instruments, and a cathode ray tube for research purposes. There were no flight controls which he could operate — in fact, he couldn't even see out of the small window, which was above helmet level! It was bad enough for the pilot, flying the U-2 in that cramped cockpit and constricting partial-pressure suit, but the backseat equipment operator flew in even more claustrophobic conditions.

At North Base, Lockheed pilots Ray Goudey and Bob Schumacher flipped a coin to determine who would go in the back for the first U-2D flight. Schumacher lost. Afterwards, he reported feeling like a dummy, stuck in the dark recess with no control of his own. The Skunk Works engineers relented, and in a second U-2D conversion they added two 'picture windows' at eye level on each side of the modified Q-bay, so that the poor occupant had at least a limited view of the outside world. After a short series of test flights, the two-seat models were turned over to the Air Force in March 1960. They were known as the 'Smokey Joe' aircraft. This was not only an acknowledgement of their role in monitoring rocket plumes. Smokey Joe was a rotund, cigar-smoking Indian character from a cartoon, and the Special Projects Branch perpetuated the now-established U-2 tradition by adopting him as their mascot. A large painting of Joe greeted visitors to the unit's small office which was tucked away in a corner of the Edwards Flight Ops building, and a smaller version appeared on the tails of the U-2 fleet.

This now numbered five aircraft, and an additional three test pilots were recruited, as the demand for U-2 test flights soared. There were about twenty maintenance personnel assigned. The unit averaged a hundred U-2 flying hours per month in the early sixties, and also controlled a B-47 and C-54 used on special test flights. Major Harry Andonian took over as chief of the Special Projects Branch in November 1960. Although the Smokey Joe appellation was supposed to apply only to the MIDAS support effort, the entire unit came to be known by this name during Andonian's tenure.

The Branch recruited three flight-rated navigators to fly in the second seat of the U-2D, and the aircraft were soon chasing the fiery rocket plumes from US missile launches at Vandenberg AFB and Cape Canaveral. When launches were planned at the Cape, the aircraft were deployed to nearby Patrick AFB or to Ramey AFB, Puerto Rico. During these events, both two-seaters would be airborne if they were serviceable. Having taken off at least an hour before the missile launch, they would climb to 60,000 feet and take up a heading in line with the planned launch azimuth of the missile, about two hundred miles from the launch pad. Sometimes both aircraft were positioned uprange; on other occasions, one would fly downrange to be underneath the missile's trajectory. Once the missile was launched, the navigator could monitor the operation of the revolving turret head, which was nicknamed the 'pickle barrel'. This could be scanned through 180-degrees horizontally, while in the vertical plane there was enough movement to follow the missile until its trajectory exceeded thirty degrees above the aircraft. On the U-2, the navigator could operate the system manually if required. On the MIDAS satellite itself, the process would have to be automatic, with the scanner operating in search mode until the missile plume was detected, whereupon it would lock-on and focus the radiant energy onto the detector. There were high hopes for the system; it was expected to distinguish different types of missile launches, and deliver staging and trajectory information to ground stations below.

The first MIDAS satellite was launched by an Atlas booster in February 1960, but the Agena second stage failed. A second launch in May made it successfully into orbit, but the data-link transmitter failed on the second day. A long pause ensued, since it became apparent from the U-2 flights that there were major problems with the detectors. They were really pushing the state-of-the-art with these systems, and there was a woeful lack of data on natural background radiation in the atmosphere, and the emission characteristics of the missile plumes themselves. Under certain conditions, the detector would fail to discriminate between a hot rocket exhaust and sunlight reflected from high clouds. Such false alarms were clearly unacceptable.

So the Smokey Joe U-2s were set to work on a major programme of solar radiation and other background measurements, using various spectrometers. Two more MIDAS satellites were sent up in 1961, and one did manage to identify the launch of a Titan missile, but this was under known and ideal test conditions. For the system to contribute to the US early warning network, absolute reliability was essential. All hopes of achieving an early operational capability were abandoned. But the U-2s were kept busy, as the laboratories came up with new radiometers which operated in a wider spectral range, or with greater sensitivity at the target wavelengths. By 1964, tests of an infra-red TV detector were being carried out on one of the U-2s. Two other aircraft in the Edwards fleet had different model spectrometers installed at this time for background measurements, while yet another carried an ACF line tracker. All four aircraft were constantly shuttling between Edwards, Patrick and Ramey to fly on the Eastern and Western Missile Ranges, not to mention occasional visits to research facilities such as L G Hanscom Field for sensor modifications.

It was trial and error all the way — especially since it had become apparent that the new generation of solid rocket propellants burned cooler than the liquid-fuelled devices, and that the radiation signatures of ascending rockets changed according to the altitude. In the event, the effort to develop a reliable space-based ICBM launch detection system lasted almost the entire sixties, but did eventually produce operational satellites designated DSP-647 in the seventies.

In the meantime, Special Projects Branch lent its support to various other space programmes. Reconnaissance cameras which had been designed for the Discoverer and SAMOS satellites were tried out first in the U-2 (they were, in any case, direct descendants of the Dragon Lady's Type B system). When NASA launched the world's first meteorological satellite — Tiros — a camera-equipped U-2 was on hand to simultaneously fly an identical path across the US and thereby check the accuracy of the photographic sensors onboard Tiros. Equipment for the polar-orbiting Nimbus weather satellites was checked out in advance on a U-2. In 1967, the Special Projects Branch started test flights of a heat-seeking guidance system made by North American Autonetics which was intended to be part of a satellite-killer that would home in on the warmth of enemy space vehicles as they whirled through space.

Meteorological research became the major activity for one particular Edwards U-2A. Serial 66701 was controlled by the Air Force Cambridge Research Laboratories (AFCRL) at Bedford, Massachusetts, and was often to be found at nearby L G Hanscom Field. Named 'The Saint', this aircraft carried a variety of instruments in the Q-bay which could measure atmospheric pressure, moisture and ozone content, and electrical fields. It flew high above thunderstorms photographing the cloud tops while other AFCRL aircraft skirted or even penetrated the build-up below. It flew over mountains to measure the effect that high ground could have on air flow: even at the U-2's high altitude, this effect was apparent. During one flight over the lee of the High Sierras, a smooth but strong updraft carried The Saint and its surprised pilot from 68,000 feet to 69,500 feet in a matter of minutes.

During missile launches from Cape Canaveral, The Saint joined the other Smokey Joe U-2s, to photo-document the cloud tops which were causing false readings on the MIDAS infra-red detector equipment. Measurements of atmospheric moisture that it took at the same time helped to prove that even minute amounts could affect the radio interferometer system used for precision tracking of rocket launches at the Cape.

Meanwhile, U-2D serial 66722 was pulled off Smokey Joe duties at the start of 1964 and reconfigured as a single-seater for Project HiCAT. This was a fresh study of high-altitude clear air turbulence sponsored by the Air Force Flight Dynamics Laboratory, which aimed to statistically define the phenomenon so that future fast and high-flying aircraft could be designed to the appropriate structural criteria. The AFFDL had studied the previous data on turbulence above 50,000 feet, which was almost entirely derived from the NASA instrumentation carried on U-2 training flights during the CIA's Operation Overflight. This time, a more sophisticated instrumentation package was designed, so that true gust velocity components encountered along the aircraft flight path could be determined. (The earlier data had been obtained by measuring only the acceleration response of the U-2 as it met turbulence.)

Lockheed designed and built a special gust probe with fixed vanes which was stuck on the nose of the U-2. It was painted in red and white stripes, and so inevitably became known as 'the barber's pole'. The vanes sensed vertical and lateral gusts, while longitudinal gusts were sensed by a pressure transducer in the aircraft's pitot-static tube. As the vanes moved in rough air, they generated electrical signals which were recorded on magnetic tape. Since the aircraft itself was bound to move as it encountered the rough air, this was measured by accelerometers and gyros placed in the Q-bay, so that the appropriate corrections could be made.

Equipped with this ingenious package, the HiCAT U-2 was test flown for the first time in February 1964. The research programme extended over the next four years, although for the first year or so the HiCAT aircraft was also tasked with other duties, so that the accumulation of data was slow, and sometimes on a random

Captain Rial Lowell from Special Projects is standing in the cockpit of the Project HiCAT U-2. At the tip of the large red-and-white stripe 'barber's pole' mounted under the nose can be seen the fixed vanes which sensed vertical and lateral gusts, and the special pitot tube for sensing longitudinal gusts. Earlier U-2 pilots would have been aghast at the prospect of deliberately flying into regions of clear-air turbulence!

basis during test flights with some other primary purpose. In October 1965 the aircraft was fitted with new digital instrumentation, and was henceforth dedicated to HiCAT. The U-2 was now deliberately flown into regions where clear air turbulence was predicted, resulting either from jetstreams, weather fronts, temperature inversions or mountain waves. An average of three flights was made each week, with the aircraft airborne for four hours each time.

Eight years earlier, the first U-2 pilots at Groom Lake would have been aghast at the idea of going looking for CAT like this — their firm impression was that the U-2 was the most fragile of aircraft, not built to last. It would be inviting structural trouble to fly in such a way! Yet serial 66722 was to survive nearly fifty-five hours of flight in high-altitude turbulence over the subsequent four years of the HiCAT project, during which time a grand total of 285 sorties and 1,221 flying hours were logged! The U-2 made HiCAT searches all over the North American continent, flying out of Edwards, Hickam, Ramey, Hanscom and Elmendorf AFB in Alaska. In June 1966 it was deployed to Christchurch, New Zealand, to sample conditions over the South Pacific, and moved on to the favourite 4080th Wing deployment base at Laverton in Australia the following month. In March 1967 the research was extended to Europe, with flights staged from the Royal Aircraft Establishment airfield near Bedford in the UK. From there, the aircraft flew mainly over the eastern Atlantic and the Scottish Highlands, but two sorties over the French and Italian Alps were made in mid-April. Then it was back to the eastern US, with 66722 deployed

Serial 66722 yet again, this time a few years later, when it was liberally adorned with photo-resolution stripes for a test programme. Major Mel Hyashi of the Edwards Flight Test Center is in the foreground.

to Barksdale, Loring, Albrook (Panama) and Patrick AFBs in succession. The project ended at Edwards in February 1968.

During long excursions away from base, the U-2s were often accompanied by the converted JB-47E bomber controlled by the Special Projects Branch. It was used as an escort to aid in navigation and communications relay, in the same way that SAC and CIA U-2s were trailed by KC-135s, especially on long overwater flights. By 1966, the bomber had given way to a more prosaic C-130, which could also transport ground equipment, the special fuel, maintenance crews and so on. Early this year, Harry Andonian completed his tour as chief of the Branch, and handed over to Lt Col J K Campbell. There was still no shortage of research tasks, but later in 1966 two of the Edwards U-2s (The Saint and one of the two remaining two-seaters) were converted to U-2C standard and re-allocated to SAC. The stock of Dragon Lady airframes was diminishing fast, and the demands of the war in south-east Asia took priority.

The research tasks at Edwards now included Project Cloudcraft for the National Academy of Sciences, and the continuing effort to perfect equipment with which to track ballistic missile launches from space. This was known as Programme 461 until the end of 1966, when it was redesignated Programme 949. Project 4076 — the infra-red tracker forming part of an anti-satellite system — continued, and there were various other highly classified projects rejoicing in such titles as 'Have Charity', 'Lariat' and 'Have Echo II'.

Special Projects was also tasked to clear new items of U-2 equipment for operational use, such as a doppler navigation system introduced on SAC aircraft in the mid-sixties. There were test flights with the big Type H LOROP (long-range oblique photography) camera before it went into service. An even bigger camera configuration known as the 'Type C' was also tested over a number of years, but was never cleared for service. This was a 100-inch double folded focal length monster developed by Hycon and Perkin Elmer. Using his drift sight, the U-2 pilot could select several targets far in advance of the aircraft and store them in the camera's memory, so that when it came within range the operation was automatic. The Type C produced fantastic results on occasions, but never worked consistently. It was just too complicated, mechanically and optically.

The workload at Special Projects remained high in the late sixties, and with the aircraft so frequently detached to Hanscom, Patrick and elsewhere, the pilots and support crews had to put up with long TDYs, just like their colleages in the SAC and CIA units. Around 1970, however, the workload began to decline as the ICBM launch warning system matured (it finally became operational in 1972) and the US abandoned development of an anti-satellite

Two aircraft continued to be based at Edwards until the late-seventies, and were sometimes flown by students from the Test Pilots' School seeking a new challenge. This is the last surviving U-2D model on display at NAS China Lake in October 1975, by which time it had been given a coat of gloss white paint with a smart red cheat line.

system. From December 1969, an alternative aircraft was available at Edwards for high-altitude research, when NASA and the Air Force began the jointly-funded operation of two YF-12 Blackbirds. In early 1971, Special Projects relinquished one of its two remaining U-2A models, but there was new work for U-2F serial 66692, which had been acquired in June 1967. This aircraft was now modified with a huge dorsal spine, which housed two tracking domes, one above the Q-bay and one about halfway to the tail. It spent the next three years with Project TRIM (Target Radiation Intensity Measurement), measuring the radiation characteristics of re-entry vehicles.

When the TRIM project finished, this aircraft was demodified to U-2C standard and re-issued to SAC's 100th SRW. This left just two U-2s at Edwards. These were U-2D 66721 and U-2A 66722, and they shared the distinction of being the last two Dragon Lady aircraft powered by the original J57 engine. There were still some research tasks, but the aircraft were now sometimes made available to students of the Test Pilots School at Edwards as part of their course. After all that they had heard about the aircraft's challenging flight characteristics, these rookie test pilots were naturally keen to have a go in the Dragon Lady. Newcomers were first flown in the back seat of the U-2D; throughout its twenty-year history, the Special Projects Branch had trained all its own U-2 pilots this way. The remaining two aircraft at Edwards were eventually retired in 1978, both having served as test aircraft throughout their twenty-one-year flying careers.

This left seven of the original-production U-2s still in military service, all of them now gathered at Beale AFB under the control of the 9th SRW. But there were a further two U-2C models still in daily use, just a couple of hours' drive from Beale. From Moffett Field on the edge of the San Francisco Bay, the NASA Ames Research Center was flying daily Dragon Lady missions to the upper atmosphere in pursuit of civilian scientific enquiry. Unlike NASA's earlier association with the U-2, when the government had used the civilian scientific agency as 'cover' for Operation Overflight, this time the scientists were fully in control.

The two NASA U-2s had first touched down at Moffett Field on 3 June 1971. Officially on indefinite loan from the Air Force, these former CIA U-2G aircraft had been in storage for a couple of years before Lockheed was asked to make them ready for service as 'Earth Resources Survey Aircraft', according to the official NASA terminology. Skunk Works engineers removed the carrier-landing modifications from Articles 348 and 349, and some other items, which reduced their zero fuel weight to 13,800 lb. They were painted in smart white and grey colours with a blue cheat line and NASA markings, and given the civilian registrations N708NA and N709NA. The pair of high-flyers were to complement three other aircraft (an RB-57, a P-3 Orion and a C-130 Hercules) which NASA had been flying since 1966 as scientific testbeds for the sensors which it soon hoped to send into space. NASA was responsible for launching the first satellites to be dedicated to the systematic study of the earth's surface. These Earth Re-

NASA 709 stands outside the hangar at Ames Research Center, NAS Moffett Field, California, during the early days of the new civilian research programme there. This U-2C was the last of the original aircraft from mid-fifties production to be retired, after a career spanning over thirty years.

sources Technology Satellites (ERTS) were due to be launched in 1972 and 1973, but the value of supplemental coverage from high-altitude aircraft was well appreciated. At the very least, the two U-2s would contribute to the fund of experience and data being built up before the ERTS system became operational.

A precedent for the allocation of the two former spyplanes to a non-defence reconnaissance role had been set in 1967. After a formal study, the CIA and the DoD agreed with a proposal (first mooted by Art Lundahl of NPIC) that some classified overhead photography be shared with other federal agencies for the purpose of environmental resource study. The Gemini and Apollo manned space missions had also demonstrated the tremendous advantages that photography from very high altitude offered to hydrology, geology, agronomy, natural resource inventory and a host of other disciplines relating to the way that Man controls (or attempts to control) the natural resources here on Earth. The sensors that had first been developed to winkle out the defence secrets of the communist world were slowly being put to use in the wider interest of mankind.

Lockheed's role in the new NASA U-2 operation was not limited to overhauling the aircraft. It secured a contract to provide all the pilots, physiological support and maintenance crews. In the latter two categories, the Skunk Works was not short of trained personnel, since it had provided similar services under contract to the CIA since the very start of Operation Overflight. As for pilots, Lockheed gained the services of three experienced U-2 drivers. Jim Barnes and Bob Ericson had just officially retired from the Air Force, having in reality spent their last fifteen years as pilots in the CIA U-2 operation. The third pilot was British; Ivor 'Chunky' Webster was one of the RAF pilots who had been sent for training on the U-2 (in 1961). Having enjoyed the experience so much, he was persuaded to resign his commission and settle permanently in California to continue flying the Dragon Lady. Completing this tremendous fund of U-2 experience, Marty Knutson — one of the first two pilots selected by the CIA for Operation Overflight — was employed by NASA as project manager for the U-2s at Ames.

There was no flying job at Moffett Field for another former CIA pilot, however. Frank Powers applied, but was told that his notoriety precluded selection. NASA management wanted to downplay the connection with the U-2's famous past as much as they possibly could.

In August 1971, the two aircraft began flying on a regular basis over five 'control' areas chosen for their particular ecological interest. Four of the areas were in California and Arizona, and therefore reached easily from the Moffett Field base, but the fifth was the Chesapeake Bay region of the eastern US seaboard, which required a U-2 deployment to the NASA airfield at Wallops Island, Virginia. For the first time, details of the U-2 payload specifications and cameras were made freely available through NASA, which needed to acquaint the wider scientific community with the research possibilities that the high-altitude aircraft now offered.

Some years earlier, military photo-interpreters had realized the reconnaissance possibilities offered by colour infra-red film. Such film is sensitive to the green, red and near infra-red portions of the spectrum. Since each object on the ground reflects and absorbs different wavelengths, such film can therefore reveal items that go unrecorded by ordinary black and white or natural colour film. It can 'see through' camouflage netting placed over objects of military interest, for instance, or determine what sort of material has been used to build a given structure. By photographing the same object at the same time through separate cameras, each containing film which is sensitive to different portions of the spectrum, much that is new about the other side's military deployments can be learnt. In civilian applications such as earth resources, the film can distinguish healthy vegetation (which appears red) from dead or dying vegetation (which appears brown or yellow), and has wide application to the analysis of land use. Since the colours displayed by the developed print bear no relation to those which we discern in the visible spectrum, this film is also known as false colour film. The technique is called multispectral photography.

In order to take advantage of these new developments in imagery collection, a range of alternative camera mounts had been devised for the U-2 Q-bay. In the A-3 configuration, for instance, three Hycon twenty-four-inch cameras were carried, all fixed at the vertical position, allowing films with different emulsions which were sensitive to different portions of the spectrum to be exposed to the target at the same time. While these cameras produced large (nine by eighteen inch) photographs for detailed resolution — especially important for military photo-interpretation — the civilian scientists often required a less detailed coverage of a broader area. So some extra configurations were offered on the NASA U-2s, using Wild-Heerbrug RC-10 cameras with six- and twelve-inch focal lengths. These provided nine-inch square negatives and could be carried in pairs, or in combination with one twenty-four-inch camera (the A-4 configuration).[2]

Many early NASA U-2s were flown with a multispectral camera which used four 100 mm lenses to record four different wavelengths simultaneously on a common film. This system emulated the main sensor on the first two ERTS satellites — since renamed Landsats. By getting the U-2s into service nearly a year before the first Landsat was launched, NASA was able to acquire preliminary imagery, and prepare the processing and analysis system for the subsequent deluge of data from the satellite. After the launch, the U-2s carrying the multispectral system staged 'underflights' along the satellite's track at 65,000 feet, to provide a means of comparison with the imagery being relayed from space.

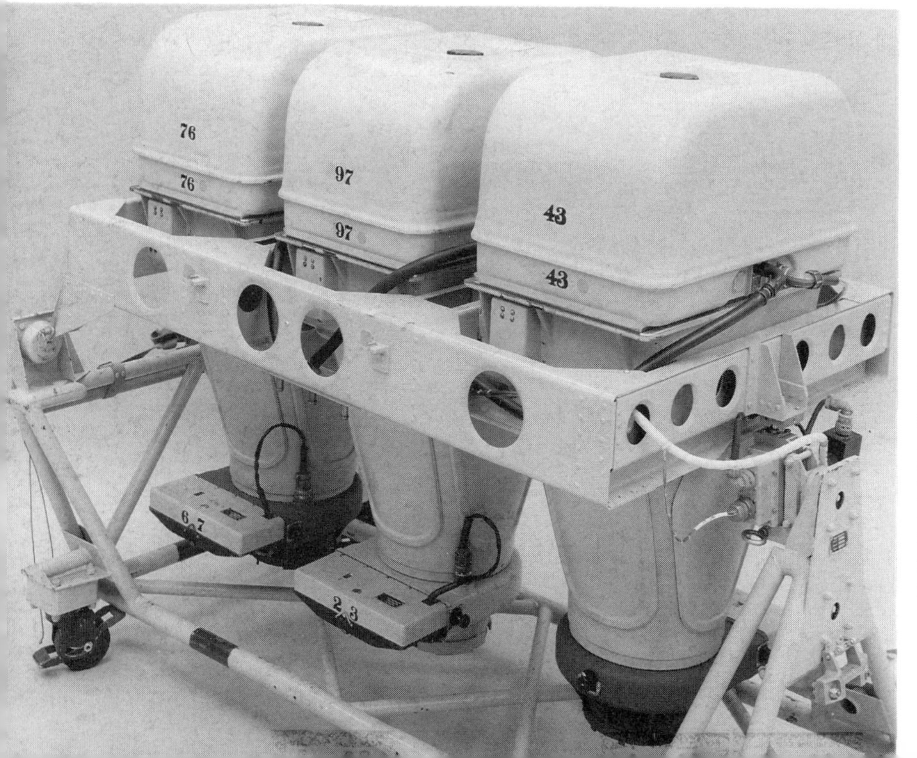

Three twenty-four-inch focal length Hycon HR-732 reconnaissance cameras are shown in the A-3 mount designed to fit in the Q-bay of the U-2. Configurations like this could take advantage of new photo-interpretation techniques using colour infra-red films sensitive to different portions of the spectrum.

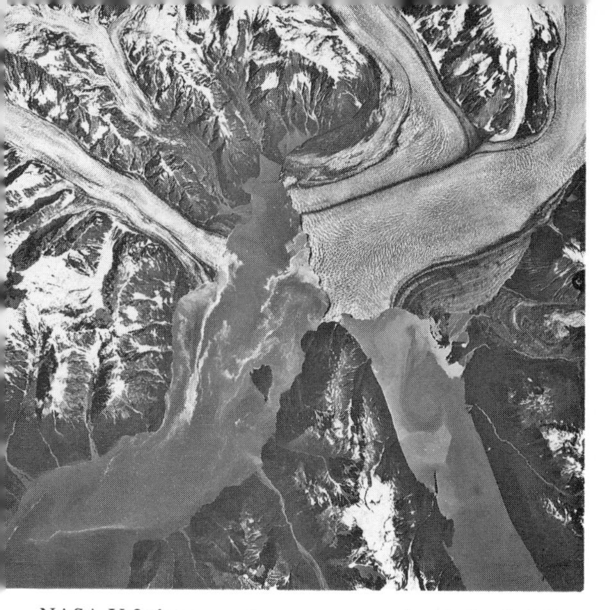

NASA U-2s have spent many summers in Alaska, building up a comprehensive image data-base. During the 1986 deployment, the Hubbard Glacier in south-east Alaska surged unexpectedly. This photograph taken by the Wild-Heerbrug RC-10 camera (six-inch lens, ground scale 1:130000) on 3 September 1986 shows that the advance of the glacier (top right) has blocked the mouth of Russell Fjord (bottom right) as it spreads into Disenchantment Bay (bottom left).

Although the first Landsat was operational in late 1972, there were problems with one of the sensors. Landsat 2 was not launched until January 1975. Long before that time, the NASA U-2 operation had been extended and placed on a more permanent basis. The High Altitude Missions Branch was established within Ames, and the U-2 operation was given the grand-sounding title of Airborne Instrumentation Research Project (AIRP). This was meant to reflect the fact that the aircraft were now also used for research purposes other than earth survey. In November 1973 they had been recruited into NASA's long-term Stratospheric Research Program. The two aircraft carried a variety of sensors to 60,000 feet or higher to measure various gases and particles, such as ozone, nitric oxide and man-made pollutants. The first of what became an annual series of deployments to Eielson AFB, Alaska took place in summer 1974. There was a month-long deployment to Hickam AFB, Hawaii later that year, followed by visits to Howard AFB, Panama Canal Zone and Loring AFB, Maine. By 1977 the two U-2s had sampled the stratosphere from the eastern US coast to 1,000 miles west of Hawaii, and from ten degrees south latitude to the North Polar region.

In both its scale and purpose, the Stratospheric Research Program could be compared with the old HASP effort conducted by the SAC U-2s a decade or more earlier. Once again, U-2 pilots were sent up on long orbits around the sky in an attempt to measure the effect of the human race's cavalier attitude to the environment. In the late fifties, man had caused radioactive particles to be flung into the atmosphere from nuclear explosions. Now he was polluting it with fluorocarbons from a million aerosols.

The sampling devices on the U-2 were more sophisticated now. There was an Ames-designed Q-bay sampler weighing 500 lb. which used chemiluminescent reactions to continuously measure gases *in situ.* An alternative Q-bay payload, also designed at Ames, consisted of four cryogenically-cooled samplers plus two whole air samplers to collect gases for laboratory analysis after landing. There was also an Ames version of the traditional filter-paper type sampler for collecting aerosol and halogen particles. Fourteen years after NASA U-2s began collecting this data, the world finally woke up to the danger, and the High Altitude Missions Branch hit the headlines as it flew the Dragon Lady into an ominous hole in the ozone layer over the South Pole.

Just as these flights served to warn of man's potential demise at his own hand, another series attempted to define how he had evolved in the first place. The results suggested that the 'Big Bang' theory of the formation of the universe needed serious modification. A team of astro-physics researchers from Berkeley designed and placed an 'Aether Drift' radiometer in the U-2, which looked upwards rather than down-wards. A special Q-bay hatch top with two sensor ports was fabricated to accommodate the ultra-sensitive device. It measured the cosmic microwave radiation which had been discovered in 1965 to be still coming from the most distant parts of space, where evidently the energy from the initial, universe-forming event was still radiating outwards. The Berkeley scientists concluded that far from being a cataclysmic explosion, this event was a very smooth, almost serene process, with matter and energy uniformly distributed and expanding at an equal rate in all directions. Not only that, they also discovered that the entire Milky Way galaxy was streaking through the universe at a velocity greater than one million miles per hour, and that the

NASA U-2C spreads its wares. Reading from the left: an infra-red scanner, three radiometers, a filter sampler, an ocean scanner, an air sampler, the aether drift radiometer (in front of aircraft), a cryogenic sampler, the original B camera, RC-10/HR-732 combination, A-3 configuration of three HR-732s, Itek KA-80A optical bar camera, dual RC-10 configuration, and RC-10/Vinten combination. Note the drop tanks, modified to contain further experiments.

universe was probably not spinning. Both these conclusions challenged previously-accepted assumptions.

The data from which the discoveries were made was obtained from eleven highly-demanding night flights over the western US flown at 65,000 feet. 'Chunky' Webster had to fly the U-2 level to within one-half a degree, because only then would the effects of the earth's atmospheric microwave radiation be cancelled out, thereby allowing the sensor to accurately detect the very-low-frequency light radiation coming from way beyond the quasars in outer space. The Aether Drift radiometer was, in effect, a camera which built up an image part by part spread over the eleven flights, but the frequency of the signal that the scientists were trying to detect was 20,000 times lower than the frequency of visible light! Thanks to plane and pilot, the flight criteria were met — an on-board measurement system indicated that average lateral displacement of the U-2 during Webster's flights was only plus or minus one-sixth of a degree!

Remote-sensing technology and applications expanded steadily throughout the seventies, thus ensuring a steady demand for the services of NASA aircraft numbers 708 and 709. More satellite sensors were 'pre-flighted' on the two U-2s: a colour scanner for the Nimbus satellite, which would measure oceanic tides, sediments and micro-organisms, and a heat capacity mapping radiometer which would provide continuous thermal mapping of the earth's surface.

In February 1977, another U-2 veteran arrived at Ames. Former CIA pilot and reconnaissance manager Jim Cherbonneaux moved from Washington to become head of the High Altitude Missions Branch. By the time he arrived, the two aircraft had acquired imagery from all fifty US states covering some thirty-five per cent of their total surface area. The imagery had proved useful in nearly all the earth science disciplines. Forest diseases and insect infestations had been detected, and timber harvesting practices improved; photographic data of watersheds had indicated where pollution needed to be tackled; large-area coverage provided 'the big picture' on population and agricultural trends to state and local government land-use planners. NASA U-2 photography had also proved invaluable in assessing the effects of natural disasters such as fires and floods. In rapid-response missions for the State of California, the aircraft were despatched to the scene of forest fires. The photographs taken showed the extent of the conflagration, possible access routes for fire-fighters, and potential firebreak locations. During the 1975-76 drought, U-2 photos helped measure water levels in rivers and reservoirs.

'What this is, is classic plowshare technology,' noted Cherbonneaux, referring to the aircraft's (and his own) military origins. 'You know what I mean? Beating swords into plowshares?'

Not everyone saw it that way. To some California citizens, the NASA U-2 was still a spy in the sky. In 1977, the state Coastal Conservation Commission commissioned a photo-survey to help its wastelands management programme. But the photos also served another purpose. Unlicensed construction in the coastal zone had recently been prohibited, and licences were hard to get. The survey was flown with the big twenty-four-inch cameras, and therefore provided high-resolution nine by eighteen photographs from which it was possible to determine whether landholders had defied the ban on construction. Citizens of the town of Bolinas got particularly vocal about it; a resident attorney summed up local feelings: 'I do have this gut reaction to this eye in the sky able to look at all the little things in people's backyards. It's a feeling of distaste, you know, that this is just one more step on the road to 1984.'

Yet all the imagery which the NASA U-2s collected was available for public inspection at Ames, via a computerized image retrieval system. Anyone could purchase full-scale reproductions of any frame from a huge data bank in South Dakota. In 1978, new ground was broken in the administrative use of the U-2 when the state of Alaska and ten federal agencies clubbed together to fund a photo survey of the entire state. They concluded that it was simply the most cost-effective way — in such a remote region — to gather all sorts of information about natural resources, land settlement claims by native eskimos, wildlife conservation, energy management, and so on. The U-2s from Ames were joined in Alaska by the WB-57F operated by NASA Houston, and in the first year of the programme they photographed 25,000 line miles of the state at a cost of $496,000. It worked out at less than $20 per data mile, which the bureaucrats reckoned was very good value compared with trekking through the tundra or hiring a lightplane and zig-zagging about at low level. During the mapping of Alaska, the U-2s flew at between 60,000 and 65,000 feet with the dual RC-10 camera configuration. The six-inch camera was loaded with black and white film and the twelve-inch one with colour infra-red film. The flights could only be conducted in the short summer season when the snows had melted and vegetation was flourishing, and so the programme took three years to complete.

More technology from the world of military reconnaissance was gradually declassified for use on the NASA aircraft. A twenty-four-inch

NASAS Lockheed U-2 pilots outside the High Altitude Missions Branch at Ames Research Center, 1987. From left to right, Jim Barnes, Dick Davies, Jerry Hoyt and Doyle Krumrey. All former Air Force Dragon Lady pilots, these four had accumulated more than 12,000 hours in the aircraft between them. Jim Barnes retired some months after this photograph was taken, having flown U-2s for over thirty years.

focal length panoramic camera which gave much greater resolution than the Itek KA-80 (albeit over a narrower strip of territory) was introduced. Two of them were mounted in the U-2 to provide convergent stereo coverage. The aircraft's utility in the firefighting or disaster assessment role was boosted by a line scan camera which provided real-time imagery to a ground receiving station by means of a data link.

When Lockheed won a contract in 1978 to restart U-2 production as the TR-1, NASA quickly became interested in the capabilities of the upgraded model. Down at NASA Houston, the three WB-57 aircraft also in use for earth resource surveys were getting old, and had consistently been out-performed by the Ames-based U-2s. The Administration decided to retire two of the WB-57s and acquire another Dragon Lady. In fact, NASA received the very first example from the reopened production line, when Marty Knutson delivered N706NA from Palmdale to Moffett Field on 10 June 1981. The aircraft was designated ER-2.

Knutson had now risen to become the chief of the Airborne Missions and Applications Division at Ames. When another promotion gave him additional responsibility for NASA's Dryden Flight Research Center at Edwards in 1984, he finally quit flying the U-2. He had logged over 3,000 hours in the aircraft over a twenty-eight-year period. Fellow NASA U-2 pilot Bob Ericson retired around the same time, leaving Jim Barnes as the sole remaining pilot from the earliest days of the U-2 programme to be still flying the aircraft. New pilots for the NASA operation were recruited from the ranks of retiring Air Force U-2 pilots. Dick Davies, Jerry Hoyt, Doyle Krumrey, Ron Williams and Jim Barrilleaux all chose to continue their association with the Dragon Lady in this way.

The ER-2 offered significant payload/range advantages over the two earlier models. The new aircraft had large superpods attached to each wing, providing extra accommodation for sensors. The other two aircraft were routinely flown with wing drop tanks which had been adapted to house various payloads, but the available volume was less than one-fifth that offered by the superpods. For the first time, a camera (typically the RC-10) could be carried in the nose or the superpods, thus freeing the Q-bay for other experiments. The ER-2 was also more comfortable — and easier — to fly.

Along with the new aircraft came yet more

NASA's first ER-2 (80-1063/NASA 706) during an early test flight in 1981. The superpods are shown to good effect as the aircraft banks left.

new sensors. The Pentagon released the Itek Iris II very high resolution optical bar panoramic camera for civilian use. This very expensive and sophisticated piece of optical technology was by now the main reconnaissance camera in use on SAC U-2s. It had also been taken to the moon by the Apollo astronauts. NASA was cleared to offer researchers a digital X-band synthetic aperture radar in the ER-2, as well as the capability to data-link to ground stations either direct or via the Tracking and Data Relay Satellite (TDRS) system. But the most significant of the new sensors was a Daedalus Thematic Mapper Simulator (TMS), a multispectral scanning radiometer which recorded radiance from the earth's surface in eleven discrete wavelength bands.[3] Such detailed measurement of light and heat could, with the aid of a computer, produce colour maps which vividly delineated even subtle variations in surface composition. Such manipulation of digital data provided capabilities which even the multispectral camera systems could not match.

Geologists, for instance, could map differing varieties in rock formation, picking out shales from sandstones, and different types of soil. Oceanographers received maps detailing water quality and thermal pollution. Agronomists could tabulate differences in health, maturity and species of vegetation. In the TMS maps, one could distinguish redwoods from fir trees, wetlands from dry pasture, and fields planted with artichokes from others planted with Brussels sprouts! Of course, colour infra-red film exposed by camera still provided greater resolution than the TMS, and the two types of sensor were regarded as complementary. They were often both used on the same surveys.

By the mid-eighties, NASA high-altitude survey aircraft had obtained over 400,000 frames of photographic imagery covering large portions of the US. This extensive collection of high-quality, large-scale negatives was housed in the Applications Aircraft Data Management Facility at Ames, right next to the High Altitude Missions Branch. All sorts of organizations were now beating a path to the facility's door to review archive film. High schools and colleges, water districts, county and state planning commissions, engineering firms, utility companies, emergency service agencies, forestry departments, oil companies . . . the list was endless.

The demand for new surveys continued. Having been well satisfied with the results of the 1978-80 mapping survey, authorities in the state of Alaska contracted for further flights each summer. There were also continuing deployments to Wallops Island in Virginia. In the summer of 1984, a U-2 flew from there to photograph more than 80,000 square miles in five eastern states, in order to survey the damage being caused to standing trees by the gypsy moth caterpillar, and assess whether spraying programmes were having the desired effect. The following year, the ER-2 flew over Florida in a major survey of the state's citrus trees for the Department of Agriculture. In both surveys, the Itek Iris camera was used, loaded with colour infra-red film. The Iris was becoming a favourite tool for regulatory agencies, thanks to its high resolution (two feet from 65,000 feet using black and white film) and wide coverage (a thirty-seven nautical mile swath beneath the aircraft). The Environmental Protection Agency used it to check up on illegal dumping of waste.

Disaster assessment flights also continued. In December 1986 they even set a controlled fire in California's San Gabriel mountains so that scientists could investigate all sorts of related phenomena. A NASA U-2 made its contribution by flying overhead to record it all on the TMS and film. Two channels of the TMS were relayed by data link to fire fighters on the ground. The readout of thermal data enabled them to track the active fire front, which was obscured from their vision by the smoke plume.

Although NASA was not keen to publicize it, there were also a number of missions being flown by the U-2/ER-2 aircraft in support of Defense Department projects. The most notable of these was Teal Ruby, a satellite which would detect and track aircraft from space by measuring their infra-red signature. It formed part of the Air Force's Air Defense Initiative (ADI) to defend US airspace against future missile and bomber threats. While these could currently be detected by radar, alternative forms of detection might be necessary if the USSR developed stealth technology. The Teal Ruby concept was made possible by advances in charged-coupled devices and cryogenic cooling of sensors. The satellite would carry a six-foot tall infra-red telescope which 'stared' down at the earth's surface and registered disruptions to the normal background signal return (caused by an aircraft) on a mosaic of thousands of focal-plane detectors.

This time without superpods, the ER-2 is shown on deployment at Mildenhall in the UK during April 1985. Visible on the wingtip skid is a particle sampler, essentially some sophisticated sticky paper which can be electrically rotated to face the slipstream. Main purpose of the UK visit was to test-fly an infra-red sensor for the forthcoming Teal Ruby air defence satellite.

It was complex technology, and one of the problems was establishing a data base of background measurements. The NASA ER-2 was enlisted to fly a similar multiwavelength infrared sensor as part of a Teal Ruby support effort codenamed 'Hi Camp' (Highly-Calibrated Airborne Measurements Program). This provided an atmospheric, terrestrial and oceanic background data base, making precise measurements of the clutter which Teal Ruby would have to deal with. The ability of the Hi Camp sensor to pick out aircraft from the ER-2's 65,000 feet cruising height was also tested. In a year-long series of highly co-ordinated flights over the western US and Europe (the latter while the ER-2 was deployed to Mildenhall airbase in the UK during the spring of 1985) a variety of US Air Force aircraft ranging in size from a T-38 to a C-5 were flown against the Hi Camp sensor. The Teal Ruby satellite itself was ready for launch by 1986, but the Challenger shuttle disaster forced a three-year postponement.

The trip to the UK with the ER-2 was the first to non-US territory by the High Altitude Missions Branch. The second came in January 1987 when the same aircraft was deployed to Darwin, northwest Australia. It was engaged on the Stratosphere-Troposphere Exchange Project (STEP), a continuation of earlier atmospheric studies by the Ames-based U-2s. NASA and the National Oceanic and Atmospheric Administration (NOAA) funded STEP to obtain yet more data on the mechanisms and rate of transfer of particles, trace gases and aerosols from the troposphere into the stratosphere. Darwin was chosen as a suitable launch point for flights into the region where the world's coldest and highest tropopause was to be found, as well as the largest and highest cumulo-nimbus clouds.

However, the most important deployment ever carried out by the NASA unit was yet to come. It also involved the most hazardous flying — long hours over inhospitable terrain from which rescue in the event of an accident might prove impossible. But the stakes were very high. In August and September 1987 the ER-2 flew twelve times out of Punta Arenas, southern Chile, across deepest Antarctica. Its mission was to take detailed measurements in the recently-discovered hole in the ozone layer that was developing over the South Pole every winter. Over 170 scientists, NASA managers and support crews descended on the desolate area at the southernmost tip of South America; in addition to the ER-2, NASA also deployed the Ames Research Center's converted DC-8 airliner. Between them, the two aircraft carried no fewer than twenty-one separate scientific payloads to measure every conceivable variable that might be linked to the alarming phenomenon of ozone depletion. The whole effort cost $10 million, but it soon became apparent that it was worth every penny. The scientists found that the hole was bigger than ever, and proved that man-made chlorofluorocarbons (CFCs) were to blame. Since the screening effect of the delicate stratospheric layer of ozone prevents harmful amounts of the sun's ultra-violet radiation reaching earth, this was significant news indeed.

CFCs are chemicals which have a variety of applications. In the form known as Freon gas, they are found in most refrigerators and air conditioners. Other CFCs are used for making foam insulation or fast-food containers, as industrial solvents, and as the propellant in aerosol-type spray cans. For years, few were willing to believe that this relatively benign family of chemicals could do any harm — until a team of scientists from the British Antarctic Survey alerted the world to measurements they had taken from the ground, which indicated a dramatic decline in ozone levels. At NASA, they quickly recomputed data from the Nimbus 7 satellite, which carried an ozone mapping spectrometer. Sure enough, the hole showed up: a forty per cent reduction in the amount of ozone over Antarctica during late winter and early spring.

The findings seemed to support a controversial theory advanced some years earlier by two California scientists. They figured that ninety-nine per cent of all CFCs were rising into the stratosphere, where their seeming stability counted for nothing. Once there, ultra-violet radiation from the sun could break down CFCs into their constituent parts — including chlorine. This very reactive chemical could then attack the easily-broken-down ozone (O_3) by detaching one oxygen atom to form chlorine monoxide (ClO) plus oxygen (O_2). The chemical industry poured scorn on their findings, but the US government was sufficiently impressed to ban the use of CFCs as a spray can propellant in 1978. The Europeans were not so impressed, even when the hole over Antarctica was discovered in 1985. They pointed out that meteorological conditions down there were unusual and extreme: very cold temperatures at all levels, and strong upcurrents from the troposphere. After all, the hole was known to disappear each November, when a vortex of winds over the pole breaks up, allowing air to flow in laterally from elsewhere in the stratosphere. Maybe the ozone was being pushed out of the stratosphere every September by a natural effect which was simply part of cyclical global weather variations. Maybe solar flares were responsible. Maybe the distinctive clouds which form at record altitudes in the polar stratosphere were somehow responsible. And so on. Some American scientists suspected that European doubts were motivated by the fact that the chemical industries there had more to lose if CFCs were banned.

NASA's mission to the Antarctic was designed to settle these questions, once and for all. While the DC-8 would fly at the lowest extremities of the hole, with some of its sensors peering upwards, the ER-2 would fly right through the very centre of it. It would carry some payloads from the recent STEP series of atmospheric sampling flights, but also some specially-designed experiments. Altogether, there would be fourteen separate sensors carried in the nose, Q-bay, and wing pods (the shorter 'Spear' type, rather than superpods, were to be used for these missions). The most important of these, if those blaming CFCs for the ozone hole were to be vindicated, was a chlorine monoxide detector designed by Harvard chemistry professor James Anderson. But, as Anderson himself pointed out, 'There isn't a single instrument on either plane that, in the long run, won't be crucial. Every chemical that can be measured must be measured, and measured precisely. If there is any ambiguity in our findings, our impact will be weakened. The burden of proof is enormous.'

Conditions at the windswept Punta Arenas airfield were rudimentary. Before the deployment, the taxiways had to be resurfaced to prevent damage to the ER-2's delicate landing gear. Even basic office accommodation had to be specially built in the draughty military hangar allocated to the NASA team. Under difficult conditions, the scientists laboured to perfect their experimental payloads, and then crossed their fingers as they watched the ER-2 soar into the mostly grey and turbulent skies over the Magellan Strait. The missions lasted well over six hours.

It was not good flying weather down there. At ground level, fronts and surface winds of up to sixty knots could develop rapidly. At altitude, the polar vortex caused winds of up to 200 knots, and temperatures as low as minus ninety-five degrees centigrade were experienced. Dual VHF radios and INS were fitted to aid with navigation and communications. The pilots took arctic survival courses before they left the US. All the same, they figured their chances of surviving an ejection onto the icecap were very slim. Unlike an earlier generation of U-2 pilots, who had flown to the North Pole on sampling flights in the early sixties, the NASA pilots did not have the comfort of an accompanying C-54 rescue aircraft with paramedics. In fact, the ER-2 had performed almost flawlessly during the six years since it had been delivered — but there was always a first time!

But the flights were uneventful, and the date take was excellent. Anderson's chlorine monoxide detector measured levels up to 500 times the normal concentration. The edge of the hole was found to have extended further north this year than ever — as far as Punta Arenas, in fact! Meteorlogical factors peculiar to the region could not be entirely discounted, but they were not themselves the cause of so much ozone depletion. Even as the scientists and airmen strived to gather this vital data, a UN convention of thirty nations in Montreal agreed to a fifty per cent reduction in CFC production by 1999. It was a compromise: the US and Scandinavian countries wanted a complete ban, and when the final results of NASA's expedition to southern Chile were published in early 1988, their case was greatly strengthened.

Could the same phenomenon of ozone depletion be occurring over the North Pole? A field investigation to the Arctic suggested it could, and so in late December 1988 the DC-8 and ER-2 were despatched from Moffett Field to Stavanger in Norway. Each aircraft flew fourteen times into the polar vortex over the next two months. The ER-2 missions lasted up to eight hours, going as far as eighty degrees North and ranging from the Barents Sea to Greenland. For safety's sake, the flight tracks were kept to within 250 nautical miles of emergency landing strips, so that the ER-2 could glide in from 65,000 feet if it encountered trouble. The precaution wasn't necessary; the ER-2 performed flawlessly once again, although the pilots had a hard time navigating and landing because of the strong prevailing winds.

When the flying was over, the NASA team called a press conference in Oslo to disclose their preliminary findings. The news was bad. There was every indication that the ozone layer over the Arctic was being eroded to a similar extent. The two aircraft had measured amounts of chlorine monoxide — an ozone destroyer — that were up to fifty times its normal concentration in the atmosphere. While the scientists went home to analyse their data and draw some firm conclusions, ministers and officials from 120 nations met in London to discuss the crisis. It was already clear that the Montreal agreement was inadequate; the US and European countries now declared they would phase out CFCs completely, and wanted others to agree to at least an eight-five per cent reduction.

While the ER-2 was ranging far and wide on these vital missions, the two NASA U-2C models back at Ames Research Center were reaching the end of a thirty-year flying career. Article 348 (NASA 708) was retired in mid-1987 when it reached 10,000 flying hours, and put on display at the Moffett Field visitors' centre. To replace it, NASA acquired a US Air Force TR-1 on loan, pending the arrival of a second purpose-built ER-2 in 1989. The other U-2C (Article 349/N709NA) made its last flight in April 1989, to a final resting place on display at Robins AFB. It was the only original U-bird still flying. Only a few months earlier, NASA pilot Jim Barnes finally hung up his pressure suit after flying the Deuce (he never called it the Dragon Lady!) for over thirty-one years. The veteran pilot had amassed a remarkable 5,760 hours in the type, a record which seems very unlikely to be matched.

1 Four of these projects are detailed in the following paragraphs. The fifth resulted in the only payload that was ever dropped from a U-2 during an operational reconnaissance mission. The details of this payload remain classified.

2 Some other camera combinations were used by military U-2s. The A-1 configuration was three mapping cameras each of six-inch focal length, plus a twenty-four-inch camera in a rocking mount to permit vertical, left and right oblique views. The A-2 configuration was three twenty-four-inch cameras fixed in position to give simultaneous vertical, left and right coverage. These were the main alternatives to the two big singly-mounted Hycon cameras — the Type B for the very largest (eighteen-inch square) negatives, or the Type H for long-range oblique photography with its sixty-six-inch folding optics. There were others, of course, such as the infra-red scanner used over China in 1964-65, and the more modern Itek twenty-four-inch optical bar panoramic camera. In fact, this last became the standard camera in use on SAC U-2s by the mid-seventies.

3 The word simulator appeared in the scanner's title since it was designed to replicate a similar system carried by Landsat-4.

Chapter Nine
Taming the Dragon

You can literally fall in love with this airplane. It's different, it's unique. There's probably a large amount of ego involved in flying it — you know you're part of a select group, and you get attention wherever you go. And while the rest of the Air Force is only training for the big event, we're out doing it already!
Major Thom Evans, 1986.

It was a bright, sunny afternoon in late September 1957, and as Colonel Jack Nole climbed his U-2 through 50,000 feet he could see the silver thread of the Rio Grande river far below him as it twisted towards the big bend. To the southwest, bluish-purple shadows were building up on the east flank of the Serranias del Burro mountains in Mexico. Nole had taken off from Laughlin AFB thirty-five minutes earlier on a routine test flight. Less than four months earlier, the commander of the 4028th SRS had led the first three U-2s into the Texas base in formation, and his unit was slowly building up experience on the unique design before it was tasked with its first operational deployment.

Colonel Jack Nole, who set a new record for the highest parachute escape from an aircraft when he bailed out of his crippled U-2 in September 1957. Nole was the first commander of the 4028th SRS, the U-2 squadron set up within SAC's 4080th SRW, initially to conduct high-altitude sampling of nuclear test debris.

As Nole later recalled, 'Suddenly, the aircraft's nose began to drop. There was no sound, no warning. I pulled the stick back into my lap, trying to bring the nose back up. But there was no control. In the far left corner of the instrument panel, the wing-flap position indicator told the story: the wing flaps were full down. I hadn't actuated the flaps to put them down — but there they were.'

It was subsequently determined that rainwater had drained into a switch, corroded the contacts, and caused a short circuit, although some of Nole's contemporaries believe he may have inadvertently lowered the flaps when selecting or deselecting gust control. Whatever the cause, Nole was immediately in big trouble as the aircraft picked up speed in a descent. Despite cutting the engine and extending the gear and dive brakes, the nose kept dropping until the U-2 was plummeting straight down. Then the tail section broke off.

Nole managed to contact his 'mobile' — the back-up pilot sitting on the ground at Laughlin who had strapped him in and seen him off, and who was now monitoring the flight from the base command post. 'Bale out! Bale out!' was the urgent advice.

By now the crippled U-2 was gyrating through

one outside loop after another. 'With each huge somersault', said Nole, 'I would fall out of my seat and my helmet would crash against the canopy. When the aircraft turned upside down for the fourth time, my helmeted head smashed the canopy assembly loose.' By now, however, Nole had managed to perform the complicated sequence of moves which released a U-2 pilot from his aircraft. First he disconnected the oxygen hoses and the electrical connection which heated the faceplate of his helmet. Then he severed the radio lead, released the seat safety belt and shoulder harness, and unlocked the canopy handle. All this in the most cramped of cockpits, with movement made doubly difficult by the constriction of the pilot's pressure suit, which would automatically inflate to compensate for the pressure drop resulting from an engine shutdown.

As Nole's head smashed against the canopy, it sailed away into space and the pilot nearly followed, but his seat pack snagged against the ledge of the cockpit, and the fierce slipstream pinned him there for what seemed like an eternity: 'There I was, bent over backwards against the fuselage, and the aircraft going end over end. I was afraid it might break up completely at any time, and I had better be free when it happened! I thrashed around, kicked and pulled, and finally came free.'

Now he faced another life or death decision. During his struggles in the cockpit, Nole had been unable to locate the 'green apple', a small ball valve which turned on the emergency oxygen supply from seat-pack to pressure suit. 'Normally, the valve was on your left side, tucked into the crease between your thigh and your hip,' he recalled. 'But the valve is green, the pressure suit is green, and the inside of the cockpit is green . . . '

With no emergency oxygen feeding the pressure suit, it was slowly beginning to deflate, and Nole detected the first signs of hypoxia as his vision began to fade. 'Dimly, I realised I had two alternatives,' he recalled. 'The first was to let myself fall until my parachute opened automatically at the preset 14,000 feet. But it would take more than two minutes to free-fall those seven and a half miles; by that time, there was a good chance I'd have suffocated. The suit's made to force oxygen into you under pressure; it's not like breathing in air on the ground. In the suit, you open your mouth, oxygen flows in, and you have to make an effort to exhale. If I was unconscious when the chute opened, I would not be able to open the face plate of my helmet and breathe naturally. The suit is made to hold air inside, not to let it in from the surrounding atmosphere!'

The second alternative was to open his parachute immediately, and hope that he could then find the emergency oxygen valve to sustain him through what would be a lengthy descent. But this would entail a terrible risk; in just a few seconds, a body dropping through the thin air at his height could accelerate to 375 mph. A parachute snapping open to arrest such a descent could be torn to shreds in the slipstream, or else deploy with such force that the body to which it was attached could be ripped apart from the shock of its opening!

If by good fortune that didn't happen, the extreme cold would pose another threat to the pilot's survival, since it would take him half-an-hour to descend all the way from 53,000 feet with the parachute deployed. 'Into my dimming vision floated the release ring of my parachute ripcord,' said Nole. He took the risk — and pulled it.

Miraculously, Nole's parachute opened gently, without a trace of shock. Accident investigators later surmised that since the pilot had been flung from the aircraft as it curved upward in an outside loop, his body had described a similar upward arc. Nole must have pulled the ripcord just as he reached the apex of this arc, and before he started to fall towards the earth at an ever-quickening pace. 'It opened so gently that I was perfectly horizontal in the air, at a level even with my chute,' Nole continued. 'Then I swung back like a giant pendulum until I was even with the chute on the other side.' The giant oscillations continued in the thin air, with Nole apparently powerless to stop them: 'Each time I swung, I was afraid that air would spill from the chute's high side and that it would collapse, dropping me like a stone. I began pulling the shroud lines at the top of each swing, in an effort to stop the oscillation.' By now, the pilot had located and pulled the 'green apple', but the exertion of trying to stop the giant swings was exhausting his emergency oxygen supply quite quickly. He would have to get outside air soon.

'Usually it's a simple matter to locate and pull the string, just below the chin, that releases the faceplate from the helmet,' said Nole. Not this time, however! Nole had to painstakingly peel the pressure glove off his right hand before he was able to get the faceplate off. He was now at 20,000 feet, and still swinging wildly from side to side. For the first time in seventeen years of military flying, the veteran pilot was hideously airsick. 'I threw up over half of Texas,' he recalled.

By now, Nole had company. Two other U-2s were airborne from Laughlin, along with a Cessna U-3 chase aircraft. Fellow pilots Dick Atkins, Warren Boyd and Dick Leavitt located the descending parachute and flew slow circles around it. Nole approached the ground, still oscillating violently. But luck was still on his side: 'I came right down over one of those small, rolling Texas hills. On the one side of it was this big, flat-topped rock. As I drifted over the hill, my body swung back, and my dangling seat pack caught on the rock, and jerked me to a stop in the air, so that I fell backwards to the ground without injury.'

The cramped nature of the original U-2 cockpit should be apparent from this view. The old-fashioned control yoke is in the stowed position, with the rubber sighting cone of the driftsight above it. Throttle wheel and flap controls are to the lower left, while communications panels were usually to be found opposite. Note the canopy thrusters on the cockpit rail, installed in the mid-sixties to aid an emergency escape.

Jack Nole had survived by far the highest parachute escape in history. It had taken twenty-two minutes; they later told him that it should have taken another eleven minutes, but those giant swings had spilled air from the chute and sped him down. Otherwise, he might not have descended to thicker and warmer air in time. 'We can only conclude that Colonel Nole survived through an act of God,' the accident board reported.

Six U-2 pilots had already lost their lives by the time that Nole made his miraculous escape. Only one other had gone 'over the side' and lived to tell the tale. This was Bob Ericson, who had unaccountably run out of oxygen while flying at 35,000 feet on a training flight from The Ranch in December 1956. He was flying an aircraft still equipped with one of the 'smoky' -37 engines, and fumes in the cockpit may also have contributed, as Ericson began to lose consciousness. Fortunately, the pilot revived as the aircraft oversped and began to go out of control. After encountering considerable difficulty in removing the canopy, Ericson baled out over Arizona, and both aircraft and pilot landed in open country.

Wilbur Rose became the first fatality when he stalled and crashed at The Ranch during the training of the first group of operational pilots. He was trying to shake off a 'hung' pogo. A few months later, another CIA recruit died at the remote desert site. Frank Grace took off on a night training flight, but apparently became disoriented and flew into a telephone pole at the end of the runway. Then a mysterious accident in Germany claimed Howard Carey, who was climbing out of Wiesbaden when his aircraft broke up in mid-air near Kaisersauten on 17 September 1956. He was 'buzzed' by two Sabre fighters belonging to the locally-based Royal Canadian Air Force group, and subsequently lost control of the aircraft.[1] Some U-2 pilots believed that the aircraft was so fragile that this one simply disintegrated in the fighter's turbulent wake. Examination of the wreckage revealed that the wing tank filler caps were missing, suggesting that some sort of fuel overpressure had led to wing failure.

When Jack Nole lived to tell the tale of just how difficult it was to get out of a high-flying U-2 that suddenly became disabled, he was only confirming what was already apparent from two fatal accidents earlier in 1957. In both cases, the unfortunate pilots had evidently made frantic efforts to get out of wildly-descending aircraft, without success. Lockheed test pilot Robert Sieker was the first of these, on 4 April, with his body discovered only yards from the aircraft wreckage. He had been followed by SAC's Lt Leo Smith, on 28 June. In the second accident to befall the newly-formed 4028th SRS within a few hours, Smith was actually dangling outside the aircraft as it hit the ground, encumbered by the various umbilical cords attached between his pressure suit and the bottom of the cockpit floor.

These tragic losses convinced the Air Force that an ejection seat really would have to be installed. It had been eliminated from the original design to save weight. While higher headquarters were pondering the issue, another attempted bale-out proved unsuccessful when Capt Lacombe crashed at night in Texas. Now the Skunk Works were authorized to develop an ejection system for the U-2, and by the middle of 1958 they had come up with a lightweight modification of a Martin-Baker seat. After consultation with the operating units, final design was undertaken, but retrofitted aircraft weren't available until late 1958. By this time, four more flyers had lost their lives in U-2 accidents, but 1959 was to prove a relatively uneventful year, with no fatalities. The first pilot to have need of the new escape system was Frank Powers on 1 May 1960 — but he didn't use it.

The CIA pilot was almost four hours into his lengthy overflight of the deepest USSR when he approached Sverdlovsk. Some thirty-five miles south-east of Sverdlovsk, he had completed a ninety-degree left turn as scheduled, and was writing down the time, altitude, speed, and engine readings on his progress chart. 'Suddenly, there was a dull thump, the aircraft jerked forward, and a tremendous orange flash lit the cockpit and sky,' he recalled. Knocked back in his seat, Powers exclaimed 'My God, I've had it now!'

As subsequently determined, his aircraft had been disabled by an SA-2 surface-to-air missile exploding nearby, perhaps after-removing part of the tail. The right wing started to drop, and Powers managed to bring it level in the normal way. But then the nose started going down, and the pilot's back pressure on the control yoke failed to bring it back up. 'Either the control cable had severed, or the tail was gone,' he said. 'I knew then that I had no control of the aircraft.' He pulled the yoke all the way back into his lap, but to no avail, and the dive accelerated. Then a violent movement shook the U-2, flinging Powers all over the cockpit as, probably, the wings were torn off.

'What was left of the aircraft began spinning, only upside down, the nose pointing upward toward the sky, the tail down toward the ground. All I could see was blue sky, spinning, spinning.' The pilot's pressure suit had inflated, indicating that cockpit pressurization had already gone. Powers prepared to eject. Despite the heavy 'G' forces, he managed to locate his emergency oxygen valve, and pulled it. (After Nole's accident, a white cross had been painted on the 'green apple' to make it easier to find.) He began groping for the destruct switches, located at shoulder height to the left of the front panel, but then changed his mind. What if he couldn't get out before the seventy-second time delay ran out and the cyclonite charge planted behind him on the Q-bay bulkhead exploded? Thrown forward and upwards as he was, getting into the proper position to eject was proving impossible. In the cramped U-2 cockpit, there was only the smallest of clearances between the ejection seat and the overhead canopy rails. Powers figured that if he were to fire the seat in his present position, the front rail would have cut off both his legs just above the knee.

'Yanking at one leg with both my hands, I succeeded in getting my heel into the stirrup on the seat. Then I did the same with the other heel. But I was still thrown forward, out of the seat, and couldn't get my torso back. Thus far I had felt no fear. Now I realized I was on the edge of panic.' He was at 34,000 feet, already halfway to the ground. Fortunately, Powers paused to think about his predicament, and realized that he could perhaps bale out instead. He reached for the canopy release handles, managed to turn them, and the canopy sailed off into space. Next he released the seat belt.

'Immediately, the centrifugal force threw me halfway out of the aircraft, with movement so quick my body hit the rear view mirror and snapped it off. I saw it fly away. That was the last thing I saw, because almost immediately my faceplate frosted over. Something was holding me connected to the aircraft; I couldn't see what. Then I remembered the oxygen hoses; I'd forgotten to unfasten them.'

With his body now wrapped around the front canopy rail, Powers tried to reach back into the cockpit to activate the destruct switches. He knew they were close to his left hand, but the 'G' forces were too great. 'Unable to see, I had no idea how fast I was falling, how close to the ground . . . And then I thought, I've just got to try and save myself now. I gave several lunges and something snapped, and I was floating free.' His parachute, which was set to open at 15,000 feet, deployed almost immediately. Powers removed the faceplate for his first, unwelcome sight of the country that was to be his enforced home for the next twenty-one months.

On 14 July 1960, Major Raleigh Myers of the 4080th SRW became the first pilot to use the U-2 ejection seat when he departed his aircraft some thirty miles north-west of Uvalde, Texas, during a low-level proficiency flight from Laughlin. The pressure suit was not worn on such flights, so Myers was not encumbered with that uncomfortable garment. Instead, the pilots wore a regular flight suit and helmet, with oxygen mask attached. Since the mask contained the pilot's radio microphone, it was kept on to ensure that air-to-ground transmissions were audible. But after completing a series of touch and go landings, Myers requested and received clearance to leave the pattern prior to setting up for his final landing. He flew off to the east, and decided to unhook his oxygen mask *to have a cigarette!* Although he turned off the oxygen valve, a Teflon seal in the valve's pressure reducer failed, and oxygen continued spilling out of the mask. The inevitable happened. Myers' cigarette ignited an uncontrollable cockpit fire, and he punched out. He was not the first U-2 pilot to be caught smoking in the cockpit; in an earlier incident, an enraged General had discovered that one flyer had even installed a makeshift ashtray!

Any confidence that the U-2 ejection system was fully proven was rudely shattered in the early part of 1962, when two accidents caused a major rethink. In the first of these, on the night of 2 January, Captain Chuck Stratton made a remarkable high-altitude escape from a SAC U-2A after an autopilot failure over the swamps of Louisiana. In the second, also at night, Captain John Campbell was not so fortunate over Edwards AFB on 1 March when he attempted to eject from a U-2F during refuelling practice. He went down with the aircraft. In both cases, it was discovered that the seat didn't work as advertised.

Stratton's aircraft suddenly pitched up and rolled to the left. He was unable to regain control before the Mach limit was exceeded. Once again, the tail section came off, and possibly the wings too. The aircraft entered an inverted flat spin. Despite the disorientation, Stratton managed to initiate the ejection sequence by pulling the D-ring between his knees. When it was all over, he recalled being ejected from the cockpit, with his arms raised above his head. He then began spinning around wildly in the thin air, unable to control the nauseating motion despite extending his arms and legs in various positions. He began to pass out and so, like Jack Nole before him, Stratton decided to take the risk and deploy the parachute while still at high altitude. He too was lucky, and got a good chute with no injury. Then came the inevitable long descent, resulting in the emergency oxygen supply running out. Of necessity, he opened the faceplate, and was relieved to discover that he was now low enough to breathe oxygen from the outside air. It was a dark night, and Stratton landed in a tree. The canopy snagged above him in the branches so that he didn't fall into the swamp below. He spent the rest of the night in the tree, talking to the rescue parties through his emergency radio. At daybreak, they arrived in a boat and helped him aboard.

Later, the rescue party gathered up the wreckage, which was spread over several miles of swampland. When they showed Stratton the cockpit section, he couldn't believe his eyes. *The ejection seat was still in it!* He felt certain he had ejected. During the accident investigation, Stratton agreed to undergo hypnosis in order to try and work out what had really happened. While the investigation proceeded, Campbell's accident shed more light on the mysterious event. The body of this unfortunate pilot was found hanging half out of the cockpit. Campbell had evidently tried to bale out, because the canopy had been manually opened. The ejection seat was still in place, and yet the initiators had fired. It seemed that the seat had somehow failed on both occasions.

The Skunk Works ran tests, firing some seats containing dummies from a spare fuselage which had been salvaged from a previous U-2 accident. The seat was supposed to fire through the canopy, and performed this function well in the tests — until they tried firing it through a canopy that had been cold-soaked to replicate the conditions of prolonged high-altitude flight. Then it rose up the tracks, slammed into the plexiglass, and got no further! The hardened canopy was equal to the force of the seat's rocket thrust.

Now they could reconstruct Stratton's escape with some confidence. He had indeed tried to eject, but after the prolonged flight at high altitude, the seat had bounced back off the canopy. Then the seat belt had released automatically, as advertised. (When the U-2 seat was first installed, the belt was severed by a striker bar as the seat rode up the rails, but this was soon modified to provide the pilot with more restraint during the ejection sequence. Thereafter, the belt was designed to be released after a time delay, just long enough for pilot and seat to rise clear of the aircraft.) Stratton was now free of restraint, but still being bounced around inside the cockpit. Somehow, he managed to pull the canopy release latch, and once the canopy was off he was thrown free of the aircraft by the 'G' forces. In Campbell's case, the seat must also have failed to penetrate the canopy, but he had not enough time (or luck) to recover from the setback and make a manual escape. In that first successful ejection two years earlier, Raleigh Myers had been lucky — most of his flight was conducted at low altitude on a hot day, and the canopy had therefore not been subjected to the cold-soak phenomenon.

So the seat had to be modified. Canopy piercers were installed on the protective rail above the pilot's head — three pointed metal spikes which could help shatter the plexiglass. The rocket charge was also increased substantially, and this entailed strengthening the cockpit floor bulkhead. Lockheed also started work on a gas-powered system which would automatically fire the canopy away when the ejection seat sequence was initiated, and this was installed on the fleet during 1964. Now the U-2 had a half-decent escape system, but pilots were still aware of its serious limitations. They were advised, for

Air Force pilot Bob Birkett reaches for a celebratory drink after clocking up 1,000 hours of flight in the U-2, August 1976. Having just climbed out of a U-2R model, he is wearing the full-pressure suit which marked a great improvement in personal comfort over the earlier partial-pressure garment.

instance, that an ejection from a spinning or diving aircraft should not be attempted below 10,000 feet, and that low-altitude ejections might not be successful unless the aircraft was in a positive climb and doing more than 120 knots.

A number of pilots continued to discount the seat's utility. 'Yeah, we had an ejection seat,' said SAC pilot Dan Schmarr in an interview many years after he retired, 'and it worked fine on the ground, or flying straight and level. But when the tail came off and you actually needed the thing, it didn't work!' Schmarr and some others were under the impression that the seat's failure to operate properly was caused by it binding in the tracks, when subjected to extreme 'G' forces. The author has not been able to confirm this as even a contributory cause in the seat's failure to penetrate the canopy in the 1962 accidents. The modified seat certainly worked for 'Deacon' Hall over California and Leo Stewart over South Vietnam in 1966, and for Sam Swart over Louisiana the following year. It also worked at 42,000 feet over Arizona for Vic

Milam on 21 May 1968, although Milam was lucky not to collide with wreckage from his disintegrating aircraft on the way down. His accident was attributed to autopilot malfunction leading to loss of control and structural failure.

When the larger R-model was introduced in 1968, U-2 pilots finally had a full-capability ejection seat designed for use in all flight regimes. This included 'zero-zero' situations, and this seat has subsequently been used successfully in a runway ejection. This was a 1984 accident at Beale AFB caused by structural failure on take-off (described further in the next chapter). On at least two other occasions, pilots who might have ejected chose to stay with their aircraft after engine failures at take-off. With his U-2 less than 100 feet into the air, Jerry Hoyt managed to retain control and land straight ahead; luckily, he was on a 12,000-foot runway at the time. Another U-2 went quiet on Dick Davies during climb-out, but he quickly pushed the manual fuel button and got a relight, enabling him to turn downwind and land. Both these pilots weren't put off by the experience, since they now fly U-2s for NASA. The later ejection seat has also been used at altitude, and unlike the original version, the pilot is retained in the seat all the way down to 15,000 feet, below a drogue chute, before the aneroid initiates his separation, and his main chute deploys.

The R-model was also able to accommodate a pilot wearing the full-pressure suit. This garment literally enclosed him from head to toe in an inflatable rubber bag, and was a good deal more comfortable to wear than the partial-pressure suit. The very earliest U-2 pilots had to wear an MC-3 suit consisting only of capstans, long tubes

Frank Powers models the MC-3A partial-pressure suit and MA-2 helmet. Life-preserving capstans run the length of his arms and legs, and the laces used to pull the suit tight against the body are clearly visible. The uncomfortable helmet neck seal is apparent; behind the faceplate is the built-in microphone. Not shown is the fire-resistant nomex garment which was worn over the pressure suit, to protect the laces and capstans and prevent them getting caught up in the cockpit.

running from the shoulders to the wrist, and down the body and legs. When inflated, oxygen compressed to five pounds per square inch held the capstans against the body, so that the occupant's body fluids would not expand if he was exposed to the low air pressures of the upper atmosphere. So the suits had to be really tight-fitting to be worth wearing at all. Despite being individually tailored to each pilot, partial-pressure suits were exceedingly uncomfortable, especially on a long flight. They stretched doubly tight across joints such as elbows and knees, creating pressure points which could become intensely irritating as the hours wore on.

The MC-3 was later modified as the MC-3A to include a bladder on the torso and upper legs (pressurized to one pound per square inch), and this allowed the suit to be somewhat less tightly laced than was necessary when the capstans were providing the only protection. The bladders also improved the air conditioning properties of the suit, since ventilation air circulated through them to rather greater effect than through the capstans. Despite this, most pilots perspired freely while wearing the pressure garment, and the long underwear that they wore beneath the suit was frequently soaking wet by the end of the flight. Some reckoned to lose a full pound of body weight this way, for each hour that they were confined within the suit.

To add to the general level of discomfort, there was a tight cork seal which fitted around the neck, over which was placed an MA-2 pressure helmet. Frank Powers described the neck seal as feeling like one was wearing a badly shrunk collar around which a tie was fastened too tight. Pilots often returned from long U-2 missions with the skin around their neck chafed red and raw. There were also special gloves with a pressure bladder stitched in, these being attached to the suit, and put on after the pilot had first donned a pair of white silk glove liners. Completing the outfit, there was an outer garment containing water flotation collar, parachute harness and seat pack, and protective coveralls on top of that. Heavy boots were worn to offset the lack of pressure protection for the feet.[2]

The modern full-pressure suit, by contrast, is relatively comfortable. The 'fishbowl' helmet is attached to it by means of a wide metal ring, thereby eliminating the maddening hermetic cork ring seal around the neck. (Eventually, the partial-pressure suit was modified to accept this ring, and those pilots still flying early-model U-2s were able to use the new and much more comfortable helmet.) The ingeniously designed gloves also attach to the suit by a metal ring-fastener.

The first suit worn by U-2R pilots was the S1010A, in which the regulator providing breathing oxygen and communications gear, parachute harness and flotation device, were all mounted integrally. It was replaced by the S1010B, in which the communications and oxygen regulator were moved to the helmet, while the flotation gear and parachute formed part of a separate harness worn over the top of the suit. In the latest suits, designated S1031, an exposure garment is built in, to provide an inch-thick layer of thermal air if the pilot should end

Major John Swanson emerges from a TR-1 at the end of an operational reconnaissance mission from Alconbury, UK. He is wearing the latest version of full-pressure suit, designated S1031, and with a number of improvements. Even so, it feels good to remove it after nine hours of confinement!

up in the water. These suits weigh about forty pounds, cost around $10,000 to make, and each pilot has two of them!

Even though flying the U-2R in the full-pressure suit is considerably less taxing than before, pilots are still given two days off after a long operational mission before they are allowed to fly again. To withstand the rigours of this type of flying, they have to be physically fit in the first place, of course. Even so, each is subjected to a medical examination before each flight, when he reports to PSD (the Physiological Support Division). The small details must not be overlooked; a tooth which has not been properly filled may contain a small amount of air. At altitude, this could expand and cause a severe toothache. There may also be post-flight problems. After so many hours of breathing 100 per cent oxygen, the inner ear tissue can absorb too much, causing headaches and earaches. Some pilots have had to be scrubbed from the programme because of such effects, which can build up over the long term.

Despite all the medical checks and precautions, accidents have been caused by physiological problems. On 28 July 1966, Captain Robert Hickman took off from Barksdale AFB for a routine monitoring flight around Cuba. He failed to make a scheduled turn over Florida. The aircraft continued south-east on the same course, while on the ground, frantic attempts were made to make radio contact. But nothing more was heard from Hickman, as the U-2 continued serenely along a constant track on autopilot. It eventually crashed on a mountainside in southern Bolivia more than 3,500 miles away when the fuel ran out! Hickman is believed to have had an embolism.

An even more remarkable incident occurred on the last day of January 1981. Captain Edward Beaumont was in the early stages of check-out at Beale AFB, having made his first trip in the U-2CT only nine days earlier. This day, he was flying one of the last single-seat U-2C models remaining in Air Force service (they were finally retired a few months later). On a bright winter's day, he performed a number of touch-and-gos, and then climbed out for some work at medium altitude. After this, he reported descending through 14,000 feet. Some time later, his mobile control officer on the ground at Beale was surprised to hear Beaumont key the mike, but make no transmission. Instead, all that could be heard was a heavy breathing sound as the U-2 pilot's transmitter remained open, but silent. The tower was alerted, and a T-37 trainer that was also flying locally was instructed to rendezvous with the errant U-2 and attract Beaumont's attention.

As the two pilots in the T-37 drew alongside, they could hardly believe their eyes. The U-2 pilot appeared to be slumped at the controls, with the aircraft in a gentle, turning descent. Beaumont had had a catatonic seizure, and was completely unconscious. With the accompanying pilots in the T-38 powerless to intervene, the U-2 floated slowly towards the Sierra foothills north of Oroville. As it neared the sloping ground, some high-voltage power transmission lines barred the way. The T-37 pilots braced themselves for a searing explosion as the black airframe flew into the 230,000-kilovolt wires.

It never came. Incredibly, the U-2 clipped the bottom two wires with a wingtip, but failed to incinerate. In fact, the contact with the power lines had the effect of rolling the aircraft into the correct attitude for a forced landing in an adjacent cow pasture. Had its wingtip not been flipped up in this way, the aircraft would have cartwheeled as it impacted the gently sloping terrain with one wing low. As the astonished T-37 pilots orbited overhead, the U-2 flopped into the muddy field and ground to a halt with the engine running. Fuel began spilling from a ruptured tank, but it ran downhill and therefore failed to ignite.

The sudden arrival on terra firma revived the stricken pilot. Although confused, he managed to shut the engine down. But the drama wasn't yet over. As the still-groggy Beaumont began to extricate himself from the aircraft, his foot slipped and caught in the D-ring of the ejection seat, which he had failed to make safe. It fired through the canopy, flinging the pilot upwards with it. Beaumont's body described a somersault, but he landed on his feet to one side of the aircraft, while the seat thudded into the ground nearby. His only injury was a chipped tooth! When the preliminary accident report was circulated, SAC generals and Lockheed managers alike thought that someone had made up the whole story as a joke. Not surprisingly, Beaumont was scrubbed from the U-2 programme on medical grounds. The U-2C which ended its flying days in a cow pasture, is now on display at Beale.

Splendid pilot's-eye view through the driftsight includes a large airfield, just below centre, and coastal areas to left and top. Operated by a hand control on the lower right panel, the tracking periscope provided two magnifications (x 1 and x 4). A glass bubble containing the scanning prism protruded slightly below the fuselage, to provide an all-round view. The sextant also used a portion of the driftsight optics, and was an alternative device for navigation.

Given the unforgiving nature of the aircraft and the unique flight regime at high-altitude, it is not surprising that problems with the autopilot have led to a number of serious incidents over the years. When he was shot down over the USSR, Frank Powers had been hand-flying his aircraft for the past thirty minutes after the autopilot had malfunctioned and the nose had pitched up. Chuck Stratton had experienced a similar pitch-up which had caught him unawares and led directly to loss of control. In other circumstances, the autopilot had a tendency to let the airspeed increase. If the aircraft was in the coffin corner regime, this could quickly become disastrous if allowed to go uncorrected, since the nose would soon tuck under, and the tailplane fail.

This was the suspected cause of the two accidents at the 4080th Wing on successive days in July 1958, which killed Chris Walker (one of the first group of British pilots to train) and Al Chapin. The wreckage pattern in both cases indicated a high-altitude break-up following loss of control, but neither pilot was the inattentive sort, so the autopilot came under suspicion. Hypoxia may also have been a factor, since it was subsequently discovered that the liquid oxygen being used at Laughlin at the time contained a minute amount of water, which might have frozen certain components within the pilots' breathing system. In Walker's case, however, he had evidently been conscious at some stage of his emergency. In a repeat of the Sieker accident the year before, Walker's body was found a short distance from the aircraft, indicating that he had baled out at the last minute, with too little time for his parachute to deploy.

The original Lear autopilot in the U-2 was replaced in the sixties by a more modern design from the same manufacturer. But even the best equipment cannot defeat Mother Nature. Since the U-2 cruises on a constant Mach schedule, and Mach number depends on temperature, even the smallest variation in outside air temperature can be significant. Actually, pilots have encountered changes of up to fifteen degrees in less than a minute. Then, says one, 'the autopilot starts lowering the nose to try and catch up with it, and suddenly you're hanging in the straps looking straight down at the ground. Then it starts going back the other way, constantly trying to chase Mach.' The autopilot's unerring adherence to the Mach hold has its advantages in other respects, however. With so little wind velocity at cruising altitude, it enables the U-2 pilot to navigate by the simple and time-honoured method of dead-reckoning.

Autopilot misbehaviour caused by a severe temperature gradient may have sabotaged one of the last planned overflights of mainland China. On 5 January 1969 Lieutenant Colonel Billy Chang of the Black Cat squadron was climbing out of Korea at the start of a photo-reconnaissance penetration when his aircraft suddenly pitched down at 55,000 feet. Chang lost control and ejected over the Yellow Sea, but did not survive. A Korean fisherman found his gruesome remains in the water some time later; the pressure suit was in tatters, and sharks had eaten most of the body.

Many pilots have remarked on the difficulties of starting down from high altitude cruise. 'The enormous lift provided by those wings is definitely a problem when one wants to go down instead of up,' reported Captain Robert Gaskin in a 1977 article for *Air Force Magazine*. 'The U-2 has to be coaxed, argued with, and finally forced to stop climbing by playing all the dirty tricks in the book to increase drag and destroy lift. The descent from altitude is slow and laborious because, initially, the engine's RPM can be only slightly reduced or flame-out will occur. Also, the indicated airspeed cannot be allowed to climb too high lest the tail separate from the fuselage.

'You move your left hand forward and put the gear handle down as the first step in the descent check list. There is no need to worry about limiting airspeed with gear down. The aircraft will come apart before the gear is overstressed. Then you manually open the bleed valves, just in case they don't open automatically at the same time, as the throttle is pulled back to a precomputed power setting. Too rapid a power reduction with the bleed valves closed will almost guarantee a [compressor surge and] flame-out. Drag devices come out of the fuselage and wings. Shuddering, shaking and groaning, the nose finally gives up and grudgingly drops to maintain descent Mach. Now at lower altitude, the power is further reduced, to increase the rate of descent. The controls now become heavy and sluggish, causing you to wrestle with the aircraft. You must lead your roll-out from turns by as much as thirty degrees.

'Before landing, it is best to get comfortable. The faceplate of the helmet comes up and you shut the oxygen off because you can now breathe ambient air. Your first breath exposes you to all the smells you have been missing inside the goldfish bowl helmet. The heater has stirred all the lingering smells — old leather, paint, electrical motor scents — that identify an aircraft to its pilot.'

Gaskin was describing a routine descent. Sometimes, however, the descents were far from routine. On 15 March 1960, Captain Roger Cooper of the 4080th Wing was on the return leg of a U-2 sampling flight from Minot AFB to the Great Bear Lake and back. He had risen to almost 70,000 feet above the desolate landscape. As he passed through fifty-six degrees north, almost above the middle of Saskatchewan, his electrical system began to malfunction. He began losing engine instruments, and so began a slow descent while reducing power to avoid an overtemp. Soon thereafter, the engine flamed out, and thick white smoke filled the cockpit, almost obscuring the instrument panel. Cooper managed to radio a warning on HF to his mobile in the command post at Minot AFB, almost 500 miles to the south-east, before he also lost battery power.

He was nearly thirteen miles above the frozen wastes. The odds were stacked against a successful ejection at such a height, and yet if he stayed with the aircraft, there was the imminent possibility of an electrical fire taking serious hold. He opened the ram air position to clear the smoke, hoping that the lack of oxygen at this height would cause the fire to go out. Luckily, it did. He continued to descend, but turned 180 degrees to his reciprocal heading since he knew the weather was better to the north. There were two possible emergency landing fields within reach to the south, but Cooper knew that there was a low overcast with blowing snow at both. Instead, he decided to try for a deadstick landing on one of the many frozen lakes beneath his flightpath.

As yet, however, he couldn't see one. There was a solid overcast beneath him, but since he still had airspeed, rate of climb, and turn and bank instruments, he felt confident that the cloud could be successfully penetrated. However, as he glided slowly towards the tops of the clouds at 25,000 feet, the turn and bank indicator stopped working! In the murk, he would have to resort to using the magnetic compass as a bank indicator. Fearing disorientation, he prepared to eject if the aircraft got out of control, but the clouds were surprisingly smooth, and he emerged into the clear at 15,000 feet with wings still level.

Now he had difficulty in seeing the ground, because the canopy and most of the windscreen were covered in frost. He tried to jettison the canopy, but it merely flipped into the open position. Now he couldn't read his maps because of the wind blast, but he did notice a frozen lake below with a sawmill on one shore. He prepared to land, by installing the ejection seat pin, tightening his seat belt and shoulder harness, and manually extending the gear. At 7,000 feet, he noticed a radio station off to his right, and decided to fly over it to attract attention. This meant selecting another lake, but there were plenty to choose from, and he eventually made an almost textbook approach, and a two-point landing in the snow. The U-2 rolled 300 feet to a gentle stop, and Cooper quickly evacuated. But there was no more sign of fire, or of any other external damage. The pilot was just beginning to unpack his emergency survival equipment when a rescue helicopter appeared, and carried him off to warm safety. The aircraft was later flown off the frozen lake, and Cooper was cited for his 'sense of responsibility and calm, professional airmanship'.

The one good thing about engine failure in the U-2 was the distance the aircraft could glide — nearly 250 nautical miles from 70,000 feet if there was no wind. This would take a full seventy minutes to accomplish. The SAC U-2 squadron set up a roll of honour in the squadron mess with the names of all those U-2 pilots who had successfully accomplished a deadstick landing. They became members of the 'Silent Birdman' club. Among those joining Roger Cooper in its ranks was Dick Callahan, who deadsticked into Hamilton, Bermuda, in 1959; Chuck Maultsby, after his unintentional excursion over Soviet territory during the Cuba Missile Crisis; Ward Graham, who made it safely into the small airfield at Flagstaff, Arizona, with only partial flaps and no drag chute in September 1965 after a fifty-minute glide; and Richard Woodhull, who made the quiet approach into Kingsley Field, Oregon, in February 1967.

Lockheed test pilot Bill Park also made a U-2 deadstick landing around this time — into the company's congested and built-up Burbank air-

Two members of the Silent Birdman club renew their acqaintance with the U-2. Pat Halloran (left) and Roger Cooper (right) both made successful deadstick landings; Cooper's epic recovery onto a frozen lake in Canada is related in this chapter. Halloran once flew a record thirteen-and-a-half hour refuelled mission in a U-2F. The two Dragon Lady veterans were participating in the 4080th SRW reunion in 1987.

field. Someone on the ground had apparently neglected to open the manual fuel shut-off valve, and Park hadn't been airborne long before he exhausted the fuel remaining in the sump tanks. Partly as a result of this incident, Kelly Johnson moved all further company U-2 test flights to the wide, open spaces of Palmdale.

At least most of these silent pilots lost their engine over land. But Captain Robert 'Deacon' Hall had the misfortune to lose his engine over the wide expanses of the Pacific Ocean. With two other U-2s flying some way ahead of him, Hall was participating in the first Crowflight deployment to Australia in October 1960.

He was flying at optimum ferry altitude — 60,000 feet — on the first leg between Laughlin and Hickam AFB, Hawaii, when he heard a dull thud and the hydraulic gauge began to wind down. Since the drive motor which boosted fuel to the engine above 40,000 feet worked off the hydraulic system, the inevitable soon happened, and Hall had a flame-out on his hands. He was almost exactly halfway between California and Hawaii. Hall declared an emergency on HF, which was picked up by a passing Pan Am airliner and relayed to Hickam. He drifted down to 35,000 feet in order to get a relight, but had already worked out that the engine would quit on him again if he attempted to rise above 40,000 feet. Hall did some rapid calculations, and soon concluded that he did not have enough fuel to reach Hickam in a medium altitude cruise. This was a long ferry leg, and there was only some half an hour of fuel reserves for a journey all the way at 60,000 feet. Now confined to a maximum 39,000 feet, Hall's U-2 was heading smack against a strong eastbound jetstream.

More calculations. Hall reckoned that he might be able to glide in from 200 miles out of Hickam, if the fuel supply lasted that long. He later described a comic exchange with the Pan Am pilot who acted as radio relay: 'I asked the guy to tell Hickam I had lost my engine, and was declaring an emergency. He asked me how many engines I had. Just one, I told him. There was a pause. "You really have got an emergency!" came the reply. Later, after I had relit, he asked me for all the usual information — souls on board, ETA, fuel remaining, and so on. I told him between two and a half and three hours fuel onboard, with ETA Hickam three to three and a half hours. Another long pause. "You're not going to make it!" he exclaimed.'

But Hall did indeed make it. The fuel finally ran out with 150 miles to go, but the aircraft arrived overhead Hickam in perfect shape, at 3,000 feet and 100 knots. During training flights, U-2 pilots practised frequently for such an engine-out approach, and Hall was now at what they termed the 'high key point'. From here, the pilot was supposed to turn into a downwind leg with flaps at twenty-five degrees and landing gear down, reaching 1,500 feet at the 'low key point' and 800 feet as he turned onto finals. Then he could also extend the speed brakes. But Hall's aircraft had no hydraulics, and although the gear had manually extended, he had no flaps or speedbrakes. Also, as he pointed out, 'we never practised with no fuel on board!' By the time he reached low key, he was going much too fast, and with not enough altitude left to bleed off speed in a turn, he had no option but to dive for the end of the runway and hope that the brake chute did its stuff. He crossed the threshold at 120 knots instead of eighty, but the chute popped, and he came to an uneventful halt. As he climbed out of the cockpit, he turned to the other two ferry pilots, who had landed more than an hour earlier, and called for the customary welcoming beer. 'We're sorry, Deke,' they replied. 'We figured you've been gone so long, that you weren't gonna make it. We've drunk it!'

Landing a flamed-out U-2 could be only marginally more difficult than landing a fully-functioning aircraft after a particularly long, tiring mission, or if one was a newcomer to the Dragon Lady. The first touchdown by a novice U-2 pilot was always a terrible moment of truth, the tension not eased by the inevitable gathering of onlookers, who awaited the entertainment with a macabre sense of anticipation. This initial qualification flight (or IQ1) began with a take-off and climb to 20,000 feet, followed by a holding pattern, a simulated penetration on instruments, and a low approach in VFR conditions. The U-2 would then be joined by a Cessna U-3 chase plane in which the trainee's instructor pilot (IP) rode as passenger. The chase plane would fly on the U-2's wing throughout the remainder of the flight, which consisted of four touch-and-go landings and a full stop. The IP would monitor the U-2's airspeed, altitude and headings, call out the necessary adjustments, and generally lend moral support to the lonely man in the hot seat. On the ground, meanwhile, another qualified pilot would be acting as mobile control officer. Driving a high-speed staff car, he would speed along the runway beside the landing aircraft, calling out the height above the ground in feet to the unsighted pilot.

As the years of U-2 operations progressed, the tension attending an IQ1 mission probably increased, since the trainee was mindful of the long history of serious landing accidents on the type. In the early 1970s, the SAC wing began filming each landing at home base for postflight analysis, and the video camera recorded some spectacular arrivals, which were duly rescreened as a warning to prospective pilots. An earlier such film was shot by accident; maintenance men were running a U-2 tracker camera on a test bench outside the hangar at Davis-Monthan at the very moment on 18 September 1964 when Major Robert Primrose stalled on finals and slammed into the ground. The tracker lens was pointing towards the runway, and so recorded the entire affair, which was apparently caused by a gusting crosswind. The death of 'Pinky' Primrose shocked the wing, for he was an experienced pilot, but the book said that you must not attempt to land a U-2 in crosswinds of more than fifteen knots, unless it was an emergency.

The aircraft's vulnerability to crosswinds has been a major operational problem over the years. To defeat the wind, landings were sometimes made on appropriately-aligned taxiways. More often, however, a diversion would be required to the nearest field where the wind was blowing down the runway. During a Crowflight deployment to Andersen AFB, Guam, in the early sixties, three U-2s were stranded by an approaching hurricane. By the time the alarm was sounded, the wind was already too high for U-2 operations. The detachment commander closed his hangar doors and battened down the hatches, while he watched all the other aircraft on base take off for safe refuge elsewhere. Unfortunately, the storm hit the base full square, and the hangar roof collapsed onto his three aircraft. The bent and battered remains were airlifted back to the US, but only one was worth repairing to fly again!

Some time after this, Lockheed pilots demonstrated landings in gusts of thirty-two knots and ninety degrees of crosswind component. The trick was to keep the downwind wing low and generate a significant yawing moment. This would keep the aircraft going straight down the runway, rather than weathercocking into wind. The demonstration was spectacular, but the Air Force decided not to increase the crosswind limits as standard procedure. It already had its work cut out ensuring that some U-2 pilots correctly performed the vital tasks preparatory to a landing in normal conditions!

The first of these preparations was to accurately compute the necessary speed over the threshold, which depended on the aircraft's weight and the amount of flap to be used. This 'T-speed' was a minimum twelve per cent above stall speed, and the pilot added thirty knots to arrive at his pattern entry speed, twenty knots for the correct downwind speed, and only ten knots for the start of finals. Ideally, this last ten knots would bleed off down the final approach so that the aircraft arrived at the threshold at T-speed, at an altitude of ten feet.

The other was to ensure that the fuel was precisely balanced in the long wings. The standard way to do this was to pause during the descent from altitude at 20,000 feet, by slowing and extending the retractable stall strips. If one wing dropped early as the speed was cut back, then fuel had to be moved across to the other by transfer pump. A final check for lateral trim was supposed to be accomplished on the downwind leg. A simple mechanical device helped the process: a pointer and plate was attached to the control yoke. Pilots noted the extent to which the pointer was off-centre during their after take-off check. It would be in the same position for landing if the fuel was correctly balanced.

A number of U-2 landing accidents were caused by incorrect fuel balance, while others were attributed to wrong T-speeds, or failure to control the aircraft during the rollout. It was a dismal catalogue. No fewer than three Chinese pilots lost their lives in landing accidents; Major Chih who perished in a fireball at the side of the runway at Taoyuan in 1962; Major Charlie Wu, whose attempt to put down a flamed-out U-2 on a small Taiwan airstrip in 1966 ended in disaster when he crashed into houses beyond the edge of the runway; and Major Denny Huang, again at Taoyuan, this time in a U-2R in 1970. Among the Americans who came to grief was Paul Haughland, the young lieutenenat killed at Laughlin in 1958 on his IQ1; Leslie White, also on his first flight, who was lucky to survive when his aircraft crashed and burned at D-M in 1966;

Captain Tony Bevacqua tucks into the regulation preflight meal of high-protein steak and eggs, with toast and coffee to follow. The arrangements for waste elimination in the pressure suit were rudimentary, so a careful diet was essential prior to the long-duration missions. Bevacqua is thought to be the youngest pilot to check out in the U-2 — he was only twenty-four when he first flew it at The Ranch in 1957.

John Cunney, killed at D-M in late 1971; and Usto Schultz, who in 1972 became the latest trainee to wipe out a landing gear, also at D-M.

Schultz was in good company. Even the renowned Ray Goudey had once done this, out at The Ranch. Tony Bevacqua, one of the early 4080th pilots who trained at the secret site, recalls doing the same thing. 'It was my first landing on the runway there, and I was flying an aircraft with a history of wing drop. I approached the threshold OK, but then the damned wing dropped, and I applied power to go around. But the mobile called to shut it down, and I hit and slewed off the tarmac at an angle. The gear sheared off when I hit a ditch. Of course, that wouldn't have happened if I had landed on the lakebed. It wasn't so much of a problem, going off to the side a bit there . . .'

On occasions such as this, pilots felt that the advice coming over the headset from mobile or the IP was a hindrance, rather than a help. At least one pilot reckoned that the airspeeds being read off the dials by the IP in the Cessna chase plane as he guided the trainee U-2 pilot into land 'were all three or four knots different from those we were reading in the U-2 cockpit'.

Having mastered the basics of landing and take-off, the newcomer would don his pressure suit for the first time and head up to high altitude. Here he was truly on his own, exploring 'coffin corner' and all the rest. No matter how good the ground school, dealing with the Dragon Lady's fickle ways was largely a matter of instinct and feel, which is why the selection criteria for those wishing to fly U-2s were (and are) so stringent. Sometimes new pilots learned their trade the hard way. Ronnie Rinehart was on an early high flight over Phoenix in 1972 when he began encountering a buffet. Presuming that the aircraft was going faster than it should, he slowed down. The buffet continued, so Rinehart eased off on the throttle by the recommended two knots. This failed to correct the condition, so he repeated the process.

The first two-cockpit U-2CT inflight over southern California. It may have looked strange, but 'the two-headed goat' certainly helped cut down on training accidents, as the instructor pilot monitored the nervous trainee from the rear seat.

The aircraft suddenly stalled, snap-rolled, and entered a high-speed spin. Somehow, the aircraft didn't come apart, and Rinehart managed to regain control. When he got down, they could tell he had survived some extraordinary manoeuvres by examining the film from the aircraft's tracker camera — every few frames, the view of the ground was replaced by one of the sky! Rinehart joined at least two other pilots (Marty Knutson and Barry Baker) who had spun and rolled in the aircraft, and yet managed to bring it home.

After the high rate of incidents forced a review of its training procedures in 1972, the Air Force accepted a Lockheed proposal to build a two-seat dual-control version for instruction. The Skunk Works took a damaged U-2C airframe and spliced in parts from other U-2s which had been written off. The result was a tandem-seater with a raised second cockpit for the instructor in place of the Q-bay. The aircraft, serial 66953, was first flown by Bill Park at Palmdale on 13 February 1973. In a smart, all-white paint scheme, it was delivered a few months later to the 100th SRW at Davis-Monthan, where its strange profile quickly earned it the nickname of 'two-headed goat'.

The advent of the U-2CT — nearly eighteen years after the type's first flight — meant that the training programme could finally dispense with the U-3 chase planes. Now the IP went along with his student for the ride, able to control the aircraft from the rear cockpit, which was fully equipped. There were still some anxious moments, however, especially on that nerve-wracking IQ1 mission. Captain Glenn Perry, who made that flight on 18 October 1974, later described it in detail for *Air Force Magazine:*

'We proceed to the end of the runway to do a high-speed taxi exercise. I am congratulating myself for exceptional performance in handling my first tail-dragger when the IP barks "Fly the wings!" and I realize that the left pogo is on the ground. At airspeeds of about ten knots the wings are flyable even though the aircraft is not. Proper technique demands that the wings be held absolutely level throughout the ground roll. That is no easy task when all other controls are being worked feverishly in order to track down the runway!

'I am trying to avoid a ground loop when we finally come to a stop. All this effort and I'm not even airborne yet! The IP takes the U-2 and demonstrates a maximum performance take-off. In only a few hundred feet we leap off the ground and climb nearly straight up. I keep waiting for the back side of a loop, but it never comes . . .

'The first touch and go — a demonstration by the IP — shows graphically what kind of learning situation I am about to face. The IP sets a fine example as he reaches the end of the runway at between five and ten feet altitude and exactly on computed threshold speed. The throttle comes to idle and he continues a steady

Ramp scene at Beale AFB in 1979 shows both U-2CT models basking in the sunshine. Although SAC retired its last single-place U-2C in 1981, these two trainers survived another six years.

descent as the Mobile vehicle dashes madly down the runway beside us, with the officer calling off our height. I suddenly realize how tough it's going to be to see out of this thing. The glare shield and instrument panel obscure everything except side vision and a tiny glimpse of the far end of the airfield. "Eight feet . . . six feet . . . four . . . two . . . one foot . . . holding one foot . . . tail's coming down . . . one foot . . . hold it off!" The U-2 shakes violently, stalls completely, and slams sharply onto the runway. I am informed over the intercom that that was a perfect textbook landing.

'The next landing is mine. My pattern is rough and crude, and I am impressed with the effort needed to fly the U-2. It is like a headstrong child: it demands constant attention to make it obey. Somehow, I manage to manoeuvre to the landing threshold at approximately the correct airspeed and altitude. I put the throttle to idle and listen for Mobile's calls. "Eight feet . . . six . . . four . . . two . . . one . . . hold it off." Sloppy, but I luck out. It is nowhere near the desired full stall, but at least it is close to two points, and the U-2 does not come back off the runway.

'But once I'm on the runway, my problems have only just begun. All my briefing and determination to the contrary, I allow the stick to go forward in an obvious gesture of relief at being on the ground in one piece. The tail breaks loose and starts to do its own thing from one side to the other. I desperately jab at the rudders and finally reduce the oscillations to something more manageable. The unavoidable impression is that any minute now this thing is going to swap ends.

' "Raise your right wing." I have completely forgotten about the wings and am rolling along on the right pogo. "Raise your left wing." Wings-level eludes me completely as I bounce from one pogo to the other. "Flaps are up. Reset trim and go when ready." One glance inside the cockpit to check the trim, and I'm headed off the runway again. I slam the rudder, add power, and breathe easier as we end that torment by becoming airborne.

'The remainder of my landings are a little better, but my control on the runway does not improve at all. It seems all I have learned is how to recover from one disastrous situation after another. Finally we make a full stop, and I realize how exhausted I am. Taxying back to the parking ramp I look inside the cockpit for an instant and the aircraft wanders off the centre line. Even at ten knots, the beast wants to destroy me!'

Introduction of the U-2CT brought a marked decline in training accidents, and so construction of another one was authorized. Like the first two-headed goat, serial 66692 was rebuilt with the second cockpit after an accident. More than a generation of U-2 pilots were trained in the two aircraft before they were finally retired in late 1987. By then, the Air Force had the use of three purpose-built TR-1B dual trainers (although one was officially designated U-2R(T)!), with a fourth on order. When it ordered the U-2 back into production in 1978 as the TR-1, this time the Air Force made sure that a trainer version was available from the outset.

Flying the larger and much-improved U-2R and TR-1 versions was somewhat easier, of course, but pilots were still painstakingly trained to cope with various emergencies. Flame-outs and electrical and autopilot malfunctions were much rarer now, but could still occur. Major Larry Faber of the 95th RS had to deadstick a TR-1 into Ramstein AFB in Germany in February 1985 after the engine vibrated and flamed out during an operational mission. Three months later, he was in trouble again over Germany with a complete electrical failure, but safely brought the aircraft down to the same base.

A remarkable save of a U-2R was recorded by Captain Jonathan George of the 9th SRW in 1986. He was flying his first operational mission out of one of the wing's overseas detachments when the autopilot disconnected and the pitch trim ran full nose down. The aircraft pitched over and exceeded allowable Mach, but George managed to grab the yoke and somehow pull the U-2 out of a steep dive before the airframe failed. But try as he might, he couldn't reset the trim. The fledgling pilot lowered the gear and started a slow descent to base, during which he was forced to 'bear-hug' the yoke, exerting up to fifty pounds back pressure to prevent the aircraft from pitching over. After an hour of this, he arrived overhead the base, but due to the tremendous fatigue, muscle cramps, and other physiological problems caused by such heavy exertion while completely enclosed in his full pressure suit, George was unable to properly configure the aircraft for landing without losing control. He therefore executed a spiralling approach and flew a perfect no-flap approach to a full-stop landing. They had to carry him from the aircraft and send him to hospital, where he was treated for severe muscle strain and exhaustion.

Flying a sampling mission out of Eielson AFB, Alaska, on 9 October 1962, Lt Col Forrest Wilson displayed similar determination to get a crippled aircraft back in one piece. The generator in his U-2A failed almost 300 miles from the nearest land. Most of the instruments failed, together with the autopilot and communications. Wilson was obliged to let down immediately, lest he inadvertently exceeded the engine limits, but he could not have gained a relight had he flamed out, since he had no electrical power. Having set up a descent, Wilson was faced with the prospect of navigating from an uncertain position to a suitable landing field, with only a stand-by magnetic compass and altimeter to guide him. In the Arctic darkness, he had to read these instruments with a hand-held flash-light, peering through the frost that was building

Back-up pilot makes sure that his colleague is properly installed in the U-2 prior to flight. Scratch marks made by the pilot's bonedome on the sunshade confirm that there was little, if any, room to spare once the canopy was closed. The pilot's green card containing pre-plotted oxygen and fuel curves is apparent; throughout the flight, he will check actual consumption against this chart. Like the flight maps prepared for him by the unit navigator, the chart is mounted on cardboard for ease of handling.

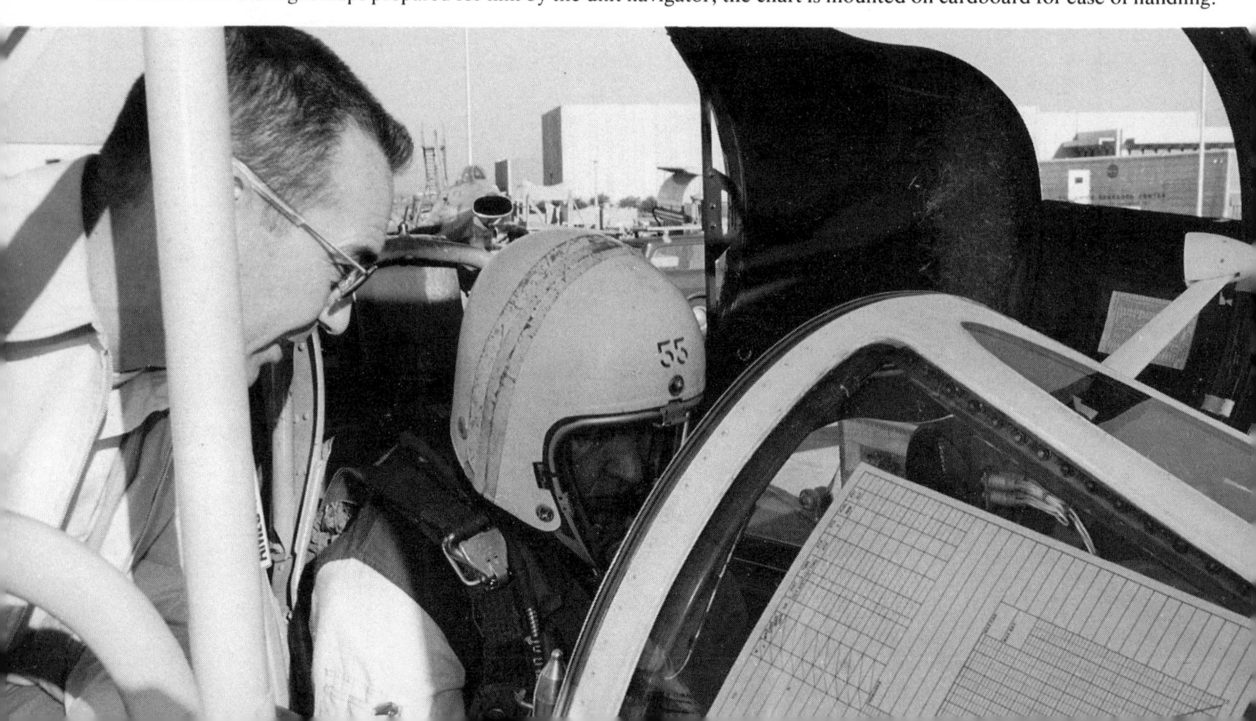

up on the faceplate of his pressure helmet after the cockpit heating failed.

Wilson set course for Kodiak airfield, his body going numb from the extreme cold which began penetrating his pressure suit. Fatigue developed as he fought to maintain pressure on the control yoke — he had no power to correct the out-of-trim condition which had developed. Wilson arrived overhead Kodiak, having reached an altitude at which he could remove his faceplate, but the runway lights weren't on and he could not attract anyone's attention down there. Rather than trying to land on an unlit field, he flew more than 200 miles further to reach Elmendorf AFB, where he made a safe landing three and a half hours after his difficulties began. Wilson won the Kolligian Trophy for his epic flight.

Emergencies like this apart, what is it like to fly twelve miles or more above the earth, completely alone, for hours on end? It can, of course, be terribly boring. On a nine-hour operational mission, mostly flown on autopilot in a long, lazy racetrack pattern, there may be only a couple of hours work for the pilot. Today, most of the sensors operate automatically, so once the aircraft has been established at altitude, the pilot's only official duties may be the routine monitoring of oxygen, fuel and other flight parameters. Many of them take reading material up with them. Novels and short stories mostly, although the dedicated will take study material, maybe connected with a college degree.

There is the radio of course, with unrivalled reception of a multitude of stations, thanks to the high altitude at which they fly. Apart from that, there are few diversions available within the cockpit, other than to raid the larder. The menu and dining arrangements are hardly of five-star quality, however. The hungry U-2 pilot squeezes his meal out of a tube and up a straw inserted through a pressure valve at the side of the helmet. For an entree, he might choose beef and gravy, with butterscotch pudding for dessert. The 'wine list' includes orange, lime, apple and tomato juices. The tubes of food can first be warmed up by a heater in the cockpit.[3]

It is the view outside the cockpit which leaves a marked impression on nearly every U-2 pilot. Shaped in the military mould, and also needing to be heavily self-reliant in the particular flying job that they do, many of them are relatively gruff and taciturn individuals. They are not given to flights of fancy, or poetic visions. Even so, few manage to avoid being affected by 'the breakaway phenomenon', a strange sort of reverie experienced by those who fly for prolonged periods at extreme altitudes. Some describe it as a sense of detachment, a shaking-loose of all earthly concerns and responsibilities, but others are affected in different ways.

'I get a personal feeling of serenity,' explains Major Thom Evans, who has been flying U-2s since 1975 and has now amassed more hours on the type than any other current Air Force pilot, 'and yet at the same time a feeling of power. You realize that you're up there by yourself, entirely alone, doing something that not many others can do.'

The view, of course, is terrific, and when he tires of gazing at the earth below, the Dragon Lady pilot need only look at the sky all around him for further amazement. 'The sky is a very brilliant blue at the horizon, and gets darker and darker as you look up until directly overhead it's almost black, even at mid-day,' notes Evans.

Major Thom Evans, high-time U-2 pilot with the US Air Force, says he gets a unique feeling of serenity from flying long periods at high altitude.

The view from the cockpit is nearly always stunning, especially on a clear day. This is the San Francisco area from 70,000 feet, with the city's airport visible near the centre, protruding into the bay. The Pacific shoreline is at the bottom. Away at the horizon, the Sierra Nevada mountains are covered in snow, and the curvature of the earth can just be made out.

According to Doyle Krumrey, a former Air Force pilot who now flies for NASA, U-2 pilots 'get to see things that a guy who has never flown the airplane can never fully appreciate or understand. I remember back in south-east Asia when we were sitting out over the Gulf of Tonkin, or running up the Chinese coast on those SIGINT sorties. We would launch after dark and come home early in the morning . . . very boring flying, with not much activity going on. But you could sit over the Gulf and look down south, and see stars in a way you had never seen before. They would twinkle, but they would also turn colour — greens, blues, reds. We were looking at them through the diffraction of the earth's atmosphere; it was wonderfully pretty to me.'

Thom Evans likes the sunsets best: 'They are unique. You see what we call a "terminator" every time. That's the line between light and dark. It starts out on the eastern horizon, which turns a strange and fuzzy grey. This brilliant line forms at the very edge of the horizon like a miniature compressed rainbow. It gradually works its way up towards the vertical and over towards the west. Behind the terminator is darkness and the stars. As it approaches the western horizon, it becomes more brilliant — oranges, reds and blues. Then all of a sudden it is night.'

Marty Knutson, who clocked up 3,800 hours in the U-2 over nearly thirty years, agrees that sunrise and sunset provide the most beautiful scenes of all, and also refers to the almost metaphysical experience which high-altitude flying can provide. 'Views like that,' he says, 'give you an entirely different perspective on life.'

1 The Canadian flight leader was interrogated twice about the incident, but refused to admit to dangerous flying. Whatever the circumstances, this Canadian group reckoned they had the U-2 squadron well-taped. Their radar regularly picked up the high-flying aircraft as they returned from training sorties or overflights, long before any US equipment, and U-2 pilots regularly observed the re-engined Sabre 6 fighters flying above 50,000 feet. 'They ruled the skies over Germany!' said one.

2 The Special Projects Branch at Edwards tried out a full-pressure suit in the early U-2 models in 1961-62. Although this suit was being successfully used in the X-15 rocket-powered research aircraft, it could not be recommended for the U-2 fleet. The oxygen supply was only good for six hours, and the suit could not be test-inflated every so often

during the flight. Dragon Lady pilots were in the habit of doing this, not only to check that the thing was working, but because it had the effect of relieving some of the fatigue by stimulating blood circulation and massaging tired muscles.

3 For the pilot clad in a pressure suit, arrangements for the elimination of body wastes are perforce rudimentary. In the early-model U-2s, two urine bottles were provided, since the area around the crotch was not protected by the partial pressure suit, and a zip could therefore be inserted. But passing water was a complicated procedure, which could take up to ten minutes to complete. With the advent of the full pressure suit, a rubber sheath was provided, connected to a hose which deposited the urine in a container beneath the seat. As for solid wastes, there is no provision made in either suit, and the U-2 pilot has to live with the consequences of his actions for the rest of the flight — a messy and unpleasant business!

For this very reason, pilots have always eaten carefully in the twenty-four hours preceding a flight. Before suiting-up, they eat a low-residue meal such as steak and eggs, high on protein but designed to discourage further bowel movements.

Chapter Ten
Battlefield Reconnaissance

To the men and women of the 17th RW — I salute you. We have come a long way in three short years trying to serve three masters: SAC, USAFE and NATO. We still have a long way to go . . . but . . . it will pay off in the strong deterrence necessary to prevent another war.
Colonel Thomas Lesan, 1985

When the Pentagon revealed its Fiscal Year 1979 budget request to Congress in the early months of 1978, there was one major surprise. Listed among the requests for Air Force aircraft was one for 'the TR-1, a new version of the Lockheed U-2, updated for tactical reconnaissance.' The military was planning a fleet of at least twenty-five TR-1s, at a cost of about $550 million, including sensors and ground support equipment. This first-year request was for $10.2 million, 'to prepare for production'. As the year wore on, details of the new U-2 slowly emerged from the Pentagon and Lockheed, but a small group of people at Boeing's Advanced Airplane Development Branch already knew what the announcement meant. It spelled the end for 'Compass Cope', a sophisticated remotely piloted vehicle (RPV) that the company was developing under a fifty-two-month contract they had

The first TR-1A during the formal roll-out ceremony at Palmdale on 15 July 1981. Long superpods extending beyond the trailing edge of the wing were relatively new, but otherwise there was little to distinguish this model from the U-2R.

secured in mid-1976. The Cope vehicle was intended to orbit at high altitude on the borders of enemy territory for many hours at a time, its various advanced sensors relaying back information on the other side's aircraft, armour and missile deployments in real-time. But the US military hierarchy needed a lot of convincing about the feasibility of a mere drone performing such a vital mission; and now the Skunk Works had performed another one of their famous moves to outflank the opposition. They had convinced those Air Force reconnaissance planners in the Pentagon basement, as well as their civilian masters upstairs, that an aircraft designed more than twenty years earlier could do the job better!

Throughout the time that the modified Ryan Firebee drones had been operating with the 100th SRW in Korea, Thailand and Vietnam, various contractors had been developing all-new RPVs designed from scratch for reconnaissance missions. The aim was to improve the performance, reliability and maintainability of such vehicles, while at the same time reducing their cost of operation. Ryan's Model 154 'Compass Arrow' had been the first of this new breed of RPVs to fly in 1969. It could carry a twenty-four-inch Itek camera and ELINT sensors to beyond 70,000 feet, by virtue of a forty-eight foot wingspan and a GE J97 turbojet. It incorporated stealth principles in its design, such as a flattened undersurface, inwards-canted vertical stabilisers, and an engine mounted in an above-fuselage position. Designated AQM-91A by the Air Force and apparently intended to replace the U-2s flying over China in a programme code-named 'Compass Arrow', this big Ryan drone was apparently never deployed.

One major drawback of Compass Arrow was that it still required the services of a C-130 Hercules 'mother-ship' for launch, and recovery crews plus helicopters to retrieve it after a mission. So the development effort turned towards RPVs that could be launched and recovered autonomously. For the 'Compass Dwell' programme, E-Systems and Martin-arietta produced rival designs with conventional undercarriages that could take off and land under their own power. They were tested at Edwards from 1970 to 1972, with a view to an eventual European deployment in the SIGINT role. Some flights demonstrated an endurance of twenty hours, but the Compass Dwell machines were powered by piston or turboprop engines, and could not rise much above 45,000 feet. The Compass Cope programme was therefore born in June 1971, when Boeing was asked to produce an autonomous, low-observability machine capable of reaching 70,000 feet on the 5,000 lb. thrust provided by the J97 turbojet. Almost a year later, Ryan was asked to develop a rival Compass Cope RPV, this time using a Garrett ATF-3 turbofan. Both companies made extensive use of composite materials in their designs, in an effort to reduce weight as well as the vehicle's radar reflectivity.

Boeing's entrant (nicknamed 'B-Gull' by the company and designated YQM-94A by the military) first flew in late July 1973, but crashed on its third flight a week later. Fortunately, a second prototype had also been funded, and this subsequently participated in flight tests at Edwards. It was joined in July 1974 by the first of two Ryan Model 235 Seagulls (designated YQM-98A). The Ryan machines were also tested on twenty-four-hour endurance missions over Cape Canaveral in 1975 (a NASA U-2 flew as chase plane for part of the time). One of them crash-landed there and was extensively damaged. Both contractors had hopes of a production contract, but were under no illusions as to the extent of the built-in resistance to drones, in an Air Force hierarchy populated almost exclusively by 'flyboys' — former jet pilots whom they perceived as having an ingrained prejudice against unmanned aircraft. The Ryan people in particular had seen it all before, during fifteen years of day-to-day involvement with the military's reconnaissance drone operation. Time and again, they had seen their RPV proposals turned down, sidewalled, or simply lost somewhere in the acquisition bureaucracy.

As the Vietnam War ended, and funds for military procurement began to decline, the Compass Cope programme therefore faced political and bureaucratic problems. But the operational requirement was not to be denied. In Europe, the Soviets were deploying tanks, missiles and aircraft like gangbusters, and despite a continuing advantage in technology, NATO was losing ground. The US intelligence community was still haunted by the 1968 Czechoslovakia fiasco, when it had been able to give little warning when the Soviets turned a military exercise into a move to depose the liberalizing government of Alexander Dubceck.

All the satellite photography in the world could not reveal a sudden movement of massed armour beneath a cloud-covered continent. The USSR and its Warsaw Pact allies now had the military might to pose a 'blitzkrieg' threat to NATO's outgunned and outnumbered divisions. They might be able to advance hundreds of miles in a surprise attack before the West had time to react properly. And Central Europe was frequently covered in cloud. There might be little or no warning of an impending Warsaw Pact advance.

There were alternative sensors available which could defeat cloud cover, and also the USSR's growing ability to camouflage its military deployments. The most prominent among these were radar and SIGINT systems — but to be effectively utilized, they needed to be flown, rather than orbited, above the threatened area. However, if they could be successfully deployed, and their information returned to NATO's ground commanders without delay, then the West would have more warning time. And since the other side would have to funnel its forces for a surprise attack through certain chokepoints, the sensors could pinpoint the most profitable areas some way behind the front line for NATO to target with its medium-range bomber and missile forces. Using this counter-attack doctrine, which has come to be known as the FOFA (Follow-On Forces Attack) concept, NATO commanders hoped to blunt any Warsaw Pact advance by denying their front-line troops the supplies and reinforcements they would surely need in order to successfully pursue the 'blitzkrieg' strategy.

As the Compass Cope evaluation proceeded, therefore, emphasis switched from its potential role as a long-range strategic reconnaissance platform to the tactical role in a NATO scenario. Boeing's design was declared the winner, and a $77 million contract for three pre-production prototype vehicles was issued in mid-1976. These would have about seventy per cent airframe commonality with the original design. The most important changes were the choice of a 6,000 lb. thrust GE TF34 turbofan engine to replace the J97 turbojet, and the addition of a second, smaller turbofan engine (probably the Williams F107 from the cruise missile programme). This would provide emergency power in the event that the main engine failed; it would be cartridge-started and would provide power for the flight controls and to restart the TF34. If the restart failed, the emergency unit could still bring the vehicle back to base, since it was capable of sustaining flight for one hour with a full payload at 9,500 feet.

There was further emphasis on reliable and safe operation, since Boeing realized that the drone vehicle could not be operated over the densely-populated areas of Central Europe unless there was absolute confidence in the system's integrity. A dual-digital flight control system would be included, with individual components of the second system ready to automatically take over if their counterpart in the primary system should fail inflight; dual automatic microwave takeoff and landing systems would be incorporated, to permit launch and recovery by day or night and in the worst possible visibility; and there would be triple redundancy in the Sperry Univac data-link through the inclusion of a dual CW/FM system and a UHF back-up. With all these precautions, Boeing engineers felt they had addressed the reliability issue full square.

But what about the cost-of-ownership issue? In 1975, Air Force Under-Secretary James Plummer complained to the annual meeting of the National Association for Remotely Piloted Vehicles about extravagant claims and cost escalations in RPV development programmes. Apparently referring to Ryan's Compass Arrow, he compared the original development estimate of $35 million with the final total system cost of more than $210 million, and complained that concurrent developments in the defensive threat had resulted in the system becoming obsolete before an operational sortie was ever flown'.

Boeing's answer was that their Cope vehicle would demonstrate a significant reduction in operating costs over existing alternatives'. They were quoting ten-year life-cycle costs of around $100 million, with direct operating costs in the order of $650 million (in 1974 prices). They reckoned these to be half those of the current SAC U-2R operation, and fuel burn by the turbofan-powered Cope vehicle would be only ten per cent of the Dragon Lady with its old-technology J75 powerplant. The vehicles themselves would cost no more than $2 million, compared with $10 million for a new U-2, the type that seemed to be emerging from the Pentagon basement as the favoured alternative.[1]

As it turned out, none of these arguments were enough to convince the decision-makers, some of them former U-2 and SR-71 pilots who had gone on to management positions at SAC headquarters and in the Pentagon. Nor was the scepticism limited to SAC old-timers. During this period, the Tactical Air Command was a candidate to operate the Compass Cope vehicle or a manned alternative (after the TR-1 was selected, it was decided that SAC would be the operator). At the next conference of the RPV industry association in mid-1977, TAC's Lieutenant General James D. Hughes admonished his audience, telling them that 'the drone community has oversold its product', and that 'demonstration capability has been sold as real operational capability'. Since he was now the man in charge of the remaining Air Force drone assets, following their transfer to TAC in 1976 when SAC's 100th SRW was broken up, this was a depressing verdict for his audience to hear.[2] A colonel in charge of drone requirements at TAC headquarters summed up the prevailing attitude succinctly when he told a reporter at the meeting: 'You and I know that in very few cases will a drone ever replace a manned capability.'

Ben Rich took over from Kelly Johnson as head of Lockheed Advanced Development Projects in 1975, and has held the top job in the Skunk Works ever since. He maintains that modern procedures for military procurement are stifling innovation and initiative. 'This is the tortoise approach,' he says. 'Proceed slowly and never stick your neck out!'

A Lockheed proposal to reopen the U-2 production line for the second time began ascending the procurement chain. Unlike Project Aquatone in 1954 or the U-2R proposal in 1966, there was no CIA sponsorship to short-circuit the system, but the U-2 plan cleared all the bureaucratic hurdles and eventually reached the desk of USAF General David C. Jones, then Chairman of the Joint Chiefs of Staff, in late 1977. He approved it with one proviso. Sensitive to the old but enduring 'spyplane' image, he declared, 'We have to get that U-2 name off the plane!' Without further ado, he rechristened it the TR-1. Perhaps he also thought that a new label would help sell the proposal to Congress. For in reality, there was virtually no difference between the new TR-1 airframe and the U-2R design from 1966. 'It will be like the Volkswagen Beetle,' remarked Colonel Theodore Freitag, a reconnaissance specialist in the Pentagon, 'just polished up a bit from year to year'.

After a hiatus of more than ten years, the U-2 production line starts rolling again. Lockheed technicians move the forward fuselage section into place during assembly of the first TR-1. Due to the unique nature of the aircraft's construction, the Skunk Works had to rehire a number of retired former employees to get things moving again.

Lockheed was equally frank about the new version. 'I'm improving the aircraft using the experience we have had with the U-2,' said Ben Rich, who was now the head of Lockheed ADP, 'but overall, its still the U-2R. We're not changing the tooling.' The only visible difference between the U-2R and the basic TR-1 airframe turned out to be in the horizontal tail. During U-2R operations, it had been discovered that sonic vibrations from the engine were causing fatigue stress in the internal ribbing, because of the way this was spaced. The answer was to add stiffeners which showed up externally on the stabilizer; on the TR-1, this addition wasn't necessary, since the internal rib spacing was adjusted to offset the resonance.

There was one significant addition to the U-2 airframe brought about by the mid-seventies TR-1 design programme, however. Large new sensor-carrying pods were designed to fit, slipper-tank style, under each wing. Known as super-pods, they were two feet eight inches wide and nearly twenty-four feet long — three times larger than the Senior Spear pods which had already put in an appearance on the U-2R. They extended aft as well as forward of the wing, and therefore interrupted the flap surface when fitted. The fore and aft sections of the new pods could be easily removed from the midbody, so that different antenna shapes tailored to specific missions could be carried. To the same end, there was also provision for a belly radome to be attached to the pod forebody, and cooling inlets and outlets could be fitted to maintain whatever delicate electronics they contained at the correct temperature. Equipment weighing as much as 800 lb. could be carried in each superpod — not forgetting the aircraft's Q-bay, which was also still available for photo or SIGINT devices. The new pods actually made their debut prior to the appearance of the TR-1, when SAC deployed the new Senior Ruby SIGINT-collection system on the U-2R in 1977.

The modular concept of the superpods was also applied to the TR-1's nose, which could be completely replaced forward of fuselage station 169 with other noses of differing shapes to accommodate new sensors and their antennae. The payload up front was a further 600 lb., which together with the Q-bay and the two small compartments fore and aft of the tailwheel (all these being pressurized), brought the total available on a superpod U-2R or TR-1 to nearly 4,000 lb. Other differences were internal, and mainly related to the new sensors which would be coming along. The DC electrical system was modified to incorporate a brushless generator, and there were various other wiring changes.

So how did man triumph over machine, and the TR-1 get selected over the attractive Compass Cope package on offer from Boeing? The maximum payload offered by the TR-1 was about one-third greater than in Boeing's latest Compass Cope design. This was a useful margin of safety if the forthcoming sensors turned out to be heavier than expected, which had often been the case in the past. Altitude capability was about the same for both types of vehicle — typically 70,000 feet — although Boeing believed that a loitering altitude some 10,000 feet lower provided the best trade-off with other cost and endurance variables. Since their machine was supposed to be stealthy, the lower altitude did not degrade its prospects of survivability, should the opposition decide to fire off a long-range SAM such as the SA-5 or send a Foxbat-type interceptor in pursuit. The TR-1, on the other hand, would rely on a variety of new electronic warning and countermeasures systems to remain in play if the going got rough. The old System 9/12/13 and 'Oscar Sierra' suite was replaced by Systems 20/27/28/29, and there was an improved warning scope in the cockpit. Even so, the TR-1 patrol might have to be set up thirty miles behind the battle line, to ensure safety. Fortunately, at the aircraft's operating altitude, the new sensors would still have sufficient range to survey the other side's territory to considerable depth.

Compass Cope seemed to have a distinct cost advantage. It was cheaper to build in the first place, and with no pilot to worry about, it offered a twenty-three-hour endurance with a 2,000 lb. payload. It would be launched from a UK base for a two-hour transit to the border area, with remote control provided via a satellite link. At least two, and probably three, TR-1 sorties would be required to provide equivalent coverage. But the cost of all the redundancy provisions was pushing the price up, and the savings to be gained from eliminating the pilot and all the attendant PSD support paraphernalia were sunk instead into the expensive remote control system.

In the final analysis, it was the presence of that man in the cockpit which was to prove decisive, and not just for sentimental reasons born of 'flyboy' prejudice. There was the unresolved question of whether the NATO allies would approve such an extensive RPV operation above their airlanes and cities. Most crucially of all, there was the nagging worry about a drone going out of control and heading across the border. The US was planning to put some of its most sensitive, state-of-the-art intelligence-gathering sensors onboard this vehicle. An errant drone might hand all this equipment to the Soviets on a plate.

So the TR-1 was selected, and Congress eventually appropriated the funds. The U-2R tooling was brought out of storage at Lockheed Palmdale and Norton AFB, and returned to the Skunk Works. A final assembly area was prepared at Palmdale. Retired F-105 and F-106 fighters were stripped of their J75 engines, which were sent back to Pratt & Whitney for modification into the -P-13B high-altitude version. Without the benefit of the fast-track, 'black-world' procurement policies which had accompanied previous U-2 production, it took three times as long to produce the finished article. This time, military bureaucracy took root with a vengeance. First, there was a power struggle between two Air Force commands as to who should run the TR-1 development. Aeronautical Systems Division (ASD) was in charge of new projects, but Air Force Logistics Command (AFLC) claimed this one as a continuation of an existing project — the U-2R. AFLC won. Then they started carping over the technical details. The U-2 had been born and bred as a 'black-world' aircraft, and many aspects of the design did not correspond to laid-down military specifications (milspecs). For instance, a bolt in the U-2 airframe might have so many threads less than the milspec in order to save weight. Now that the production line was to be re-opened, the Air Force wanted everything to conform with the rules. Eventually, a compromise was reached whereby the Skunk Works was allowed to work to the intention, rather than the letter, of milspecs. Even so, volumes of new technical data had to be produced, so that rookie airmen on their first tours in the maintenance shop could figure out how everything worked.

Thanks to all these complications, the first TR-1 did not roll out at Palmdale until 15 July 1981. Unlike its predecessors Article 066 emerged in a blaze of publicity, with press photographers on hand and a high-powered reception committee drawn up outside the hangar. Those in attendance included Kelly Johnson and Tony LeVier, the designer and first pilot of the original U-2 from twenty-six years earlier, both now retired but still on the Lockheed payroll as consultants. There was also an impressive number of their colleagues from those early days, who were still employed at the Skunk Works. Men like Ben Rich, the small but energetic engineer whose pioneering work on thermodynamics in the 1950s had ensured the success of the Blackbird series. He had taken over as head of Advanced Development Projects (ADP) when Johnson retired in 1975. Also Fred Cavanaugh, who had worked at Burbank since the early 1950s, and was now TR-1 programme manager. Similarly Bob Anderson, the senior engineering project manager for the TR-1, had been in the Skunk Works since 1956. And so on down the line to the technical reps in the field, some of whom had been employed on the U-2 or SR-71 projects for the best part of their working lives. It was a remarkable record of continuity. As one

Fred Cavanaugh, grizzled veteran of over thirty years in the Skunk Works. He became U-2 project manager in 1962, and was intimately associated with the aircraft through till 1985, when a domestic accident put him in a wheelchair and forced early retirement. He was responsible for many innovations in the later years of the U-2 programme.

SAC's reconnaissance stablemates captured together inflight during a rare formation sortie. Many people presumed the SR-71 had replaced the U-2, but this was far from the truth. Both aircraft had unique characteristics which suited each for different roles in airborne intelligence-gathering.

of them said, 'No one was interested in leaving for more money. We loved Kelly, and we loved the work.'

The TR-1 made its first flight on 1 August 1981, with Lockheed test pilot Ken Weir at the controls. It was delivered to Beale AFB the following month, the first of at least twenty-five TR-1 models required by the USAF. Two of these would be twin-cockpit, dual control TR-1B trainer versions. Prior to the public roll-out, Lockheed had actually already completed and delivered one new-production U-2. This was an ER-2 (ER for Earth Resources) version for the NASA Ames Research Center, which would boost the U-2 fleet at Moffett Field to three machines and provide extra capacity for an ever-lengthening list of scientists anxious to utilize the aircraft's unique operating qualities in a variety of experiments. The ER-2 was first flown by Art Peterson on 11 May 1981, and delivered on the 10th of the following month.

At Beale AFB, in the rolling foothills below the Sierra Nevada mountain range, the new TR-1 joined the dissimilar team of Dragon Lady and Blackbird aircraft which had been created when the U-2s moved up from the Arizona desert in 1976. At that time, Colonel John Storrie was the 9th SRW commander. 'I think SAC's reconnaissance operations will become much more efficient with the transfer,' he said. 'The move puts all SAC recon units, including the RC-135s at Offutt, under the 14th Air Division. It makes sense from the standpoint of command and control.'

Chuck Stratton, now a Colonel and just about to complete his second tour with the 100th SRW as its last commander, pointed out that, 'The U-2 and the SR-71 are uniquely different systems which complement each other. One often hears that the SR-71 is a follow-on or replacement for the U-2, but this is not the case. SAC will have a continuing requirement for the specialized mission capabilities of both aircraft.'

Yes, there was plenty of work for the 99th

SRS, as the Dragon Lady squadron was now designated. While the Blackbird had been the main platform for photo missions for some years now, they were awfully expensive to operate — it made no sense at all to fly them on a regular basis over what SAC called 'permissive environments'. By this, it meant areas of interest that were not defended with SAMs or high-flying interceptors. The Blackbirds, for instance, flew across North Korea from the 9th's Detachment 1 at Kadena AB, and attracted enemy missile attacks on numerous occasions in the process. (None were ever hit.) The U-2s flew serenely across Nicaragua, Somalia, and other countries which had not been equipped with Communist Bloc missile defences. If a direct overflight was not possible for military or political reasons, there was always the LOROP camera which could be mounted in the Q-bay and programmed to produce high-quality imagery up to a hundred miles off to one side.

There were also other, less sensitive, photo missions for the U-2 flyers. During training flights across the US, the photographic 'take' was geared to requests for coverage by other government agencies, such as the Department of Agriculture, the Bureau of Land Management, and the US Corps of Engineers. At the request of the US Agency for International Development, they flew over Guatemala in February 1976 when that country was devastated by an earthquake and took damage assessment pictures. In October that same year, shortly after the move to Beale, Major David Hahn flew a photo mission over the Pacific in a last-ditch attempt to locate a shipwrecked sailor. He was thought to have been adrift on the high seas for three weeks in an orange liferaft, but search and rescue authorities were having difficulty locating him in the wide ocean spaces, far removed from land bases. When the U-2 returned to Beale and its film was processed, the liferaft showed up as a tiny bright speck in an otherwise endless series of black exposures of the sea surface. Cued to the precise location by the U-2 picture, rescuers reached the sailor, who could not have survived for much longer without food and water.

Another mission now undertaken by the 99th SRS was that of high-altitude air sampling. The sampling equipment was much the same as in the old Project Crowflight days. Since 1965, with the U-2s otherwise engaged in south-east Asia, the Air Force had been using big-wing RB-57F models for all of this work, which was declining in importance now that even the French and the Chinese were conducting all their nuclear tests underground. Nevertheless, the high-altitude sampling mission had to be retained in the inventory, and when the last RB-57F was retired with fatigue problems in 1973, it reverted to the U-2 squadron.[3]

However, by far the greatest proportion of U-2 flying hours was now dedicated to the SIGINT mission. With the proliferation of SAMs and electronic warfare devices on the other side of the Iron Curtain, the US military's need for up-to-date electronic order-of-battle information about the Warsaw Pact became even more acute. Exotic new receivers were developed for all three types in the SAC reconnaissance inventory. Much of it was placed on the RC-135, since it was the largest airframe and had the most room for sensors, operators and the accompanying extra power and cooling facilities. The Blackbird was also fitted with ELINT receivers, but the U-2 was in some respects an ideal ELINT-gathering aircraft. Although it could not carry a fraction of the weight of equipment onboard the RC-135s, it flew 30,000 feet higher, and could therefore 'listen' further over the horizon from an orbit in friendly territory. Furthermore, the presence of a snooping RC-135 with its large radar cross section (RCS) was easily detectable to the other side, which would shut down most of its emitters to avoid giving any secrets away. Although the U-2 was not a true stealth aircraft, its operating height and smaller RCS sometimes enabled it to avoid detection and creep close to the border for an uninterrupted eavesdropping session. As for the lack of accommodation for onboard ELINT operators, the Senior Book U-2 operation had already proved that with modern data-link technology, airborne receivers could be operated and monitored from ground stations some hundreds of miles away.

Admittedly, the Blackbird flew even higher than the U-2, but it also flew so much faster, which often prevented it from capturing the required signals; as it flashed past an enemy communications emitter, for instance, it might not be within range for long enough to pick up the entire message being transmitted. Eavesdropping satellites had similar problems, and their orbits were even more predictable than the RC-135's excursions. If instead they were parked in geostationary orbit, the laws of gravity demanded that they be placed thousands of miles above the area of interest, and therefore out-of-range for certain transmissions. Regarding the ELINT role, the nature of some signals of interest meant that the collecting aircraft needed to orbit precisely and within a confined space in order to satisfactorily capture them on tape. With its fifty-mile turning circle at full speed, the SR-71 was hardly suitable in these circumstances.

This was why the new ELINT sensor developed by E-Systems and codenamed 'Senior Ruby' was added to the U-2R in the mid-1970s. By housing Senior Ruby in the new superpods, its multiple recorders and antennae could be integrated with the existing Senior Spear COMINT sensor so that the aircraft could simultaneously collect both types of SIGINT and relay them via the Sperry data-link housed under the tail. A Transportable Ground Intercept Facility (TGIF) also designed by E-Systems controlled the sensor operation, received the downlinked signals in near-real-time, and disseminated them to the 'customers'. Elements of this new Remote Tactical Airborne SIGINT System (RTASS) made their first appearance in Europe during 1978, when the 99th SRS deployed a U-2R with superpods to the US airbase at Mildenhall in the UK for the first time. From here, the aircraft flew off on nine-hour patrols along the boundaries of the Warsaw Pact in Central Europe. Following earlier visits by 9th SRW Blackbirds and U-2s, Mildenhall had now become a permanent deployment for the SAC flyers, as Detachment 4 of the 9th SRW. The wing's other permanent detachments were now at Kadena, Okinawa (Det 1 — SR-71); Osan, Korea (Det 2 — U-2R); Akrotiri, Cyprus (Det 3 — U-2R); and Patrick AFB, Florida (Det 5 — U-2R).

'Det 3' at RAF Akrotiri twice made the headlines in the late 1970s. On 7 December 1977 it was the scene of a tragic accident when Captain Robert Henderson apparently stalled his U-2R on takeoff and crashed into a nearby operations building. The fully-fuelled plane burst into flames, and set the building alight. Henderson was killed, along with four Greek-Cypriot civilians working for the RAF. Two other Cypriots, along with four British airmen, were injured.

In 1979, the U-2 operation on Cyprus became caught up in the great debate over ratification of the SALT 2 treaty between the superpowers. Many US observers were worried that the Soviets would cheat on the treaty, and demanded assurances about their country's ability to verify that its provisions were being followed. With the recent overthrow of the Shah of Iran, the US had lost the use of its SIGINT intercept stations in the mountains north of Tehran and Mashhad. These played a vital role in watching for signs of

A U-2R in SIGINT-gathering configuration comes into land at Mildenhall, UK, in 1980 after a nine-hour flight along NATO's borders with the Warsaw Pact. A single aircraft was based there with Det 4, 9th SRW, until the mission was re-allocated to the newly-formed 17th RW in 1983.

an imminent Soviet ballistic missile test from Kapustin Yar or Tyuratam, and in picking up telemetry relayed back from the missiles once launched. Their loss cast further doubts on the verifiability of SALT 2. The Carter administration suggested that Cyprus-based U-2s could help close the intelligence gap, by flying over Turkey and the Black Sea. They could be equipped with TELINT sensors which would pick up data from the missiles once they had climbed to an altitude of about ninety miles. SALT sceptics in the Congress were briefed on the plan in early April, and details were leaked to the newspapers. A request for overflight rights was sent to the Turkish government. Istanbul, still smarting from a three-year cutoff in US military assistance following its invasion of Northern Cyprus, reacted coolly and told Washington that it would consider the matter, but only if the USSR raised no objections! A high-level delegation from the Pentagon flew to Istanbul and persuaded the Turks to grant unconditional overflight rights for US spyplanes, but the U-2 scheme was never implemented. Instead, the US deployed a new TELINT satellite, and was able to open two new ground stations in Sinkiang to monitor Soviet missile tests, thanks to the Chinese government.

On 12 February 1983, the first TR-1 to be deployed overseas landed at RAF Alconbury, just outside Huntingdon in the English Midlands. This US airbase, already home to a wing of RF-4C Phantom reconnaissance jets, had been chosen to house the new battlefield reconnaissance aircraft, since there was not enough room at Mildenhall, fifty miles down the road. 'Det 4', 9th SRW remained at Mildenhall, however, since the Blackbird was now in permanent residence there. At Alconbury, a new wing and squadron structure would operate the new Dragon Lady. SAC reactivated its dormant 17th and 95th unit numbers, as the 17th Reconnaissance Wing and 95th Reconnaissance Squadron. It seemed a strange choice; neither formation had any reconnaissance heritage whatsoever. Apparently, that didn't matter to the official SAC historians; this wing and squadron had a long lineage and it was deemed important that their colours should not be laid up forever. The 17th insignia did feature what looked at first glance like a small dragon perched above a gold shield, but this turned out to be a griffon instead. The wing motto in French, '*Toujours au Danger*' somehow didn't have the same cachet as 'Toward the Unknown', As for the 95th, its patch consisted of a donkey

When the TR-1 arrived at Alconbury in 1983, it was co-located with an RF-4C Phantom photo-reconnaissance squadron which was already resident. When USAFE was forced to choose between priorities a few years later, the Phantoms were withdrawn. The TR-1 shown here, serial 01069, was later involved in a ground collision with a bus full of security policemen!

Learning from earlier experience, the Air Force ordered a two-seat training version of the TR-1 to be completed at an early stage of the new production contract. Two TR-1B models were, in fact, the fourth and fifth aircraft to emerge. They were finished overall in a gloss white at first, but have since been repainted in the standard black coat of the single-seat version.

supported on its front legs while the hindquarters prepared to kick backwards. The U-2 pilots at Alconbury were henceforth obliged to wear this ridiculous emblem on the distinctive dayglo overalls that they wore during routine, non-flying duty. The Alconbury squadron soon acquired the unofficial nickname "Ass Kickers"! Worse was to come: in 1985 SAC decreed that the dayglo suits be replaced by the standard drab olive outfits worn by everyone else in the Air Force. This was another blow to some pilots' self-esteem.

By now, however, U-2 pilots were a dime-a-dozen. Back in the early days, a pilot didn't apply to join SAC's Dragon Lady squadron — he was asked. Even through the wartime years in south-east Asia, the outfit remained a select, closely-knit group. But since the operation had been established at Beale, they had turned out at least ten new pilots each year. They even advertised for them nowadays in Air Force publications! SAC now required only 1,500 hours prior flying experience of its U-2 applicants — in the old days, no one with less than 2,500 hours got on board the programme. A separate squadron was established at Beale in 1981 to train U-2 pilots. Designated the 4029th SRTS, it called itself 'The Dragon Tamers' until a change of wing policy resulted in the squadron also being assigned the SR-71 training mission. The first two twin-cockpit TR-1B trainers were delivered to the 4029th SRTS in 1983, although the old U-2CT models were not finally phased out until the end of 1987, after a third two-seat TR-1B arrived at Beale.[4]

The new unit at Alconbury immediately took over the SIGINT flights across Central Europe which 'Det 4' had been running on a regular basis since 1979 with a single U-2R. These operations were codenamed 'Creek Spectre'. Since the expected new sensors were not yet ready, the 17th Wing expanded only gradually in its first three years, as most of the new TR-1 models were retained at Beale and Palmdale for development and training purposes. Throughout its first two years, the wing had only three aircraft and eight or nine pilots assigned. Even so, it was a struggle to build a viable new organisation from scratch, often using makeshift facilities while purpose-built accommodation was built.

For instance, there was only one hangar available for the aircraft, and those assigned on a mission had to be preflighted in the open. In the cold and frequently wet atmosphere of a British winter, this was no joke for the maintenance people, especially those trying to keep delicate SIGINT and avionics equipment serviceable. It could even represent a hazard to the aircraft — one U-2 was nearly lost when water entered the elevator section while the aircraft was parked on an exposed apron, and subsequently froze when it ascended to high altitude. Eventually, in late 1985, the first of five weather shelters were commissioned at Alconbury, and in 1987 construction of large-span semi-hardened shelters for the TR-1 fleet began.

The British weather posed another serious problem for TR-1 operations out of Alconbury. The base had only a single runway, aligned roughly south-east/north-west. The trouble was, this frequently didn't match the wind direction, and with the Dragon Lady's low crosswind tolerance, many planned sorties were lost to this single weather factor. Others were lost when fog rolled in. Tired pilots returning from nine-hour flights over Europe were faced with frequent diversions. This was, and remains, the Achilles' heel as far as TR-1 operations in Europe are concerned. The wing lobbied to have an old wartime cross-runway reopened, but various buildings had gone up within the obstacle clearance limits for this strip in the meantime. It also tried to develop dispersed operations out of two nearby bases which had different runway alignments, but both had poorly-maintained aprons and taxiways which caused damage to the aircraft's delicate tailgear.

In March 1985, the size of the 17th RW fleet doubled with the arrival of three more aircraft. Within days, the wing also received the first of the long-awaited new sensors — the Hughes Advanced Synthetic Aperture Radar System-2 [System (or ASARS-2)]. For the first time in a nearly thirty years, operational U-2s would be carrying an actively-radiating system aloft as part of their reconnaissance sensor suite.

The history of ASARS-2 went back eight years, to the time in 1977 when the Hughes Radar Systems Group at El Segundo, just south of Los Angeles International Airport, received a follow-on contract from an earlier radar development programme designated UPD-X. In the UPD-X programme, the idea was to devise a high-resolution, side-looking airborne radar (SLAR) which could be tied by a jam-resistant air-to-ground data link to ground-based processing equipment. If carried by a high-flying platform like the U-2, such a radar could peer a hundred miles into enemy territory and identify rear-echelon targets in all weather conditions — thus providing a significant advantage over photographic sensors, which were defeated by haze and cloud cover. The photo method was only useful at night if lower-resolution infra-red film was used, whereas this imaging radar would be a true, twenty-four-hours-a-day sensor. Moreover, the long wavelengths at which SLAR radars operated (e.g. X-band, around 10 GHz) enabled the sensor to detect targets concealed by camouflage, such as foliage or man-made screening.

There were a number of advantages to be gained from down-linking the radar returns for processing on the ground. The weight and power requirement for the airborne portion of the sensor could be kept down; sophisticated computers could be employed to process and present a high-quality finished product; and ground commanders would have vital, up-to-the-minute intelligence on the disposition of the other side's forces. They wouldn't have to wait for a reconnaissance aircraft to return and down-load its film for processing at an airfield which could be some distance removed from the front-line command post.

Such radar technology was not new; similar radars had first been developed in the mid-1950s, and an early Westinghouse system had been carried by one model of the RB-57D — the U-2's cousin in strategic reconnaissance at that time. As the aircraft flew along, the SLAR transmitter directed short pulses of microwave energy in a narrow, fan-shaped beam at right angles to the flight path. The radar energy which was returned via the same antenna to a sensitive receiver produced electronic images of the earth's surface on either or both sides of the aircraft, in a relatively narrow swath. The image was recorded on film or in an electronic memory, from which it could be recalled and viewed on a cathode-ray tube. As the aircraft moved forward, new swaths were illuminated so that, over a period of time, a continuous image of the earth's surface could be obtained. By collecting imagery of a border region over a period of time, changes in the other side's military deployments could be noted.

In these early systems with their relatively low resolution, only large-scale changes could be identified; the construction of new roads or airfields, for instance. The only way to improve resolution was to build a bigger (e.g. longer) antenna. Hughes built a fifty-foot long antenna which was hung in a large pod underneath a supersonic B-58 bomber for trials in 1959. The pod was so bulky that the aircraft was limited to subsonic speeds. A breakthrough was made when the synthetic-aperture radar (SAR) technique was developed around 1960. In this approach, a much smaller antenna was made to function as a very large one by recording the successive returns from the ground in the radar's waveguide network as the aircraft flew along the track. The returns would subsequently be combined by a data-processing system to offer much better resolution, and the length of the 'synthetic' antenna (or 'aperture') was limited only by the distance along the track for which a given target was within the illumination beam of the actual (or 'real') antenna. The data-processor was the key element in a synthetic aperture radar (SAR); it had to provide the proper amplitude and phase weights to the stored returned pulses, and sum them to obtain a correct image of the scene.

Goodyear achieved an early prominence in SAR technology, and its APQ-102 system went into service on USAF and USMC RF-4 Phantoms in the mid-1960s. In the early 1970s, a high-resolution Goodyear SAR was tested on a U-2R in a programme codenamed 'Senior Lance'. Since the radar antenna was too large to fit inside the Q-bay, the Skunk Works designed an inflatable radome made of Kevlar. The Q-bay hatch was removed and the new material faired round the area. It was inflated by bleed air, and access was by means of two zip-fasteners! It was yet another example of 'can-do' Skunk Works ingenuity.

Two highly classified radars for operational reconnaissance were spawned by the Senior

A unique inflatable radome was designed by the Skunk Works to accommodate the experimental Senior Lance imaging radar. Made of Kevlar, it was zippered at the rear and down the middle for access, and attached to the base of the Q-bay.

Lance/UPD-X trials. Goodyear developed ASARS-1 for the supersonic SR-71 Blackbird, while Hughes clinched the contract for ASARS-2, which ended up on the TR-1.

The task for Hughes from 1977 was to refine the SAR concept in order to improve resolution even further. The days of the fast, low-level penetrating reconnaissance aircraft seemed numbered; for a combination of political and military reasons, RF-4 missions over North Vietnam had become virtually impossible in the later stages of the war. Since then, the Soviets had developed more sophisticated SAMs and anti-aircraft guns such as the SA-6 and the ZSU-23, and their interceptors would soon be getting look-down, shoot-down radars. If the imagery from synthetic aperture radar could approach that provided by ordinary cameras, then the need to mount hazardous, unarmed flights across heavily-defended targets would be much reduced. You could merely orbit an aircraft outside the enemy's territory or defensive missile radius, and take 'pictures' at your leisure. In the ASARS programme, Hughes was aiming for images equivalent to infra-red photo quality at a one hundred mile range. Resolutions of a few feet were on the cards, irrespective of the distance of the target from the aircraft. This was because SAR technology allowed the resolution to be made independent of range, simply by increasing the length of the synthetic array in direct proportion to the area to be mapped.

There were important reasons for choosing a high-altitude aircraft such as the Compass Cope RPV or (as it turned out) the TR-1 as the radar platform. SLAR returns were processed to present a plan, rather than an oblique, view of the imaged terrain. At first glance, such an SLAR image reproduced on ordinary film could look just like one taken by an ordinary camera from directly over the target. There were significant differences, however, such as that caused by terrain masking when the side-looking sensor painted a hill. The far side of the hill would receive no radar illumination, and would thus appear as a black shadow. For this reason, unless the SLAR platform was a high-flyer, an unacceptably large proportion of the image would be in shadow. Since it flew above the tropospheric region of disturbance, a high-flying platform would also provide the stability necessary to achieve good results (although even the slightest deviation from a perfectly straight, constant-speed course would still have to be measured, either by accelerometers or an inertial navigation sensor, and a compensatory phase correction made to the received signals).

Even with the high-flying TR-1, some shadows representing uncharted territory would remain on the image, especially at longer ranges, when the look angle from the aircraft decreased. The scientists at Hughes came up with what they called the 'squintable antenna' feature to further reduce the hidden area. By training the beam of the real antenna forwards (or backwards) and making an appropriate focusing correction and co-ordinate rotation, the synthetic array could be 'squinted' well ahead of (or behind) the aircraft. The radar could thus be made to 'anticipate' terrain which would become masked, and look around it to eliminate the potential gap in coverage. The 'spotlight mode' was another advanced feature of ASARS. By gradually changing the look angle of the real antenna as the aircraft advanced, and making appropriate phase corrections, the radar could be made to repeatedly map an area of particular interest. By doing this, an even better radar map could be obtained.

This was not all. There were plans to 'cue' the radar's spotlight mode by feeding in collateral data, in particular from the U-2's existing SIGINT sensors. If a mobile missile site or command post came on the air and was detected, for instance, the radar could be 'told' where to look to provide confirmation and additional intelligence. The combination of the TR-1, the ground station, and these passive and active methods of target detection was named the Tactical Reconnaissance System (TRS).

With all the attendant advantages mentioned earlier, plus some clever extras which modern computing power could provide in the ground station, the ASARS/TR-1 concept held great promise. The extras included a variety of ways in which the radar image could be manipulated to suit the particular purposes of detection or interpretation. Since metallic objects such as tanks or other vehicles were better radar reflectors than surrounding natural materials, returns from the latter could be filtered out in a process known as 'thresholding', thus immediately highlighting the real objects of interest in the image. In another useful process known as 'change detection', a newly-received image could be electronically compared with a previous one

Lockheed designed a new nose section to accommodate the ASARS radar sensor. Here it is test-flown on a U-2R model. These were brought up to date with many TR-1 features, and the only visible difference between the two was on the horizontal tail. Lengthwise rib stiffeners are visible on the leading edge of this aircraft's tail — on the TR-1, this precaution against fatigue was incorporated internally during construction.

retained as a reference. And since SAR imagery was processed to show targets in an overhead view, a computer could more easily classify them based on their outlines. This move towards automated target recognition was long overdue; interpreters had frequently been overwhelmed in the past by the sheer weight of data brought back by the proliferating reconnaissance sensors. Now, instead of waiting for great chunks of raw material to land on his desk for analysis, the interpreter could sit comfortably in front of a screen in a facility near the front line, and have his attention drawn automatically to changes such as the appearance of enemy armour in a particular area. And he would be watching this at virtually the same moment as it happened, thanks to the data-link.

The prototype ASARS-2 was flying on a U-2R from Palmdale by 1982. The antenna, transmitter and receiver/exciter were carried in a new nose section, thirty-two inches longer than the original 'slick' U-2R/TR-1 nose. A heat exchanger protruded from the front end in a distinctive fairing. Another two ASARS black boxes were carried in the Q-bay. The data-link remained below the tail, linked to the system by a fibre optic cable. To provide the precise aircraft positional data required by the sensor and the ground station, an astro-inertial system,

A Hughes technician prepares to load the ASARS-II radar antenna into the nose section of a TR-1. This side-looking synthetic aperture radar provides very high definition images of the terrain, and incorporates some highly-classified ECCM features.

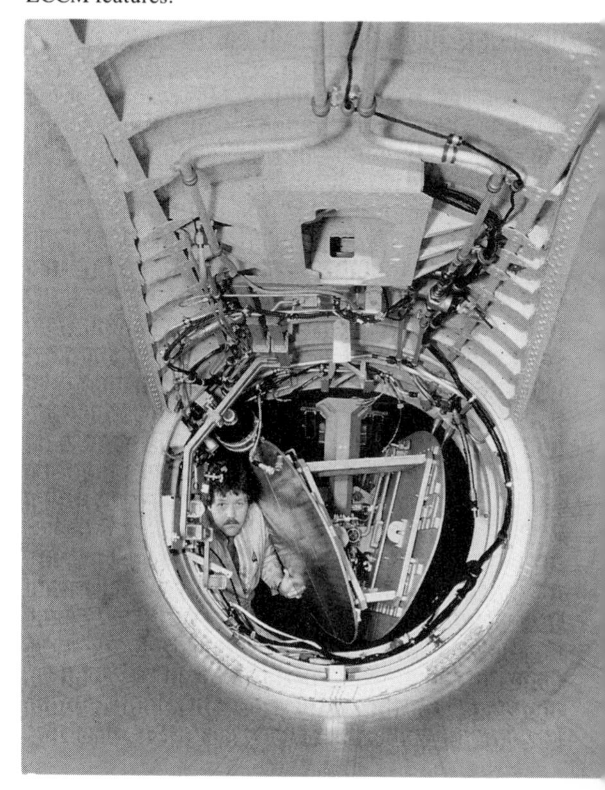

manufactured by Northrop and similar to that carried by the SR-71, replaced the Litton LN-33 INS fitted on previous aircraft. Production of the TR-1 was forging ahead, so the first of the new models were deployed to Europe without ASARS. But it was not long before Air Force top brass were waxing lyrical about the new sensor. 'It provides pictures of near-photo quality at remarkable stand-off ranges,' said Lieutenant General Thomas McMullen, the boss of Aeronautical Systems Division in 1983. 'The combined increase in range, resolution and area coverage represents a quantum jump over currently operational systems,' he went on. Nevertheless, ASARS underwent a further two years of testing in California before it was deemed ready for deployment overseas. Part of the delay was caused by a lag in development of the ground equipment, contracted to Ford Aerospace. The ground stations would not only process and present the downlinked data, but would also provide the means for the interpreters to prepare exploitation reports and send them rapidly out to the field commanders.

Lieutenant Colonel John Sanders, commanding officer of the 95th RS, flew the first operational ASARS-2 sortie from Alconbury on 9 July 1985. A week later, the first TR-1 to carry both radar and SIGINT sensors was airborne and heading for the central front. At a secret location twelve miles below, codenamed 'Metrotango', the TR-1 Exploitation Development System (TREDS), more commonly known as the TR-1 ground station, had been set up in a collection of mobile trailers. It was the start of what proved to be a protracted Initial Operational Test and Evaluation (IOT&E). Although a lot of money had already been spent, there was still only one prototype ASARS in Europe, and the Air Force wanted full testing there for both radar and ground station before committing more funds to production.

Even though it was taking a long time to get ASARS fully up and running, the programme was in far better shape than the other major new sensor being developed for the TR-1. This was PLSS — short for Precision Location and Strike System. It was essentially a sophisticated ELINT system which provided immediate intelligence on the opposition's air defence network to a ground station. If hostilities broke out, PLSS could additionally provide airborne relayed

The PLSS concept of three TR-1s operating in a racetrack pattern with their ELINT sensors providing bearings from threat emitters. This data is relayed to a ground-based Central Processing System (CPS) where it is cross-triangulated and then passed to a Tactical Air Control Centre (TACC). From here, strike aircraft could be despatched, and were guided to the weapons release point by the TR-1 triad.

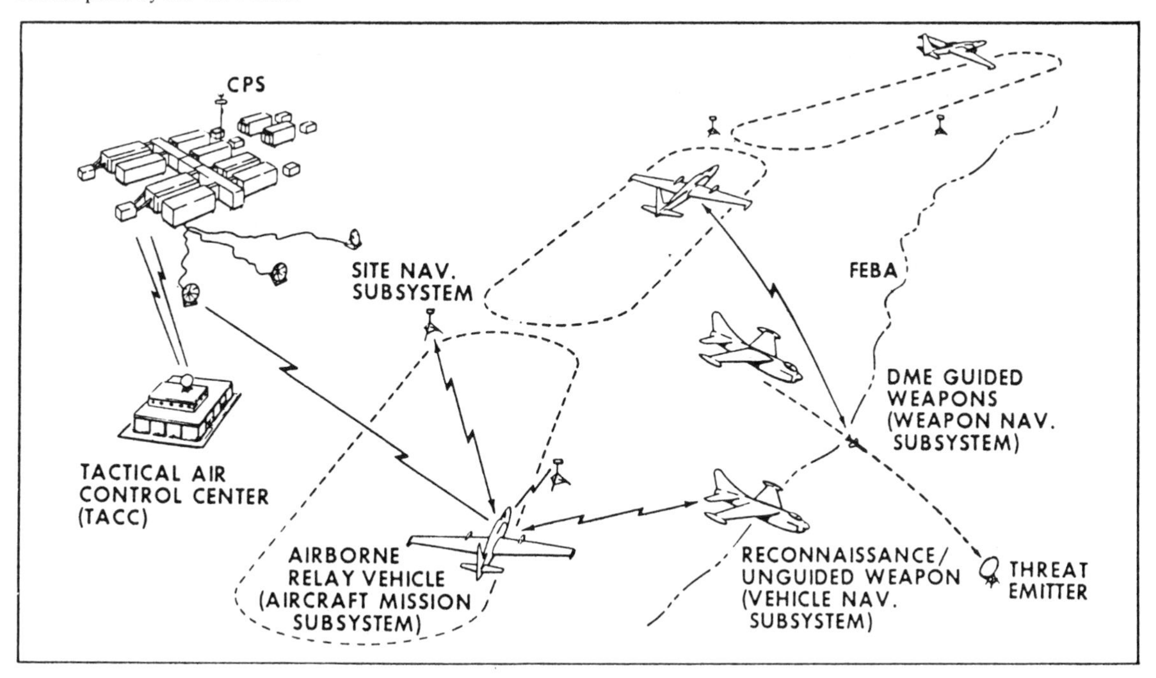

guidance to bring attack aircraft to within striking distance of the enemy radars and SAMs.

The trouble with conventional airborne ELINT systems was that, while they could determine the basic parameters of the targeted signal within seconds (antenna scan, transmitter power, pulse repetition frequency and so on), they needed to stay tuned to the emitter for much longer while the ELINT aircraft continued its progress, so that different bearings on the emitter could be taken and its exact location determined by a triangulation process. This having been done, the site could be attacked if neccessary with ordinary dumb bombs, but the *modus operandi* which had been established in Vietnam was for a dedicated air defence suppression aircraft (the famous 'Wild Weasels') to carry an ELINT sensor and radar-seeking missiles, so that SAMs could be silenced as soon as they came on the air and the direction-finding process had fixed their position. Unfortunately, the other side learned to foil this tactic by shutting down their radar before its position could be fixed or the anti-radiation missile launched.

If this shut-down meant that a SAM went unfired against a strike aircraft then the US commanders were happy. Unfortunately, this was not always the case, and as North Vietnam's air defence network grew in mobility, sophistication and redundancy, different solutions to the suppression of enemy air defences were sought. In early 1972, the Air Force defined a new SAM suppression programme codenamed 'Pave Onyx'. Task Four of Pave Onyx was an Advanced Location and Strike System (ALSS). ALSS would employ a 'triad' of high-flying aircraft each equipped with data-links, and ELINT sensors with a range of 200 nautical miles. The initial candidate aircraft was the RB-57F, with the Compass Cope RPV seen as an alternative in the long-term. The aircraft would establish non-overlapping racetrack patterns in friendly airspace. When a threat emitter came on the air, the ELINT sensors would each detect the signal and relay it to a ground station. There, the minute difference in the time taken for the signal to reach each aircraft would be measured. By locating the aircraft precisely with respect to one another, and to their distance to the ground station, the emitter's position could be accurately determined, even if it had only come on the air for a short while, and then shut down again. This was called the time-difference-of-arrival (TDOA) technique. The emitter's position would be displayed on a screen in the ground station. Strike aircraft could then be directed to attack it, with one member of the airborne triad providing a relay service via data-link, to guide the pilot of the strike aircraft to a continuously computed weapons release point. Ordinary free-fall or glide bombs could now be used against the site. Better still, if the strike aircraft carried 'smart' bombs guided by Distance Measuring Equipment (DME), these could be the receiving point for the relayed signals. Before launch, the signals would be fed to the host aircraft's cockpit as bearing and range-to-go; after launch, ALSS could continue to direct the weapon all the way down to impact, through its autopilot. Using this technique, the attacking aircraft could release the weapon without coming within range of the threat site; either way, the attack did not rely on anti-radiation missiles, which had to be launched within tightly prescribed firing envelopes and even then could be confounded if the SAM radar was shut down.

Study contracts for ALSS were let to teams led by IBM and Lockheed, and IBM was selected to develop the system in March 1972, with a view to deployment early the following year. But the RB-57F fleet was in bad shape, with yet another round of wing cracks, so another aircraft was needed to form the 'triad.' The call went out to Lockheed and the 100th SRW. Some of the wing's old U-2C/F models were taken out of storage, and the refuelling modification was deleted from the F-models. The eighteen small ELINT antennae for ALSS were mounted on a rack which fitted low down in the Q-bay; the Skunk Works designed a new dilectric Q-bay lower hatch, which bulged out below the fuselage in order to house the data-link antenna. Initial flight tests took place in October 1972, but development problems ensued. In the meantime, the air war in south-east Asia had begun to wind down, but ALSS obviously had potential in other theatres, such as Europe. The Air Force therefore decided to proceed with an Initial Operational Test and Evaluation (IOT&E) to serve as guidance for future development efforts.

The various elements of ALSS were set up on the White Sands Missile Range in New Mexico. In addition to the ground station containing the computers and communications gear, there were two ground relay beacon stations and a number of remote beacon stations. These served as accurate reference points to fix the position in space of the U-2 triad and the strike aircraft by distance measuring techniques. Eight emitters representative of Soviet Bloc radars were also made ready. Between May and September 1973, forty-two ALSS test missions were scheduled, with the U-2s launching from Davis-Monthan to fly designated triad orbits at various ranges from the emitters. A number of F-4 Phantoms would operate from Holloman AFB, New Mexico, as the strike aircraft.

The evaluation was a mixed success. The U-2 seriously blotted its copybook. The fleet was grounded twice in May for accident investigations, causing the loss of ten scheduled ALSS missions. On another six occasions, the 100th was unable to get three aircraft launched. Of the twenty-six missions actually flown, the U-2s were late arriving on station twelve times, despite the use of spare aircraft kept on standby status whenever a mission was scheduled. Pressurization, oxygen, autopilot and hydraulic system failures were all to blame. Six of the eight target radars were located to within seventy-five feet (and the other two to within a hundred feet), but the system really only worked well when detecting S-band emissions. Rockwell Mk 84 glide bombs with DME guidance were launched towards the targets by the F-4 Phantoms. An average miss distance of seventy-four feet was recorded: ALSS seemed promising, but the overall reliability of the system was low. There were many sophisticated components making up the ALSS, and they all had to work properly all of the time. They didn't.

Despite this, the Air Force thought that the concept was worth pursuing. ALSS became part of the Pave Strike programme to improve the Air Force's ability to perform precision air-ground strikes. The old U-2C models were kept on flying status (they were also used for training), and the ground-based elements of ALSS remained at White Sands. Some tests were flown to determine how well ALSS could direct DME-equipped RPVs to a target. The Air Force also began to define the parameters of an improved system, which was ultimately designated a Precision Location and Strike System (PLSS). It also decided to try ALSS out in Europe, in a 'dense threat emitter environment', where it could also be tested for integration into an existing command and control network. The deployment was codenamed 'Exercise Constant Treat'.

In early May 1975, the 100th SRW sent five U-2C models configured for ALSS to the US airbase at Wethersfield in eastern England. Before they set out, Lockheed resprayed all the aircraft in a two-tone grey camouflage pattern. The new colour scheme had nothing to do with avoiding enemy interception; the British Labour government had requested that the regular black paint be removed before deployment. Some bureaucrat or politician in London was obviously

A forerunner of PLSS was tested in Europe in the early summer of 1975. For Exercise Constant Treat, five U-2C models were deployed to the little-used airfield at Wethersfield in the UK. Following a British request, the aircraft were repainted from black to a more innocuous-looking two-tone grey prior to the deployment. In this picture, groundcrew and back-up pilots are preparing for launch of the triad. The ALSS equipment was mounted in the Q-bay, on a slightly-bulged lower hatch.

nervous about sanctioning a mass deployment of the famous spyplane. Somehow, a repaint job was supposed to allay this concern.

The ground elements of ALSS were set up in West Germany; the ground station at Sembach AB and five remote beacon sites scattered from Bremerhaven in the north to Neu-Ulm in the south. From mid-May until early July, eighteen ALSS missions were flown. The U-2 triad would launch at ten-minute intervals from the rural English base and set course for Germany, followed by an airborne spare. If it was serviceable, the fifth aircraft would be preflighted and ready to go in case one or more of the others aborted. These precautions proved necessary; thanks to various malfunctions, the spare aircraft had to be called forward into the triad on ten out of eighteen missions, and the fifth aircraft was launched five times. The U-2s generally flew for five hours on racetrack patterns parallel to the East German and Czechoslovakian border; although special test transmitters had been set up on West German territory below, the aircraft also tuned in to the real transmissions emanating from the other side of the German border. F-4 Phantoms were again used as the strike aircraft (from the 36th TFW at Bitburg AB), but only unguided ordnance was dropped this time, on two bombing ranges in West Germany. The Phantoms picked up the ALSS data by means of a DME system housed in a converted fuel tank below the right wing.

Official British nervousness over the ALSS deployment was increased on 29 May, when one of the U-2s crashed during the fourth test mission. Captain Robert Rendleman ejected to safety and the U-2 pancaked onto a track in the forest-covered hills of the Hunsruck near Winterberg in West Germany. Air Force spokesmen stressed that the plane was unarmed and carrying no cameras; it was one of five engaged on tests of 'a precision high altitude navigational system', — which was an approximation of the truth! The last of SAC's U-2Cs was hurriedly flown over to replace the crashed aircraft, and the ALSS flights continued.

When it was all over, the verdict was much the same as before: ALSS was too unreliable. Data-links failed, receivers and transmitters went off line, connectors failed to connect. Even during an English summer, crosswinds at Wethersfield were above U-2C limits on four occasions, forcing two missions to be delayed and another two postponed overnight. When everything was up and running, it could work. Unguided bombs struck within 150 feet of their emitting targets on the German ranges. However, within the group of Warsaw Pact emitters which it was expected to locate during the European deployment, ALSS had only picked up a third of those which had been reported active by other means of detection.

Seemingly undeterred, the Air Force pressed ahead with plans for the follow-on development of PLSS. Not only would PLSS have to be more reliable; it would have to accomplish a whole lot more when it was working. Since ALSS could only pick up pulsed signals, it was of no use whatsoever against the new breed of mobile Soviet SAMs such as the SA-6 which employed continuous-wave radar illumination. It could also be spoofed by frequency-agile radars. Neither could ALSS be justified as an ELINT system alone. Unlike Senior Ruby and similar exotic collection systems, it could only acquire frequency and pulse repetition interval (PRI) data on the emitters it picked up. In order to reprogramme ECM self-protection systems on its fighters and bombers, the Air Force also needed to know the pulse width, scan type, polarity and scan rate of the other side's radars.

So the PLSS requirements included capability against CW as well as pulsed signals; against frequency and PRI-agile radars; and ability to determine more signal parameters. There would be wider frequency coverage, aimed at targeting VHF and UHF communications emitters as well as radars; and better signal sorting and data processing, to include the computation of bearing information from the three airborne sensors as well as TDOA data. The vulnerability of the data-link to jamming would be reduced. There were even more goodies on the list: PLSS would be able to detect emitters even if they had only broadcast for five seconds. It would be able to direct up to thirty weapons against targets at a time — triple the capacity of the computers and data-links used in ALSS!

Preliminary design study contracts for the improved system were let to two teams for six months in 1976. This time, IBM joined forces with Lockheed, E-Systems and Collins, and this team beat another comprising Boeing, Hughes, RCA and McDonnell Douglas Electronics to the full-scale $120 million development contract, awarded on 1 July 1977. PLSS was under way, and a prototype was expected to be ready for flight tests in 1981.

Like ASARS, the PLSS development started while the Compass Cope RPV was still in contention as the airborne platform, but within a year the TR-1 had been chosen, and there was talk of needing another thirty-odd aircraft in addition to those already assigned to deploy with the radar sensor. The Lockheed company which was leading the PLSS effort was not, however, the Lockheed-California Company of which the Skunk Works was a part. It was Lockheed Missiles and Space Company (LMSC) up at Sunnyvale near San Francisco. LMSC was the prime PLSS contractor, responsible to the Air Force for design, development and integration of all the various subsystems, such as the ELINT sensors (E-Systems); DME equipment (IBM and Harris Corp); data-link (Sperry); ground communications (Collins) and ground station computers (Control Data). At first, LMSC had no formal relationship with the Skunk Works over PLSS; it was dealing with one set of people in Air Force Systems Command (AFSC) at Wright-Patterson AFB, while Burbank dealt with another group in Air Force Logistics Command (AFLC) at Robins AFB. Strictly speaking, as far as LMSC was concerned, the PLSS airframe was 'government-furnished equipment'.

Structural complications such as this bedevilled the PLSS programme almost from the start. Despite what they had learned from ALSS, both the government and LMSC badly underestimated the difficulty of integrating the various PLSS subsystems. LMSC had thirteen different sub-contractors working for it at various times during the development. Co-ordinating all the black box requirements and interfaces became a nightmare. A year-long cutback in funding during 1980-81 didn't help, and neither did LMSC's decision in August 1981 to uproot its PLSS workers and move them to a new division in Austin, Texas. (At the same time, it also moved another troubled programme, the US Army's Aquila RPV, to the same location.)

LMSC claimed that it had gathered together at Austin 'a dynamic and personal team of technical and operational problem solvers'. Nevertheless, the PLSS schedule slipped badly, and cost growth became a major problem. In an effort to make up lost time, there was little attempt to use ground simulation methods to aid in system integration. The result was that when the equipment was finally readied for flight tests (initially on one aircraft only) in December 1983, there were a large number of 'bugs' still remaining to be ironed out. There was a serious problem with spare parts — none had been provided! This caused more delays.

The first TR-1 modified to carry PLSS was Article 074, and it could not be mistaken for any other aircraft, thanks to the peculiar configuration of the nose. The phased array antennas of the E-Systems ELINT intercept receiver faced both ways behind a long, flat, slab-sided radome which immediately gained for the aircraft the nickname 'Pinnochio', after the Disney cartoon character whose nose grew and grew! DME equipment from Harris was housed in the two superpods.

Flight tests of the complete triad were rescheduled for September 1984, and then postponed again. The production decision was put off. In July 1985, the TR-1 triad was finally airborne. The aircraft flew weekly PLSS missions from Beale to the China Lake weapons range, where the emitters simulating Soviet systems were situated. The ground station was at Sunnyvale, LMSC's headquarters. In between the tests, LMSC system engineers tried to assimilate the results and get everything working properly. Despite optimistic claims by LMSC within days of the first three-aircraft test, it was another six months before consistently satisfactory results were being reported.[5] By now, the customer had lost patience, and the PLSS concept was in danger of being overtaken by events.

It was, after all, an extraordinarily complicated and expensive way of neutralizing the enemy's air defences. Apart from the Central Processing Subsystem (e.g. ground station), which would have to be housed in a hardened bunker if the system ever went operational in Europe, there were up to nine other widely strung-out ground sites each with a DME beacon, these being necessary to fix the position of each TR-1. Then there was the TR-1 triad itself (plus ground spare aircraft), its ELINT sensors, data-link and an atomic clock for the precise TDOA measurement, and the receivers to go on the strike aircraft and/or weapons. Since PLSS was supposed to cover frequencies from VHF all the way up to K-band, at ranges as great as 300 nautical miles, the antennae picked up millions of pulses per second, which the software had to filter and

The peculiar slab-sided nose of the PLSS-equipped TR-1 is accentuated to the almost-grotesque in this wide-angle shot, taken at Beale AFB in 1985. Additional PLSS equipment was carried in the modified superpods. Despite some occasional successes, the system never demonstrated enough reliability to preserve funding, which was eventually cut off in 1987, after massive resources had been expended to no avail over a fifteen-year period.

prioritize through complex number-crunching. And if a twenty-four-hour operation was demanded, twelve TR-1 aircraft would have to be dedicated to the mission, making a massive demand on the operating unit's resources.

The Wild Weasel community was, not surprisingly, dead set against it. TAC and USAFE commanders also had their doubts. The primary strike aircraft envisaged to work with PLSS was the F-16, but the Air Force was stuffing so many extra black boxes into its small airframe that there wasn't enough room left for the PLSS DME/data-link equipment. This would have to be housed in a pod, which would take up a valuable wing station. Of more significance, the forward march of defence technology had now produced capable new anti-radiation sensors which could be combined with a new generation of deadly missiles. The High-Speed Anti-Radiation Missile (HARM) was joining the Wild Weasel squadrons, representing a great improvement over their old Shrike and Standard ARM weapons. And in the 'black' world, Northrop was working on a new, stealthy, but low-cost, vehicle codenamed 'Tacit Rainbow' which could be launched from the ground or from a strike aircraft and made to loiter in the vicinity of enemy radars until they were turned on, whereupon it would dive down and destroy them if necessary.

In 1985, the Air Force was still telling the world (and, more particularly, the legislators on Capitol Hill) that PLSS was 'our best defence suppression engagement system . . . the capability is critical'. It admitted, however, that 'affordability is a vital concern'. After the years of plenty during the Reagan defence build-up, the military budget was coming under increased pressure from a Congress anxious to cut the massive US deficit. In mid-1986, some of the top brass were still trying to finesse the system through to production procurement. 'While it would appear that PLSS does the same thing as the Wild Weasel,' AFSC Colonel Edward Fincher told *Air Force Magazine,* 'the systems are complementary. The Weasel accompanies a strike force and attacks radar emitters as they are encountered. PLSS directs F-16s to do this as well. But the large number of such targets in the Warsaw Pact area means that there will be more than enough work for Weasels and F-16 PLSS strike aircraft, which will always be limited in number.' (This was said before the existence of 'Tacit Rainbow' was admitted.)

In reality, the Air Force now wanted to ditch PLSS. Having spent over $600 million on the system, with only a prototype to show for it, the Air Force decided after a major programme review in early 1986 to schedule more tests, but postpone a production decision. By June, it had decided to cancel the production programme entirely. The decision was opposed by some senior Pentagon officials, including Donald Latham, Assistant Secretary for C^{3}I, who told journalists that PLSS could 'detect, locate and identify an enemy radar to within fifty feet accuracy from 300 miles away . . . if I can keep the Air Force from doing something foolish — they're trying to cancel it'. The Air Force position was restated by Chief of Staff General Charles Gabriel: 'PLSS development has been painfully slow,' he said, and although it had demonstrated some success, the new pressures on the defence budget meant that it was no longer enough of a priority to survive. 'If they can't tell me we have a cost effective system to do the job for us, I would rather walk away from it than pour a lot of money down the hole.' Even PLSS supporters like Latham admitted that another $600 million would probably be needed to buy a sufficient quantity of production systems, including additional TR-1 airframes.

As the debate over PLSS rumbled on, production of the TR-1 at Palmdale slowed, and the original plan to produce thirty-five TR-1 aircraft was revised, so that this total also now included procurement of a few additional U-2R models to supplement those remaining in service with the 9th SRW. (As we have already seen, the difference between the two types was essentially limited to the sensor fit.) The Air Force was planning to keep the R-model in service well into the 1990s, and wanted to ensure it would not run out of airframes.

Ever since the U-2s had moved to Beale in 1976, there had been three far-flung but permanent U-2R detachments, all within easy flying time of politically unstable areas which the US wanted to keep an eye on. There was still a single aircraft kept at the British base of Akrotiri on Cyprus and flown occasionally over the eastern Mediterranean in an operation codenamed 'Olympic Harvest'. Another U-2R flew out of Patrick AFB, Florida, on 'Olympic Fire' missions around the Caribbean and Latin America. The third detachment was a two-plane operation codenamed 'Olympic Torch' out of Osan AB, South Korea, which kept an almost daily eye on the bellicose and unpredictable North Koreans.

On 22 May 1984, Captain David Bonsi was climbing out of Osan AB for a routine flight when there was an explosion behind him and the entire tail section separated. He ejected safely. An investigation started, but less than two months later a similar incident occurred at Beale AFB. This time, the aircraft had hardly left the runway when the tail section crumpled. Again, the pilot survived through ejection — at only a hundred feet. Operations were allowed to continue, but on 8 October yet another U-2 was lost, when it broke up in mid-flight over Korea. A third pilot was saved by the ejection seat. Now the entire TR-1 and U-2R fleet was grounded.

The cause of the first two accidents was pinned down shortly thereafter. The long exhaust duct on the U-2R and TR-1 was made up of two parts; a forward adaptor section of approximately four feet which was bolted to the J75 engine, and to which the long tailpipe fixed in turn. This adaptor/tailpipe link, consisting of a U-shaped clamp and two quarter-inch bolts, had come loose and caused the tailpipe to move out of alignment. As the jet blast increased, the tailpipe crumpled and the trapped exhaust blew the tail off at its mounting points just aft of the speedbrake. The remedy was relatively simple: a new type of clamp was devised, and more bolts were added. As for what had caused this problem after so many years of trouble-free operations, it was thought that the various attempts to tweak the J75 for better fuel consumption had led to more compressor surges putting a strain on the connection. Faulty maintenance procedures may also have been to blame. (The third accident was unrelated. It was attributed to loss of control in IFR conditions, leading to possible overspeed, upon which the aircraft broke up.)

The fix soon reached the field, and the grounding was lifted on 21 November. The Dragon Lady had been down long enough for senior US commanders in Europe and the Pacific to miss its valuable day-to-day contribution to the intelligence picture. Referring to the grounding, General Charles Gabriel later told Congress, 'The TR-1 gives a pretty good picture along the front line more continuously than the SR-71 aircraft, and that is the reason

General Rogers (Supreme Allied Commander Europe) likes it so much.'

In the early 1980s, Lockheed had entertained hopes of selling the TR-1 to the allies abroad. The Skunk Works was cleared in 1982 to make presentations to the British and West German governments; the third production TR-1A was exhibited at the Farnborough Air Show in the UK that September. The sales effort came to naught, however. The British didn't have any money, and the Germans were happy for the time being with the converted Atlantic maritime patrol planes they used for SIGINT missions.

For a while, it seemed as if TR-1 production would be extended in conjunction with the JSTARS (Joint Surveillance and Target Attack Radar System) programme. This was yet another new airborne battlefield surveillance sensor, similar to ASARS and PLSS in a number of respects. Like ASARS, JSTARS also featured a side-looking radar, offering very high resolution so that enemy forces could be detected at long stand-off range. Like PLSS, the intelligence would be processed and distributed in real time so that strike aircraft (or missiles) could be directed to the targets by the airborne platform. There were two essential differences between JSTARS and ASARS. JSTARS had been designed virtually from scratch to serve two masters (the Army and Air Force), with both services participating in system definition and development contracting. And the JSTARS radar would be multi-mode, featuring in addition to SAR a sophisticated Moving Target Indicator (MTI) so as to pinpoint enemy armour and mobile missiles on the move.

While requests for proposals for JSTARS were being solicited from US industry in 1984, an argument broke out over the choice of an airborne platform for it. The Army suggested its existing low-level reconnaissance plane (the OV-1 Mohawk); the paymasters in Congress favoured the TR-1. For a while, both aircraft were envisaged as JSTARS platforms by the military. But then the Air Force proposed instead to use a converted Boeing 707, designated C-18. They argued that the larger plane would be able to process, analyse, and disseminate the radar data onboard, rather than having to relay it all to a ground station first. No doubt the Air Force was by now growing wary of the complexities and cost escalations that were plaguing PLSS and, to a certain extent, the ground portion of the ASARS set-up. Relying on ground stations also meant that the system was not self-deployable; even if the ground equipment was mobile, it would pose considerable demands on airlift resources if it ever needed to be deployed quickly to a potential or actual battlefield.

The Army's plan to use the OV-1 faded out in mid-1984, when the two services signed a new memorandum of agreement over JSTARS and other joint projects. The brown-suiters would get the ground stations they wanted, which would process raw radar data which had been downlinked, and assign targets to the artillery units along the front. In return, they agreed to support the C-18 as the JSTARS airborne platform, so that the blue-suiters could do their processing onboard, and send the details on those targets selected for air strikes directly to the fighter-bombers. Despite this agreement, some Congressmen continued to champion the TR-1 for JSTARS, charging that the lower-flying C-18 was too vulnerable to enemy air defences. The Air Force countered that the C-18 'lends itself to some self-defence capability, whereas providing the TR-1 with a measure of self-defence would not be possible with the present state of technology'. Since the TR-1 already had a very capable electronic warfare suite, it wasn't entirely clear what they meant.

The programme faltered for a while as doubts were raised about its cost and feasibility. Some 'nice-to-have' features were deleted. Finally, in September 1985 a team led by Grumman and also including Norden (for the radar) and Boeing (for the C-18 airborne platform) was selected and received a $657 million contract for full-scale development of JSTARS. It would be a five-and-a-half-year programme; despite strong backing from the House Armed Services Committee, support for the TR-1 as the JSTARS platform faded away.

By now, Lockheed had run out of potential further applications for the Dragon Lady; even the once-serious interest expressed by the US Navy in adapting the U-2R as an ocean surveillance tool had evaporated. The admirals had begun to look closely at the U-2 during the early 1970s, as the threat to the fleet from long-range cruise missiles launched from over-the-horizon increased. What they wanted was a long-endurance platform with sensors that could detect potentially hostile surface ships at very long distances. It could be a satellite, or a very

The U-2EPX with Navy titles returns from an ocean patrol during 1973. This was the first of two configurations used in the EPX trials, and features yet another nose contour housing an RCA X-band radar. The aircraft achieved good results, but the Navy chose not to proceed with a production system.

high-flying aircraft. The sensors could be active, or passive, or a combination of both.

The Skunk Works came up with a three-sensor package on the U-2R, and the Navy funded it as the EPX (Electronics Patrol Experimental) programme. A heavily-modified RCA X-band weather radar was mounted in the nose. Wing pods of somewhat longer design than those developed for the Senior Spear sensor were also carried. The left pod contained an RCA return beam vidicon camera, while the right one housed a cut-down version of the UTL ALQ-110 ELINT receiver. An astro-inertial system was carried in the Q-bay, and a data-link system in the E-bay, with the antenna beneath the tail. The standard T-35 tracker camera was displaced from the nose to the right pod. The CIA unit loaned two aircraft for flight trials, which began in February 1973 out of Edwards.

The trials were moderately successful. The U-2 was launched from land bases on all flights; the option to launch or recover on a ship was there, but it would cause some disruption to normal carrier air wing operations, and in any case the aircraft's endurance was such that useful times-on-station could be achieved without recourse to it. Data from the EPX was downlinked and processed by surface warships, and in an early flight the radar detected a submarine which had surfaced 150 miles away. This wasn't bad for what was always viewed as an interim sensor, so in early 1974 the Navy awarded Lockheed another $3.4 million contract for a second series of flight tests using the more powerful Texas Instruments APS-116 radar. They were already familiar with this sensor at Burbank; it was being fitted to the S-3A Viking ASW aircraft now being produced there for the Navy in large numbers, and was claimed to be capable of detecting periscope-type targets in high seas. To do this, it had a large, rapid-scan parabolic antenna. Unlike the RCA radar used in the first EPX trials, there was no way this would fit in the nose of the U-2R. So the Skunk Works revived the inflatable radome idea that they had earlier come up with for the Senior Lance programme. This protruded some twenty-eight inches below the aircraft, and provided the APS-116 antenna with a 360-degree field of view. The nose reverted to the standard 'slick' type.

Test flights of this 'preproduction' U-2 EPX configuration were conducted in 1975. Whatever results were obtained (and none were made public), the Navy did not proceed further with EPX.[6] Space-based solutions to the requirement were beckoning; in April 1976 the prototype 'Classic Wizard' satellite system was launched. Through the deployment of three satellites carrying UTL's ELINT sensors and working in tandem, fairly accurate DF bearings could be returned to ground stations and triangulated to determine the position of hostile warships. The Soviets also used this method, and had gone one stage further by orbiting nuclear-powered radars for ship detection as well.

Of course, these satellites were necessarily positioned hundreds of miles above the oceans, rather than the ten miles or so of a high-flying platform like the U-2. The Navy's interest in such a platform revived in the 1980s, but they were now thinking in terms of a new generation of RPVs. 'Recent and potential advances in critical technologies such as aerodynamics, artificial intelligence, fusion and very high speed integrated circuits lend themselves for application in an advanced technology HALE RPV,' they told industry in late 1986. HALE stood for High Altitude Long Endurance.

By the mid-1980s, drones were in favour again. The Israelis had once again proved their worth in tactical engagements, during the 1982 clashes with Syria in the Bekaa Valley. Even the US Air Force was beginning to show renewed interest. A number of short and medium term programes were under way, and the Defence Advanced Research Projects Agency (DARPA) launched the 'Teal Cameo' project to define a high-altitude theatre unmanned vehicle as a potential successor to the TR-1. Such a vehicle might also, they said, serve as a platform if the PLSS technology was salvaged. And they were talking about vehicles with an endurance of days, rather than hours.

In the face of all this, the prospects of yet another round of U-2 production appeared slender indeed. In most respects, the Skunk Works was still Number One with the Air Force R&D community, thanks to their pioneering work on a stealthy fighter, which had matured in great secrecy over a ten-year period into what was reputed to be a limited operational capability, held in reserve at Groom Lake. But even the brilliant brains in ADP at Burbank and the new Kelly Johnson Advanced Research Center at nearby Rye Canyon could blot their copybook on occasions. In 1986 the Lockheed-California Company was accused by the government of slack security in the safeguarding and classification of top-secret documents. Hundreds had gone 'missing'. Progress payments on the SR-71 and TR-1, and on at least one classified programme, were held up. The company was obliged to house-clean and tighten up its procedures. The directive did not go down well with a group of people whose leitmotif was to avoid uneccessary paperwork. As far as the Skunk Works was concerned, security was maintained in a different way, by strictly controlling the number of people with access to the programmes, and vetting them thoroughly. This was much more effective than employing an army of clerks to stamp bits of paper with the correct security grading.

This was, after all, how they had produced the U-2, the Blackbird, and the stealth plane in double-quick time and without any unauthorized leaks of information. Lockheed Chairman Roy Anderson had summed the philosophy up nicely in a 1984 speech decrying increased interference by government and legislature. 'In the Skunk Works', he said, 'we have developed articles on a classified basis for about sixty per cent of the cost of developing a system on a non-classified basis, and in one-half to two-thirds the time. Same company. Same technologies. The only difference is the management method and freedom from queries that too many times tend to be politically motivated rather than founded in any abuse of the trust the nation places in our industry.'

U-2R from Det 2, 9th SRW, returns to Osan airbase in South Korea in late 1987 following an operational mission. ASARS-radar nose and full SIGINT fit is evident. A black cat emblem adorns the tail, similar to that adopted by the Chinese air force 35th Squadron in the sixties. Det 2 began using the emblem in the late-seventies, and actually went so far as to fly a black cat named Oscar to justify their action!

Even as production of the TR-1 ground slowly to a halt, the Skunk Works was coming up with innovative new ideas in the way of sensor fits and other capabilities. They developed a large radome on a pylon to fit on top of the fuselage aft of the E-bay. In 1986, *Jane's All The World's Aircraft* obtained a picture of the modified aircraft during flight tests at Palmdale, taken from outside the base. From this distance, the radome appeared to be an AWACS-type rotodome, but the authoritative reference book had got it wrong; it would have been impossible for the Dragon Lady to provide the payload and power required for such an installation. It was, in fact, a satellite antenna with which the aircraft could relay digital data acquired by its sensors to an orbiting platform in space. From there, the intelligence 'take' could be downlinked to virtually any command post with the necessary satellite dish. It suggested the possibility of the NSA, the Pentagon, or even the White House situation room participating in a U-2 mission, as it actually took place, even if the flight was on the other side of the world.

There were other advances as well. New electronic cameras had been developed for airborne reconnaissance, which did not actually use film at all. The film plane was replaced by a charged coupled device (CCD) array, and the 'image' captured theron could be immediately transmitted to the ground on a video data link. There, digital signal processing could manipulate the 'image' in all sorts of advantageous ways. It could be viewed on a high-resolution TV screen, or converted into 'hard copy' film. Such ground-exposed film actually cut through haze, smoke or light mist better than the old method of airborne exposure. This was great for long-range imaging of the stand-off type routinely carried out by the U-2 (and SR-71, for that matter), where the 'look angle' is such that the camera has to contend with a large amount of atmospheric attenuation. Another advantage of using the CCD design was that it reduced the installed weight of these hitherto bulky cameras by half — to around 230 lb. in the case of a sixty-six-inch focal length system. Itek developed an electronic camera which Lockheed adapted for use on the U-2R in a most ingenious installation.

The Generals loved it. No matter how good the radar or SIGINT sensors were, there was nothing like a real honest picture to convince their political masters. In this respect, little had changed since the days when President Eisenhower had crawled over the floor in the Oval Office, magnifying glass in hand, examining with fascination the blown-up prints from those early U-2 flights over the Soviet Union which Allen Dulles and Richard Bissell had spread out there for him. In most other respects, things were vastly different from those far-off days — socially, diplomatically, and militarily. Except, of course, that the Dragon Lady was still up there, bringing back vital information from its lonely, lofty patrol.

1 There were others. One was a proposal from General Dynamics to revamp its 1963-vintage RB-57F, itself an extensively remanufactured version of the original Martin RB-57D, with an even larger wing and four engines, two big TF33 turbofans and two auxiliary J60 turbojets in underwing pods. Twenty-one RB-57Fs had served the USAF in high-altitude sampling, photo and SIGINT roles, but all except three had now been retired. These were all flying with NASA at Houston, where their ability to carry heavier payloads than the U-2Cs at Ames was deemed useful, but the civilian agency found — like the Air Force before them — that the RB-57F's altitude performance left a lot to be desired. General Dynamics suggested airframe and engine modifications to boost the aircraft's ceiling to 68,000 feet, and quoted $98 million in total programme costs for the reactivation of fifteen stored RB-57F aircraft and the conversion of a further ten stock B-57 airframes.

2 Despite receiving two new versions of the Ryan Model 147 drones, one with provision for interchangeable payloads and one dedicated to electronic warfare, the Air Force's only operational RPV unit was disbanded in 1979.

3 The 9th SRW was reported to have deployed a U-2R equipped with the Q-bay sampling hatch to RAF Alconbury in the UK in late April 1986. This was shortly after the Soviet nuclear reactor at Chernobyl blew up. The aircraft was said to have flown over the Baltic in an attempt to monitor the spread of contamination; at a lower level, Air Force WC-135Bs with a similar sampling system were doing the same thing.

4 Strictly, this third trainer was designated U-2(T), since it was procured with U-2R, as opposed to TR-1, funds. The 4029th SRTS was redesignated 5th SRTS in 1986.

5 In late 1986, Air Force Chief of Staff General Charles Gabriel appeared to contradict this report when he told *Armed Forces Journal International* that 'only about fifteen minutes of triad data' had been obtained in the entire flight test programme.

6 In a supplementary proposal, Lockheed suggested arming the U-2EPX with two long-range Rockwell Condor anti-ship missiles, which would have been carried on underwing pylons. This idea, too, was never taken up, and the Condor programme itself was cancelled in 1976.

Update

In December 1987, an event which had seemed impossible only four years earlier was acted out in Washington before the gaze of a bemused world. The US President, who had castigated the Soviet Union as 'an evil empire', sat down with its formidable leader and signed the first-ever nuclear arms reduction agreement. As part of the strict verification provisions of the INF treaty, both sides handed over precise details of their current medium-range missile deployments.

From background briefings, journalists learned that the data received from Moscow had caused US intelligence agencies to review their conclusions on Soviet missile strength. The analysts had apparently over-estimated the number of intermediate-range SS-20 missiles, but under-estimated the number of shorter-range SS-23 systems that had been deployed in Eastern Europe. A Soviet ground-launched cruise missile had been produced in considerably larger numbers than the US thought possible, although Moscow said it had not yet been deployed. Conversely, about 200 missiles covered by the treaty which the US had thought operational were now declared to be for 'training only', many of them being filled with concrete.

The discrepancies proved once again that no matter how good a country's technical intelligence collection capability was, it could never be omniscient. As Amrom Katz, who had a long and distinguished career in the technical development of overhead reconnaissance systems, once teased, 'We have never found anything that the Soviets have successfully hidden!' Washington

A full suite of defensive electronic equipment helps protect the U-2R/TR-1 series from enemy counter-measures. This picture shows the wingtip antennas for System 27, a digital radar homing and warning system.

sources now reported that US reconnaissance satellites had *never* imaged an SS-20 on operational deployment. The Soviets had managed to camouflage its movements so successfully that the bean-counters at NPIC had been obliged to estimate numbers by counting the garages which sheltered the missile's transporter/erector/launcher. This apparent Soviet success in the art of 'maskirovka' (concealment) only served to underline the value of alternative methods of detection, of the kind carried out by the TR-1 flights across Central Europe. If the Warsaw Pact's mobile missiles cannot be located by satellite photography, then alternative sensors such as SIGINT or radar must come into play.

Using the SIGINT sensor portion of its Tactical Reconnaissance System (TRS), the TR-1 can 'hear' for up to 350 miles. This brings virtually all of Czechoslovakia, East Germany, Hungary and Poland within eavesdropping distance, to which may be added large portions of European Russia if a patrol is flown over the Baltic Sea. The 'look' from the TR-1's ASARS radar is somewhat shorter-range, but still enough to provide good coverage of any threatening armoured movements towards NATO territory. TRS was supposed to combine the 'take' from these two sensor elements in order to provide confirmed target identification — an example of new intelligence techniques known as data fusion which are made possible by the power of modern computing. This proved more difficult than anticipated, however.

On 16 July 1985, the 17th RW flew the first full-up TRS mission with both SIGINT sensors and ASARS functioning. The flight marked the beginning of the system's Initial Operational Test and Evaluation (IOT&E). This phase lasted throughout 1986 and into the early months of 1987, and included a demonstration in September 1986 of the wing's ability to provide continuous coverage of the Central Front for forty-eight hours. However, in early 1987, US Defense Secretary Caspar Weinberger reported to Congress that the IOT&E had 'shown the need for some system modifications. Incorporation of these, as well as some expansion of planned future ground-station capabilities, have delayed ultimate program completion by several years'.

At that time, it was planned to build a permanent, hardened ground station in Europe near the spot where the temporary ground station was located. That would enable this demonstration equipment, currently housed in mobile trailers, to be upgraded and relocated to a second hardened site in NATO, although the Air Force said it was considering acquiring a purpose-built mobile station which 'we think will have more survivability'.

In the same report to Congress, Weinberger said that PLSS had been dropped because 'the complex task of processing and analyzing the vast number of signals picked up during fast-paced combat operations has proven to be more difficult than anticipated.' But the programme struggled on using some limited additional funds for testing, reluctantly doled out by a disbelieving Air Force hierarchy. In spring 1987, the triad was exercised in one of TAC's Green Flag electronic warfare mock combats on the Nellis Range. According to Lockheed, PLSS worked very well this time. A test in Europe was then planned for Fiscal Year 1988, but there wasn't enough money allocated to fund the deployment from the $15 million allocated for PLSS in the Fiscal 1988 budget. There was talk of a new platform and a substitute programme named Signal Location and Targeting System (SLATS) which would cost less and be of worldwide application.

Although TRS was encountering integration difficulties, the Hughes ASARS portion of it was producing outstanding results. It approved for production, and Hughes declared in early 1988 that ASARS 'produces long-range images *superior* to those delivered by photo techniques'. (Author's emphasis.)

Meanwhile, Grumman and its JSTARS subcontractors justified their multi-billion dollar programme in terms which suggested that ASARS didn't even exist! Among JSTARS attributes, they cited its 'unprecedented' fixed target capability, although it seemed unlikely that the Norden radar would match ASARS in high-resolution coverage. However, JSTARS did have an SAR mode which could be selected by the operators to replace the primary Moving Target Indicator (MTI) mode when a target vehicle was travelling below the minimum detectable velocity threshold of the MTI. Interestingly, Hughes was in the process of adding a Moving Target Indicator (MTI) to ASARS. Lockheed people believed that ASARS on the TR-1 was already doing eighty per cent of the JSTARS job —adding an MTI to the Hughes radar would raise that percentage still further.

With the US defence budget coming under pressure, and Secretary Gorbachev hinting at Soviet conventional force reductions in Europe, was the estimated $7.7 billion cost of twenty-two JSTARS aircraft plus 100 ground stations justified? Almost inevitably, the JSTARS contractors encountered software integration problems, and the programme fell behind schedule. Was this another PLSS-type fiasco in the making? A decision on full-rate production would not now be taken until late in 1991. The JSTARS slowdown came as no great disappointment to senior NATO commanders, who feared a 'data deluge' from all the new battlefield intelligence systems, unless more thought was given to the correlation, presentation and dissemination of their product.

There were still rumblings of discontent over the choice of the Boeing 707 airframe (now designated E-8A) as the JSTARS platform, on vulnerability grounds. The Air Force maintained that the threat was overstated. If the E-8 was vulnerable, then so was AWACS, and even C-5/C-141 transports delivering reinforcements to the front line. In any case, NATO would be attacking Soviet airfields and SAM sites if fighting broke out, to neutralize the air defence threat. Besides, the E-8 could also operate successfully from further behind the front line if forced to retreat, they said. The same holds true for the TR-1, of course, except that the likelihood of the Dragon Lady being intercepted by fighters is more remote.

Even so, concern over Soviet ability to intercept large radiating airborne platforms like JSTARS and AWACS was growing. In 1983 the Soviets had deployed the long-range SA-5 SAM in Czechoslovakia, East Germany and Hungary. This large two-stage missile with four strap-on solid boosters had never been seen outside the USSR's borders before. Its performance had been the subject of debate within the US intelligence community for over twenty years; on an optimistic view, its antiquated semi-active guidance system could be easily countered. But there were now suspicions that the SA-5 had been fitted with a new anti-radiation seeker. 'The enemy has, or will soon have, the ability to actively locate any large emitting target and either successfully jam or engage these targets with anti-radiation missiles,' said AFSC Brigadier General Charles Winters in early 1987. There was also the SA-12a to contend with, another new long-range Soviet SAM which looked like replacing the ageing SA-4 as a mobile, forward-deployed counter to high-flying aircraft. Faced with such a threat, one advantage that the TR-1/TRS has over the E-8/JSTARS is that the former can shut down its active radiating sensor (ASARS) and still perform in passive mode using the SIGINT systems.

Meanwhile, Lockheed embarked on new

TR-1 belonging to the 17th RW is prepared for its next sortie inside one of five specially-constructed individual hangars at Alconbury. In 1986, construction began on twelve fully-hardened concrete shelters for the wing's aircraft.

efforts to protect the TR-1. Funding for an unspecified 'advanced defense system' for the aircraft costing $33 million was requested. The proposed upgrade was 'now lead time away from the threat capability', said the Pentagon. Other TR-1 improvements were also in the pipeline. In a weight reduction programme, composite material was being substituted on some parts of the airframe, such as the speed brakes and lift spoilers. And an upgrade to the J75 powerplant was in prospect.

As the US Air Force grew more serious about deploying an unmanned aircraft to supplant the RF-4C in the low-altitude penetrating reconnaissance role, another new Dragon Lady mission beckoned. The plan was to use the orbiting TR-1 as a relay platform for the imaging sensor data being collected from behind enemy lines by the drone. The aircraft's operational future was secure, unlike that of its illustrious stablemate, the SR-71 Blackbird. Faced in the late 1980s with an impending budget crunch, Pentagon planners began to think the unthinkable: the 9th SRW's SR-71 squadron was threatened with deactivation. Keeping the fleet of ten Blackbirds active was rumoured to be costing $300 million per annum. A significant portion of their mission could be accomplished by satellites and other means, it was said. The Blackbirds escaped the axe in 1988, largely thanks to pressure from Congress. But it was only a temporary reprieve, and plans to retire the aircraft in the Fiscal Year starting October were made public in January 1989. The top US Air Force commanders reluctantly supported the decision.

No cutbacks were made in U-2R and TR-1 operations, however, and here was an ironic turn of events. The Blackburn had been brought into service twenty-five years earlier as a replacement for the U-2! It was really just like Milt Caniff's character in that old 'Terry and the Pirates' cartoon strip. The Dragon Lady was still outlasting her rivals, and acquiring the air of immortality.

Epilogue

The man who, in the view of gain thinks of righteousness; who, in the view of danger is prepared to give up his life; and who does not forget an old agreement, however far back it extends: — such a man may be reckoned a complete man. Old Confucian saying.

The Pan American Boeing 747 had just landed at Los Angeles International Airport after the long flight from Hong Kong. As the passengers began to disembark up the jetway, the man who had told airport officials he was from the State Department moved forward. Unusually for a US government employee on official business, he was dressed casually. On the front of his well-worn leather jacket was sewn a large red patch. There was no legend on the patch — only the face of a black cat.

Two middle-aged Chinese men stepped off the aircraft, both looking disoriented and more than a little anxious. Then one of them spotted the man in the leather jacket, and rushed over to embrace him. His companion quickly followed suit; both seemed almost overcome by emotion. The mysterious federal agent hustled the strange pair away from the gaze of curious onlookers, and within a short while had cleared them through customs and immigration formalities and put them onto another airliner headed for Washington D.C.

The American's name was John R . . . ., and he was a CIA agent, not a State Department official. Twenty years earlier he had been a security officer at Taoyuan airbase on Taiwan with the joint US/Chinese 35th squadron. The two Chinese men he had come to meet were 'Chappie' Yei Chang Yi and 'Jack' Chang Li Yi, former U-2 pilots with the squadron, who had been shot down over mainland China in 1963 and 1965 and given up for dead. In fact, they had both survived the destruction of their aircraft by baling out. Following their capture by the communists, they had been detained for two long and painful decades until the government in Peking finally decided to release them in late 1983.

Now they had come to settle in the US. It was not their first choice of residence. Naturally enough, both wanted to return to Taiwan, but in an act of unconscionable meanness and ingratitude, the nationalist government had refused to have them back! Thanks to the loyal intervention of some of their former colleagues, both Chinese and American, the CIA had quietly arranged for their admission to the US instead. The poignant reunion at Los Angeles airport recalled a similar scene on the Glienicker Bridge in Berlin twenty-one years earlier, when Frank Powers had been met by Colonel Leo Geary and walked to freedom, in exchange for Colonel Rudolf Abel, the Soviet spy.

When the two former Chinese U-2 pilots

'Chappie' Yei (left) and Jack Chiang, the two U-2 pilots from Taiwan who seemingly came back from the dead. Few, if any, in the West knew that they had survived being shot down, until Peking revealed their existence in 1983.

reached Washington, they related their incredible story to the Agency's debriefers. Both had fallen victim to the SA-2 SAM, but like Powers, had managed to escape as their aircraft broke up and spun towards the ground. Yei escaped his downing on 1 November 1963 near Shanghai without serious injury, but Chang was not so lucky when he was shot down during that fateful mission to the Paotow nuclear reactor on the night of 10 January 1965. Although his parachute successfully deployed, he hit the ground so hard that he broke both his legs. The temperature was minus twenty degrees that night, and being virtually immobilized, he suffered severe frostbite. At daybreak, he managed to drag himself into a nearby village and received medical attention which saved his feet from amputation.

Both pilots were sentenced to ten-year prison terms including hard labour on collective farms. Yei tried to commit suicide in prison, and had his sentence extended by three years. Mercifully, both escaped the attention of the Red Guards during the upheavals of the Cultural Revolution. In the mid-seventies they were released from prison, but were made to settle in their native towns. (Both came from families which had fled the mainland with Chiang Kai Shek during the Communist Revolution in 1948.) Yei taught English at a college in Wuhan while Chang worked as a mechanical engineer at a factory in Nanking. Neither was aware that the other had survived until a joyous day in October 1983 when both were called to Peking and told that they were free to leave the country — together!

They were provided with documents which were good for a journey as far as Hong Kong, and told that arrangements for onward travel from there were entirely their own responsibility. When they arrived in the British crown colony, they were put up in the Imperial Hotel at Peking's expense. The People's Republic issued a short statement detailing the release. Word reached Taiwan, where their former U-2 squadron commander 'Gimo' Yang was now a senior Boeing 747 captain with China Airlines. Yang flew to Hong Kong and met his two long-lost comrades, who told him they wanted to return to Taiwan. At the time of their disappearance, both were married, and Chang had fathered two boys and a girl. Their former squadron boss had to tell them that both wives had re-married, since the two pilots had long been presumed dead.

Yang returned to Taiwan and passed on the pair's request for re-entry to senior government officials. He was quickly told to drop the matter. Some sources suggest that the government in Taipei was prepared to court-martial them if they set foot again on nationalist soil. The Taiwan Central Intelligence Agency evidently thought that the two had been 'turned' by the communists into double-agents. Paranoia still reigned supreme in the higher counsels of state in Taipei, even though it had been apparent for years that the mainland no longer posed a military (or now even an economic) threat to the increasingly prosperous offshore island. Moreover, it was hard to imagine what useful espionage could have been conducted in Taiwan by Yei and Chang, given the publicity which Peking had accorded their release. The communist government surely did not expect the pair to be readily re-admitted to the nationalist air force!

Maybe the nationalists were also embarrassed; there was certainly an element of calculation in Peking's decision to release the two pilots. It was a key objective of communist diplomacy to make Taiwan renounce its claim to sovereignty over the mainland, and acknowledge the legitimacy of the government in Peking. The nationalists had steadfastly refused to do this ever since their defeat in 1948. Indeed, they still maintained the fiction of electing deputies to the National Assembly on Taiwan who 'represented' the mainland provinces. As part of this hardline policy, Taiwan refused all contact with the mainland, and prohibited its citizens from travelling there, although many still had close family ties. By releasing the two pilots, Peking may have hoped for a breakthrough in their quest to reunite Taiwan with the mainland by peaceful means.

Taiwan was still governed under martial law by Chiang Kai Shek's son, seventy-three year-old President Chiang Ching Kuo. It is not known whether the decision to refuse permission for Yei and Chang to return was taken by him personally. However, the President was well aware of the Black Cat squadron: he had been the head of the National Security Agency on Taiwan when the U-2 project was set up. He was widely credited by the American side with ensuring that the operation at Taoyuan went smoothly. He made visits to the squadron and posed with the pilots for photographs.

There may have been another reason behind Taiwan's refusal to welcome back the pair of U-2 pilots. As in the case of Frank Powers, there was a feeling in certain official quarters that Yei and Chang had no right to be alive. Rather than being captured by the communists, they should have taken their own lives. As was the case with Powers, however, such an instruction was never officially issued, or even discussed, at least not by the CIA. The preflight intelligence briefing conducted at Taoyuan before each overflight of mainland China was tape-recorded, and a copy sent to project HQ in the US. The record shows that the Chinese pilots were told what to do in the event that they were captured: 'You are to say that you are a pilot of the 6th group on a routine patrol.'

It may be, however, that the nationalist government privately asked them not to be taken alive. Whatever the truth, the nationalist propaganda machine certainly perpetuated the notion. When the very first Chinese U-2 was shot down in September 1962, they made a hero out of the pilot, Chen Huai Sheng. Taiwan said that he successfully force-landed the plane, but then shot himself with the hunting pistol from his survival kit, when it became apparent that he would be captured. The government named a high school in Taipei after Chen, and a memorial hall and statue of him was erected on the playing fields of a village a few miles north of the capital.

Chen may have been a hero — as indeed were all the other Chinese U-2 pilots — but the word which eventually filtered back to the Black Cat squadron was that he had not committed suicide. He had been severely injured during the shoot-down, but was rescued and taken to hospital by his captors. He reportedly died there overnight.

Gimo Yang relayed the unwelcome news from Taipei to his two former pilots, still languishing in the Imperial Hotel in Hong Kong. But he also relayed the story to Bob Ericson, the former CIA U-2 pilot now flying for NASA at the Ames Research Center in California. Ericson had spent more time at Taoyuan than any of the other US pilots, and knew both Yei and Chang. In turn, he contacted the CIA's former Director of Administration for the U-2 project in Washington, Jim Cunningham. From there, the wheels started to turn. Two weeks later, Cunningham called Gimo Yang in Taipei and asked him to return to Hong Kong. There Yang interviewed the two pilots again, this time in the presence of a CIA polygraph operator. Like Powers before them, these returning U-2 pilots had to submit to a lie-detector test before they could be fully cleared. Yei and Chang were then offered the chance to emigrate to the US. They took it.

'Chappie' Yei now lives in Texas, where he manages a supermarket with his brother. Jack Chang has settled on the east coast, where he runs an apartment block for senior citizens. Both maintain a low profile, which is understandable in the circumstances. Neither of them has re-married. The CIA arranged for them both to be paid $200,000 in compensation and back-pay. Those Americans who had served with the U-2 project in Taiwan heard the news of their resettlement with great satisfaction. 'I'm very proud that the Agency has treated these long-lost drivers like members of its own "family",' said one. 'I don't think it had any legal obligation to these guys, but it recognized a moral obligation.' As for the money, another Taiwan U-2 veteran pointed out that it was a smaller sum than that routinely handed out by the Taiwan nationalists themselves, as a reward to communist Chinese pilots who defect from the mainland!

★ ★ ★ ★

3 May 1986. On the flight deck of the China Airlines Boeing 747 freighter, the three-man crew prepared to start their descent into Hong Kong. It was a routine stop on flight CI334, a regular cargo schedule from Bangkok to Taipei. In the cabin below there were no passengers, but ninety-one tonnes of electronics, garments, fish and fruits. Captain Wang suddenly turned to his fellow crewmen, and dropped a bombshell. Instead of landing in Hong Kong, Wang informed them that he was going to fly the jet to nearby Canton. Schooled in years of obedience to the pilot-in-command, the startled co-pilot and engineer chose to accede to the captain's demand. The 747 made a safe landing at Canton airport, and Captain Wang announced that he was defecting to the People's Republic of China.

Although no China Airlines jet had ever flown into mainland Chinese airspace before, Captain Wang had done so on a number of occasions twenty years earlier. For this errant 747 pilot was none other than Major Johnny Wang Shi Chuen, former U-2 pilot and veteran of ten hazardous overflights of the mainland! But unlike his earlier excursions to Lanchow and Paotow, this time the communists didn't try to shoot Wang down. Instead, the official Xinhua news agency got the story out within hours, including the fact that the 747 captain was a former U-2 pilot. They could have added that Wang even rose to command the Black Cat squadron. After completing the prescribed ten overflights, he had been promoted to Lieutenant Colonel and chosen to succeed Gimo Yang as commander of the unit.

According to Xinhua, fifty-six year-old Wang wanted to be reunited with his ageing father in Sichuan province. They neglected to mention that he had left behind a wife and two children in Taiwan. In common with all the other ROCAF pilots selected for the U-2 squadron prior to 1965, Wang came from a staunch nationalist family which had retreated across the Taiwan Strait with Chiang Kai Shek in 1948. In common with members of other families with a similar history, he was prevented from ever visiting relatives that had been left behind on the mainland. According to some who met him in the final few months before his dramatic defection, Wang appeared to be disillusioned with life, despite his superior status and salary. He apparently did not get along well with his colleagues in China Airlines. Wang did not talk openly about politics, but his growing disenchantment with old-guard nationalist rhetoric and actions must have received a boost when the Taiwan government refused to readmit his former squadron mates Yei and Chang a couple of years earlier.

Whatever embarrassment that event had caused was but nothing compared with the dilemma that Wang's action now presented in Taipei. There was an $80 million airliner sitting on the mainland, not to mention the loyal co-pilot and flight engineer, and the communists weren't going to release aircraft or crew unless the nationalists engaged in direct negotiations! Nearly four decades of principled dissociation thus came to an end as officials from Taipei sat down in Hong Kong with officials from Peking. After four days of talks, the two sides agreed the release of the 747 and two crew.

Johnny Wang stayed behind. Within a short while, he was working in a senior position for the state airline CAAC, and was even elected a member of the People's Congress. Had he been working for the communists all along? It seems unlikely, but he probably made contact with the mainland authorities a good while before his defection. Whether engineered by Peking or not, Wang's action marked a turning point in relations between the two rival Chinese states. Some months before he died in January 1988, President Chiang began a series of far-reaching reforms. He ended martial law, lifted exchange controls, allowed opposition political parties to form, and relaxed the ban on Taiwan citizens travelling to the mainland.

24 May 1987. Under an unseasonably grey sky, a distinguished group of military veterans gathered just inside the main gate at Laughlin AFB, Del Rio, Texas. A dozen retired military aircraft lined the road which led from the gate into the base, but this group was interested in only one of them. Facing U-2C serial 66707, over two hundred members of the 4080th/100th SRW

Brigadier-General (retd) Austin Russell addresses the former members of the 4080th SRW at the 1987 reunion. Russell commanded the wing from 1957-58, and checked out in the U-2.

Reunion Association and their wives stood to attention as a memorial service was conducted. They had plenty of memories to share, of an outfit which was unique in the annals of the US Air Force. Some of them had served ten years or more in the closely-knit unit, and had sacrificed much 'for the mission and the welfare of mankind', as the roll of honour in wing headquarters used to proclaim.

Among the many former wing and squadron commanders present that day at Laughlin, one of the most senior was retired general Austin J Russell. During his short but effective tenure as 4080th Wing Commander, he had gained a reputation as a tough disciplinarian in the same mould as LeMay and others in the SAC hierarchy. Now a frail (but still upright) man in his seventies, Russell rose to make a short speech. He reminded his audience of the history they had forged and the places they had been . . . Argentina, Alaska, Cuba, Vietnam. He spoke of the medals, campaign honours and unit citations that had been bestowed on the wing in unprecedented numbers. He talked of the discomforts and frustrations that arose from the nature of the work, especially when so many of the objectives were shrouded in secrecy. Then he started to remind them of those comrades — mainly pilots — who had given their lives in the course of duty. Their thoughts turned to Anderson, of course, but also to many others, including such popular and capable men as Pinky Primrose, John Campbell and Alfred Chapin. And there in the audience, as Russell glanced up from his notes, was Chapin's widow. The wife of the pilot that had checked out the doughty general in the U-2 nineteen years earlier, shortly before he was killed in a high-altitude accident. Russell stumbled over his words, shuffled his notes, and ground to a halt. The occasion was proving too much for even this tough old warrior. Overcome with emotion, he stammered an apology and abruptly sat down.

19 September 1987. Amidst the glitzy surroundings of the Frontier Hotel and Casino in Las Vegas, another group of airborne reconnaissance veterans were holding their biennial reunion. They called themselves the Roadrunners Internationale, and the criteria for membership was that you must have served with one of the 'black'

Reunion scene at Laughlin AFB in May 1987, as former members of the 4080th SRW gather in front of retired U-2C, serial 66707, for a memorial service.

programmes out at The Ranch. By definition, therefore, their ranks consisted largely of U-2 and SR-71 people from the Air Force, the CIA and Lockheed.

Again there were speeches, and memories, and tributes. Retired Air Force Brigadier General Leo Geary stood up. The former manager of the military side of Operation Overflight had an important announcement to make. For many years, Geary had been troubled by what he and many others considered to be an important piece of unfinished business. Now he had finally gained the authority from Washington to discharge an obligation.

From the speaker's platform, he called for Sue Powers to come forward. Two hundred pairs of eyes followed the second wife and widow of the most famous U-2 pilot of all, as she rose and made her way towards him. Geary presented her with a citation made out to Captain Francis Gary Powers, which specifically commended him for his flight of 1 May 1960 'into denied territory'. There was also a military medal. It was the Distinguished Flying Cross, awarded to all of Powers' contemporaries in 1957-58, but withheld from him alone for nearly twenty years!

★ ★ ★ ★

17 April 1989. There wasn't much time left. Ben Rich and his colleagues at the Skunk Works had always known that the U-2C model with the J75 engine could smash the world time-to-climb and sustained altitude records for its weight category. But the aircraft's maximum performance had always been shrouded in official secrecy, and the US Air Force had never allowed them to make the attempt. Now the very last of the original U-2 models — Article 349/NASA 709 — was due to make its final flight in ten days' time.

The authorities had finally sanctioned the record-breaking flights, and the aircraft had been flown from NASA Ames to Edwards AFB. As dawn broke, the mid-fifties vintage U-2 was positioned close to the south runway at Edwards. Despite its age, NASA 709 was still gleaming in its smart white and grey colours with the blue trim. It had been fuelled with only 395 gallons, to give a gross take-off weight of 15,989 lb.

With observers from the Fédération Aèronautique Internationale (FAI), the press and television looking on, NASA pilot Jerry Hoyt started up and moved to the threshold. Shortly after 0700, the last of the original U-birds soared into the blue sky. Those who were unfamiliar with a high-performance U-2 take-off gasped as Hoyt pointed the nose almost vertical and climbed away.

From the control room at NASA's Dryden Research Facility at the edge of the lakebed, Hoyt's progress was tracked by radar. The old time-to-climb records in the FAI's C-1F Group III weight category were all held by a Lear Jet, the hottest of the small business jets. Predictably enough, the U-2C broke them by a very wide margin. It reached three kilometres altitude (9,842 feet) in fifty-two seconds, compared with one minute forty-eight seconds for the Lear Jet. At six, nine and twelve kilometres, the performance gap widened still further, and the Dragon Lady reached fifteen kilometres (49,212 feet) in six minutes fifteen seconds, almost three times faster than the previous record. The Lear Jet had eventually reached 54,370 feet but Hoyt and NASA 709 had a way to go yet. They passed twenty kilometres (65,617 feet) in twelve minutes thirteen seconds, and finally rounded out at 73,700 feet after just sixteen minutes.

Hoyt brought the U-2 back down and taxied up to the NASA hangar. Ben Rich and most of the Skunk Works managers and test pilots were on hand to greet him, including Tony LeVier. The spritely seventy-six-year-old who had made the first flight in a U-2 almost thirty-four years earlier was, as usual, in a voluble mood. 'It's a hell of an airplane,' he told reporters. 'That damn plane may have kept us out of World War III.' Rich said the record flight would pay tribute to all those who had been associated with the aircraft over the years.

Then next day, they added a full fuel load to NASA 709, giving a gross take-off weight of 20,900 lb. Ron Williams took it up again to set new records in the higher weight C-1G category, which were previously held by the Canadair Challenger and the Dassault Falcon 900 business jet. Then he flew the short distance to Palmdale, where Lockheed repainted the aircraft black at the request of its new owners-to-be, the air museum at Robins AFB in Georgia. On 26 April 1989, Doyle Krumrey had the honour of piloting the last of the original U-2s to its final resting place.

Bibliography

Chapter One

Beschloss, Michael R., *Mayday,* Harper and Row, (New York, 1986)
Boyne, Walt, 'The Most Elusive Hughes . . . the D-2/F-11 Recon Bomber', *Wings,* June 1977
Burrows, William E., *Deep Black,* Random House, (New York, 1986)
Godard, George W., *Overview,* Doubleday, (New York, 1969)
Holm, Skip, 'Article Airborne', *Air Progress Aviation Review,* vol. III, 1986
Johnson, Clarence 'Kelly', *More Than My Share of It All,* Smithsonian Institution Press, (Washington DC, 1985)
Katz, Amrom, (ed.), 'Selected Readings in Aerial Reconnaissance', *Rand Corporation Report,* P-2762, August 1963
Killian, James R.,*Sputnik, Scientists, and Eisenhower,* The MIT Press, (Cambridge, Mass., 1977)
Powers, Thomas, *The Man Who Kept The Secrets,* Alfred Knopf, (New York, 1979)
Richelson, Jeffrey, *American Espionage and the Soviet Target*, William Morrow, (New York, 1987)
Sloop, John L., 'Liquid Hydrogen as a Propulsion Fuel', *NASA History Series,* SP-4404, 1978
Wise, David and Ross, Thomas B., *The U-2 Affair,* Random House, (New York, 1962)
York, Herbert F., and Greb, G. Allen, 'Strategic Reconnaissance', *Bulletin of the Atomic Scientists,* April 1977

Chapter Two

Beschloss, op.cit.
Burrows, op.cit.
Johnson, op.cit.
Prados, John, *The Soviet Estimate,* The Dial Press, (New York, 1981)
Richelson, op.cit.
Rostow, Walt W., *Open Skies,* University of Texas Press, (Austin, 1982)
Rubin, Ronald, 'A Day at The Ranch', *Gung Ho Magazine,* July 1983
'The Trial of the U-2' — transcript of the court proceedings against Francis Gary Powers, Translation World Publishers, (Chicago, 1960)

Chapter Three

Aviation Week and *Space Technoloy* 16, 23 and 30 May 1960.
Bamford, James, *The Puzzle Palace,* Houghton Mifflin Co, (Boston, 1982)
Beschloss, op.cit.
Brugioni, Dino, 'The Tyuratam Enigma', *Air Force Magazine,* March 1984
Dulles, Allen, *The Craft of Intelligence,* Harper and Row, (New York, 1963)
Eisenhower, Dwight D., *Waging Peace: The White House Years 1956-1961*, Doubleday and Co, (New York, 1965)

'How U.S. Taps Soviet Missile Secrets', *Aviation Week,* 21, October 1957
Johnson, op.cit.
Moseley, Leonard, *Dulles,* Hodder & Stoughton, (London, 1978)
Peebles, Curtis, 'Tests of the SS-6 Sapwood ICBM', *Spaceflight,* Nov-Dec 1980.
Penkovsky, Oleg, *The Penkovsky Papers,* Collins, (London, 1965)
Powers, Francis Gary, *Operation Overflight,* Holt, Rinehart & Winston, (New York, 1970)
Powers, Thomas, *The Man Who Kept The Secrets,* Alfred Knopf, (New York, 1979)
Prados, op.cit.
Richelson, op.cit.
Rühle, Hans, 'Gorbachev's Star Wars', *NATO Review,* August 1985
'The Trial of the U-2' op.cit.
United States Congress, *Hearings before the Senate Committee on Foreign Relations: May, June 1960. 86th Congress, Second Session,* US Government Printing Office 1960 (uncensored portion) and 1982 (declassified portion)
Wise and Ross, op.cit.

Chapter Four

Arnold, Lorna, *A Very Special Relationship,* HMSO, (London, 1987)
Bamford, op.cit.
Beschloss, op.cit.
Glasstone, Samuel and Dolan, Philip, *The Effects of Nuclear Weapons,* Castle House Publications, (Tunbridge Wells, 1980)
Richelson, op.cit.
Dedicated To Peace — a History of the 4080th Strategic Wing, 1956-1966, published privately 1966
Strategic Reconnaissance 1956–1976 — a History of the 4080th/100th SRW, Taylor Publishing Co/100th SRW, (1976)
'HASP — Purpose and Methods', *Isotopes Inc/Defense Atomic Support Agency,* report no. 1300, 31 August 1961
Del Rio Morning Herald: transcripts of microfilm archives, provided by The Laughlin Heritage Foundation

Chapter Five

Anderton, David, *Strategic Air Command,* Charles Scribner's Sons, (New York, 1976)
Burrows, op.cit.
Infield, Glenn B., *Unarmed and Unafraid,* Macmillan, (New York, 1970)
Lavalle, Major A.J.C., (ed.) *The Battle For The Skies Over North Vietnam,* USAF South-east Asia Monograph Series, USGPO, (Washington, 1976)
Momyer, General William W., *Air Power in Three Wars,* USGPO, (Washington, 1978)
Moser, Don, 'The Time of the Angel: The U-2, Cuba and the CIA', *American Heritage*, October 1977
Prados, op.cit.
Wagner, William, *Lightning Bugs, And Other Reconnaissance Drones,* Armed Forces Journal/Aero Publishers, (Fallbrook, Ca. 1982)
Futrell, Robert F., *The USAF in Southeast Asia — The Advisory Years to 1965,* USGPO, (Washington, 1981)
'Dedicated To Peace', op.cit.
'Strategic Reconnaissance 1956-1976', op.cit.
Aviation Week and Space Technology, 29 October and 5 November 1962, and 12 August 1963
United States Government, 'Records of the National Security Council, Executive Committee, October/November 1962', reprinted in *Declassified Documents Quarterly Catalogue,* Research Publications

Chapter Six

Frank, Lewis A., 'Nuclear Weapons Development in China', *Bulletin of the Atomic Scientists,* January 1966
Humphries, Orin, 'High Flight', *Wings,* June 1983
Inglis, David R., 'The Chinese Bombshell', *Bulletin of the Atomic Scientists,* February 1965
Ryan, William and Summerlin, Sam, *The China Cloud,* Hutchinson, (London, 1969)
'Chinese U-2s Seek Nuclear, IRBM Data', *Aviation Week,* p29, 17 September 1962
'Strategic Reconnaissance' op.cit.
'Loyal Martyrs – The Air Force Roll of Honour', HQ ROCAF, Taipei, January 1982

Chapter Seven

Bailey, Bruce, 'The View From The Top', *Warplane,* Issue 69
Crickmore, Paul F., *Lockheed SR-71 Blackbird,* Osprey, (London, 1986)
Godard, op.cit.
Hollingworth, Claire, *Mao And The Men Against Him,* Jonathan Cape, (London, 1985)
Johnson, op.cit.
Miller, Barry, 'Lockheed to Flight Test U-2 For Navy Surveillance Role', *Aviation Week and Space Technology,* 29 January 1973
Momyer, op.cit.
Strategic Reconnaissance, 1956-1976, op.cit.
Wagner, op.cit.

Chapter Eight

Beschloss, op.cit.
Brugioni, Dino, 'New Roles for Recce', *Air Force Magazine,* October 1985
Burrows, op.cit.
Crooks, Hoblit, Mitchell *et al.,* 'Project HiCat', *Air Force Flight Dynamics Laboratory (AFFDL) Technical Report,* November 1968, 68-127
Dighton, Ralph, 'Former sky spy takes on research tasks', *The Bakersfield Californian,* 20 January 1964
Groves, Patricia T., 'The Second Coming', *American Aircraft Modeller,* July 1972
Keeshan, Walt, 'The U-2 Today Scores Short-Cuts to Space With Its Long, Lonely Flights', *Western Aviation, Missiles and Space,* (date?)
Miles, Marvin, 'The U-2 Spy Planes Still Flying Labs', *Los Angeles Times* (date?)
Patty, Stanton, 'High-flying "spy" planes mapping vast resources', *The Seattle Times,* 12 November 1978
Stanley, Don, 'Recycling the U-2', *California Living Magazine,* 5 June 1977
Yaffee, Michael, 'AF Research Aircraft Zero in on Tracking System Noise', *Aviation Week,* 11 November 1963
'U-2s Gathering Data on Missile Exhausts', *Aviation Week,* 18 November 1963
'Solo flights into the ozone hole reveal its causes', *Smithsonian Magazine,* (date?)
'High Altitude Perspective', *NASA SP-427,* January 1978
'Airborne Antarctic Ozone Experiment', *NASA MS 245-5,* July 1987

Chapter Nine

Powers, op.cit.
'U-2 Pilot Bares Tale of Fantastic Parachute Jump', *Del Rio News Herald,* 26 September 1963
'I Baled Out Ten Miles Up!', Col Jack Nole, *Reader's Digest,* September 1964

'Flameout at 60,000 Feet', Maj Ward Graham, *Pilots For Christ International,* Parkesburg, PA
'Learning to Land the U-2', Capt Glenn Perry II, *Air Force Magazine,* January 1976
'Flying the U2', Capt Robert Gaskin, *Air Force Magazine,* April 1977
'High Flight', Orin Humphries, Part 1, *Wings,* June 1983; Part 2, *Airpower,* July 1983

Chapter Ten

Boutacoff, David, 'Army Banks on Joint STARS for Airland Battle Management', *Defence Electronics,* August 1986
Elson, Benjamin, 'USAF Picks Lockheed Team to Develop Targeting System', *Aviation Week,* 11 July 1977
Eppers, Dr. William, 'Air Force Sensors', *National Defense,* November/December 1976
Fink, Donald, 'U-2s, SR-71s Merged in One Wing', *Aviation Week,* 10 May 1976
Ganley, Michael and Goodman, Glenn, 'Air Force Tries a New Spanner as its Budget Fuel Drips', *Armed Forces Journal,* January 1987
Graham, L., 'Exploitation of Synthetic Aperture Radar Imagery', SPIE (Society of Photo-Optical Instrumentation Engineers), vol. 79, 1976
Hiltzelberger, Major Ron and Jonkoff, Captain Vic, 'Systems Integration: An Air Force Perspective', *Defense Electronics,* November 1986
Jensen, Homer *et al.,* 'Side Looking Airborne Radar', *Scientific American,* October 1977
Johnson, op.cit.
Klass, Philip, 'Air Force to Test Enemy Radar Locator', *Aviation Week,* 23 July 1984
Klass, Philip, 'Army Emerges as Lead RPV Proponent', *Aviation Week,* 20 June 1977
Lawson, Bob, 'Et Tu, U-2?', *The Hook,* Summer 1982
Miller, Barry, 'USAF Widens Unmanned Aircraft Effort', *Aviation Week,* 9 November 1970
Miller, Jay, *Convair B-58,* Aerofax Inc., (Arlington, 1985)
Myers, Willis, 'How Real is Real Time?', SPIE, op.cit.
O'Brien, D. and Schweizer, P., 'High Altitude RPVs', *National Defense,* July/August 1974
O'Lone, Richard, 'Compass Cope Seen in Anti-Tank Role', *Aviation Week,* 13 June 1977
Prados, op.cit.
Richelson, Jeffrey T., *The US Intelligence Community,* Ballinger, (Cambridge, Mass., 1985)
Schemmer, Benjamin, 'Electronic Cameras with Instantaneous Ground Read-Out Now Make Real-time, Precision Tactical Targeting Operationally Feasible', *Armed Forces Journal,* November 1982
Stimson, George W., *Introduction to Airborne Radar,* Hughes Aircraft Company, (El Segundo, Ca., 1983)
Ulsamer, Edgar, 'ASD's Efficient New Weapons', *Air Force Magazine,* June 1977
Ulsamer, Edgar, 'Winning the Electronic War', *Air Force Magazine,* July 1983
Wiley, James, 'Side-Looking Airborne Radars See Their Market Growing', *Defense Electronics,* October 1984
'Anti-radar Weaponry Development Pushed', *Aviation Week,* 27 January 1975
'Lockheed Gears for TR-1 Version of U-2', *Aviation Week,* 28 August 1978
'Temper RPV Enthusiasm', *Aviation Week,* 23 June 1975
'TR-1, "son of U-2", meets the people', *Lockheed Life,* August 1981
'A Management Study of Cost, Schedule and Performance on EW Programs', Association of Old Crows, National Security Affairs Committee, (1986)
Unclassified Chronology and Brief History of the 17th Reconnaissance Wing, (1984, 1985, 1986)
Hearings before the House and Senate Armed Services and Appropriations Committees, United States Congress, (1984-1987)

Appendix 1
The Dragon Lady Pilots

LAC Lockheed test pilot
RAF Royal Air Force
USN United States Navy
CAF Republic of China Air Force

* 'Be it known to all men that those men whose names appear thus have made the supreme sacrifice for the mission and the welfare of mankind.'

1955 Tony Levier, LAC
Bob Matye, LAC
Ray Goudey, LAC
Bob Schumacher, LAC
Robert Sieker, LAC *
Pete Everst
Lewis Garvin
Hank Meirdierck
Robert Mullin
Lewis Setter
Bill Yancey

1956 Howard Carey *
Glenn Dunnaway
Marty Knutson
Jake Kratt
Carl Overstreet
Wilbur Rose *
Hervey Stockman
Carmen Vito
Jim Allison
Tom Birkhead
Jim Cherbonneaux
Buster Edens *
Bill Hall
Dad McMurry
Frank Powers
Sammy Snyder
Bill Strickland
Barry Baker
Jim Barnes
Tom Cruel
Bob Ericson
Frank Grace *
Russ Kemp
Al Rand
Lyle Rudd
Al Smiley
Jack Nole
Joe Jackson
Floyd Herbert
Hank Nevett
Howard Cody

1957 Dick Atkins
Warren Boyd
Ray Haupt
Joe King
Steve Heyser
Mike Styer
Dick Leavitt
Bennie Lacombe *
Skip Alison
Tony Bevacqua
Jack Graves
Ed Emerling
Richard McGraw
John Campbell *
Ken Alderman
Leo Smith *
Ford Lowcock *
Buck Lee
Alfred Chapin *
Jim Sala
Scott Smith
Jim Qualls
Roger Cooper
Pat Halloran
Frank Stuart
Jim Black
Ed Perdue
Roger Herman
Bobbie Gardiner
Marv Doering
Nat Adams
Buzz Curry
Forrest Wilson
Roy St Martin
Snake Bedford
Rudy Anderson *
John McElveen
Ed Dixon
Harry Cords
Bob Pine
Earl Lewis
Wes Mcfadden
Cozy Kline
Ted Limmer, LAC

1958 Don James
Austin Russell
Adrian Acebado
John MacArthur, RAF
David Dowling, RAF
Mike Bradley, RAF
Chris Walker, RAF *
Bill Rodenbach
Bob Wood
Bob Ginther
Paul Haughland *
Ken Van Zandt
Robbie Robinson, RAF
Bo Reeves
Dick Callahan
Buddy Brown
John Boynton

1959 Andrew Bratton
Ron Hedrick
Harold Melbratten
Gerry McIlmoyle
Bill Stickman
Dick Rauch
Floyd Kifer
Gimo Yang Shih Chu, CAF
Chen Huai Sheng, CAF *
Hsu Chung Kuei, CAF
Tiger Wang Tai Yu, CAF
Chih Yao Hua, CAF *
Hua Hsi Chun, CAF
TJ Jackson Jr
Chuck Stratton
Bob Schueler

Bob Powell
Robert 'Deke' Hall
Kenneth McCaslin
Jack Carr
Elsworth Powell
J. B. Reed
Raleigh Myers

1960 Bob Wilke
Bob Spencer
Don Crowe
Rex Knaak
Tony Martinez
Hank McManus
Dave Gammons
William Wilcox
Leo Stewart
Pinky Primrose *
Ed Hill
Joe Hyde
Dave Ray
Cliff Beeler

1961 Chuck Maultsby
Chunky Webster, RAF
Taffy Taylor, RAF
Chuck Kern
Willie Lawson
George Bull
Don Webster
Eddie Dunagan
John DesPortes
Dan Schmarr
John Wall
Dick Bouchard

1962 Art Leatherwood
Jim Rogers
Don McClain
Clair McCombs
Ed Smart

1963 Vic Milam
Ward Graham
Yei Chang Yi, CAF
Terry Lee Nan Ping, CAF *
Bookie Baughn
Ron Stromberg
Liang Teh Pei, CAF *
Johnny Wang Shi Chuen, CAF
Ken Somers
Gene O'Sullivan

1964 Jack Chang Li Yi, CAF
Yang Hui Chia, CAF
Martin Bee, RAF
Basil Dodd, RAF
Pete Wang Chen Wen, CAF *
Steve Sheng Shih Hi, CAF
Charlie Wu Tse Shi, CAF *
Jack Fenimore

1965 Terry Liu Jet Chuang, CAF
Mickey Yu Ching Chang, CAF *
Spike Chuang Jen Liang, CAF
Tom McMurtry, USN
Ken Diehl
Jerry Davis
George Worley
Arnie Strasheim
Bill Copeman
Keith Spaulding
Earle Smith Jr
Les Powell
John Amundson
Ed Rose
Don Wright

1966 Bob Birkett
Bob Hickman *
Harold Swanson
Jim Hoover
Andy Fan Hung Ti, CAF
Billy Chang Hseih, CAF *
Yang Erh Ping, CAF
Lonnie Liss
Jim Whitehead
Dave Patton
Sam Swart
Don Aitro
Les White
Marion Mixson
Richard Woodhull

1967 Tom Hwang Lung Pei, CAF *
Dick Cloke, RAF
Harry Drew, RAF
Jerry Chipman
Lash Larue
Frank Ott
Roy Burcham
Dale Kellam

1968 Johnn Shen Chung Li, CAF
Tom Wang Tao, CAF
Frederick Banks
George Freese
Jim Phenix
Curt Behrend
Doyle Krumrey
Ron Williams
Ken Chisholm
Ray Samay
Jerry Wagnon
Stan Lawrence
David Lee Pao Wei, CAF
Denny Huang, CAF *
Ben Higgins
Don Wright

1969 Thomas Block
Willie Horton

1970 Dick Davies
Jim Terry
Jim Wrenn

1971 Chen Wei, CAF
Chu Chien, CAF
Phil Daisher
Sid Head
Jerry Sinclair
John Cunney

1972 Gary Hawes
Fuzzy Furr
Don Schreiber
Usto Schultz
Terry Nelson
Dan Riggs
Ronnie Rinehart

1973 Richard Rice
Art Saboski
Robert Armstrong
Charles Smyth
Jerry Hoyt
Rich Drake
Al Henderson *
John Kent
Dale Hudler
Howard Bayne
Jim Pinson
John Sander
Ken Stanford
Chuck Crabb
Tom Doubek
John Dale
Denny Gagen

1974 Jack Stebe
Don Hahn
John Cantwell
Ho Ho Hoenninger
Dick Whitaker
Denny Thisius
Mike Lemmons
James Martin
Glenn Perry

Terry Rendleman
Snake Pierce
Muff Heckert
Dave Dickerson

1975 Stanley Rauch
Jim Barrilleaux
Bob Gaskin
James Wilson
Bill Koplin
Jim Madsen
John Little
Michael Phillips
James Evans III
Don Hatten
Richard Boyer
Dick Keylor
Thom Evans
Dave Kantrud
Ron Friesz
Jimmy Myer
Kit Busching
John Swanson
Anthony McGarvey

1976 James Winans
Larry Driskill
Ken Bassett
Chuck Voxland
Mike Kidder
Dave Bateman

1977 Mike Kelly
Richard Fossum
Fred Kishler
John Storrie

1978 Jimmy Carter
Raymond Wilson
Bob McCrarey
Mark Fischer
Bill Williams
Bob Munger
Paul Roberts
Mick Uramkin

Doug Morin
Bill Collette

1979 Dale Smith
Mike Musholt
Rick Bishop
Cecil Snyder
Butch Hinkle
Rich Snow
Bob Johnson
Bill Burke
Steve Brown
Don Feld
Mike Danielle
Dave Ebersole
Pete Balzli
Stormy Boudreaux
Dave Bechtol
Kurt Lindeman
Jimmy Mclean
Bob Ray
Tim Lyle

1980 Paul Cross
Jan Nystrom
Edward Beaumont
Mark Spencer
Larry Faber
Steve Barber
James Nicol
Cleve Wallace
John Petersen
Grant Gordon
Bruce Cucuel

1981 Bill Earnest
Dee Porter
Lou Campbell
Ken Broda
Dan House
Bill Kemmer
Ken Womack
Ron Blatt
Jim Kippert
Glenn Johnston
Bobby Fairless

1982 Rob Bateson
Marty Decker
Bill Gilbert
Bubba Lloyd
Bruce Carmichael
Sam Ryals
James O'Neal
Ashton Lafferty
Bruce Jinneman
Joe Crawley
Steve Randle
Robert Heath
Dan David
Tim Cox
Marty Gutierrez

1983 Jimmy Milligan
Dave Hensley
Dan Kelly
Mark Benda
William Walker
Alan Kopf
Howard Johnson
Dave Bonsi
Joe Mudd
Sonny Sonnhalter
Alan Popwell
Dewayne Rudd
Ken Sasine

1984 Dave Pinsky
Tom Dettmer
Joe Fusco
Todd Hubbard
Steve Benningfield
Nap Napolitano
Bob Uebelacker
James Perkins
Jim Bob Roberts
Don Merritt
Robert Johnson
Pete Lemaire
Robert Dunn
James Burger
John Zermer
Al Crawford

The above list is thought to be complete, with the following exceptions:

— a few supervisory and training officers assigned to Operation Overflight, who flew no operational missions.

— those trained by the US Air Force since 1984, whose names have been withheld for security reasons.

— Lockheed test pilots other than the initial cadre. These have subsequently included Bill Park, Darryl Greenamyer, Art Peterson, Ken Weir, Dave Kerzie and Skip Holm.

— those assigned to the Air Force Flight Test Centre, Edwards AFB. Only a partial listing of these is available, correct to mid-1964. Pilots: Harry Andonian, Weldon Armstrong, Fred Cuthill, Loren Davis, Donald Evans, Norvin Evans, Mervin Evenson, Henry Gordon, Norris Hanks, Pat Hunerwadel *, Robert Jacobson, James King, Budd Knapp, Rial Lowell, John Ludwig, Lachlan Macleay, Charles Rosburg, Wendell Shawler, Philip Smith, Donald Sorlie and Robert Titus. Observers (in U-2D): James Eastlund, William Frazier, Bobby Lynn, Charles Manske, Charles Nyquist, Raymond Oglukian and James Williams.

Appendix 2

Article/Serial Number Reference List

Original production

Article Number	Serial Number	Location of surviving airframes (all others written off in accidents)
341	–	
342	56-6675	
343	56-6676	
344	56-6677	
345	56-6678	
346	56-6679	
347	56-6680	U-2C displayed at NASM, Washington DC.
348	N708NA/56-6681	U-2C displayed at NASA Ames Research Center
349	N709NA/56-6682	U-2C displayed at Robins AFB Museum
350	56-6683	
351	56-6684	
352	56-6685	
353	56-6686	
354	56-6687	
355	56-6688	
356	56-6689	
357	56-6690	
358	56-6691	U-2C wreckage at PRC Military Museum, Beijing
359	56-6692	U-2CT battle damage repair trainer, RAF Alconbury
360	56-6693	
361	56-6694	
362	56-6695	
363	56-6696	
364	56-6697	
365	56-6698	
366	56-6699	
367	56-6700	
368	56-6701	U-2C displayed at SAC Museum, Offutt AFB
369	56-6702	
370	56-6703	
371	56-6704	
372	56-6705	
373	56-6706	
374	56-6707	U-2C displayed at Laughlin AFB
375	56-6708	
376	56-6709	
377	56-6710	
378	56-6711	

379	56-6712	
380	56-6713	
381	56-6714	U-2C displayed at Beale AFB
382	56-6715	
383	56-6716	U-2C displayed at Davis-Monthan AFB
384	56-6717	
385	56-6718	
386	56-6719	
387	56-6720	
388	56-6721	U-2D displayed at March AFB
389	56-6722	U-2A displayed at USAF Museum, Dayton, Ohio
390	–	

Supplementary production

1959-60		
391	56-6951	
392	56-6952	
393	56-6953	U-2CT displayed at Edwards AFB
394	56-6954	
395	56-6955	
1967-68		
051	68-10329	U-2R 9SRW Beale AFB
052	68-10330	
053	68-10331	U-2R 9SRW Beale AFB
054	68-10332	U-2R 9SRW Beale AFB
055	68-10333	U-2R 9SRW Beale AFB
056	68-10334	
057	68-10335	
058	68-10336	
059	68-10337	U-2R 9SRW Beale AFB
060	68-10338	U-2R 9SRW Beale AFB
061	68-10339	U-2R 9SRW Beale AFB
062	68-10340	U-2R 9SRW Beale AFB
1980-89		
063	N706NA/80-1063	ER-2 NASA Ames Research Center
064	80-1064	TR-1B 9SRW Beale AFB
065	80-1065	TR-1B 9SRW Beale AFB
066	80-1066	U-2R 9SRW Beale AFB *
067	80-1067	TR-1A LASC Palmdale
068	80-1068	U-2R 9SRW Beale AFB *
069	N708NA/80-1069	TR-1A NASA Ames Research Center on loan from USAF
070	80-1070	U-2R 9SRW Beale AFB *
071	80-1071	U-2R 9SRW Beale AFB
072	80-1072	
073	80-1073	TR-1A 9SRW Beale AFB
074	80-1074	TR-1A 9SRW Beale AFB
075	80-1075	
076	80-1076	U-2R 9SRW Beale AFB
077	80-1077	TR-1A 17RW RAF Alconbury
078	80-1078	TR-1A 17RW RAF Alconbury

079	80-1079	TR-1A 17RW RAF Alconbury
080	80-1080	TR-1A 9SRW Beale AFB
081	80-1081	TR-1A 17RW RAF Alconbury
082	80-1082	TR-1A 9SRW Beale AFB
083	80-1083	TR-1A 17RW RAF Alconbury
084	80-1084	TR-1A 9SRW Beale AFB
085	80-1085	TR-1A 17RW RAF Alconbury
086	80-1086	TR-1A 17RW RAF Alconbury
087	80-1087	TR-1A 9SRW Beale AFB
088	80-1088	TR-1A 17RW RAF Alconbury
089	80-1089	U-2R 9SRW Beale AFB
090	80-1090	TR-1A LASC Palmdale
091	80-1091	U-2RT 9SRW Beale AFB
092	80-1092	TR-1A 17RW RAF Alconbury
093	80-1093	TR-1A 17RW RAF Alconbury
094	80-1094	TR-1A 17RW RAF Alconbury
095	80-1095	U-2R 9SRW Beale AFB
096	80-1096	U-2R 9SRW Beale AFB
097	N709NA/80-1097	E-R2 NASA Ames Research Center
098	80-1098	U-2R 9SRW Beale AFB
099	80-1099	TR-1A 17RW RAF Alconbury

* originally manufactured as a TR-1A

Photograph and Illustration Credits

Aireview: page 44

via Author: pages 86, 95, 118, 179

Author's collection: pages 137, 140, 150, 154, 161, 172, 173, 181, 197, 198

Jim Ballard Photography: page 169

Bell Aircraft via René Francillon: page 9

Harry F Carter: pages 132, 159

CIA via René Francillon: pages 28, 36

Peter R Foster: page 188

David Foust: page 184

Hughes Radar Systems Group: page 178

Lockheed: pages 1, 4, 13, 14, 15, 16, 19 (both), 21, 30, 33, 59, 77, 98, 105, 114, 117, 119, 138, 149, 158, 160, 164, 167 (both), 170, 174, 176, 178, 187, 190

Jay Miller: pages 11, 42, 134, 145, 148

via Jay Miller: pages 61, 65, 70, 103, 107, 108, 111, 112

NASA: pages 25, 133, 135, 136, 152, 162

New York Times: page 49

Novosti: pages 26, 53, 84

via *Pilot Press*: pages 31, 40

Popperfoto: page 116

Pratt & Whitney: page 6

The Press Association: pages 8, 75

Jimmy Stewart: page 192

Tass: pages 37, 52

USAF: pages 50, 60, 78, 79, 123, 125, 127, 130, 131

USAF via Tony Bevacqua: pages 63, 69, 83, 157

USAF via Don Webster: pages 81, 87

via Don Webster: page 72

S Wright & R Waxman via Author: page 101

Xinhua: page 194

via 4080SRW/100SRW: pages 5, 46, 67, 80, 90, 143

Index